Aristarchus of Samos: *On the Sizes and Distances of the Sun and Moon*

This book offers the Greek text and an English translation of Aristarchus of Samos's *On the Sizes and Distances of the Sun and Moon*, accompanied by a full introduction, detailed commentary, and relevant scholia.

Aristarchus of Samos was active in the third century BC. He was one of the first Greek astronomers to apply geometry to the solution of astronomical problems as we can see in his only extant text, *On the Sizes and Distances of the Sun and Moon*. Alongside the Greek text and new English translation, the book offers readers the Latin text and English translation of Commandino's notes on the text. Readers will also benefit from a comprehensive introductory study explaining the value of Aristarchus's calculations and methodology throughout history, as well as detailed analyses of each part of the treatise.

This volume will be of interest to students and scholars working on ancient science and astronomy and the general reader interested in the history of science.

Christián C. Carman is Professor and Researcher at the Universidad Nacional de Quilmes, Argentina, and Research Member of the National Research Council of Argentina (CONICET). He works on topics related to philosophy of science as well as history of ancient astronomy.

Rodolfo P. Buzón is Emeritus Professor of Greek Philology at the Universidad de Buenos Aires and Research Member of the National Research Council of Argentina (CONICET). He has published several works related to classical philology.

Scientific Writings from the Ancient and Medieval World

Series editor: John Steele

Brown University, USA

Scientific texts provide our main source for understanding the history of science in the ancient and medieval world. The aim of this series is to provide clear and accurate English translations of key scientific texts accompanied by up-to-date commentaries dealing with both textual and scientific aspects of the works and accessible contextual introductions setting the works within the broader history of ancient science. In doing so, the series makes these works accessible to scholars and students in a variety of disciplines including history of science, the sciences, and history (including Classics, Assyriology, East Asian Studies, Near Eastern Studies and Indology).

Texts will be included from all branches of early science including astronomy, mathematics, medicine, biology, and physics, and which are written in a range of languages including Akkadian, Arabic, Chinese, Greek, Latin, and Sanskrit.

The Medicina Plinii
Latin Text, Translation, and Commentary
Yvette Hunt

Learning With Spheres
The *golādhyāya* of Nityānanda's *Sarvasiddhāntarāja*
Anuj Misra

The *Arithmetica* of Diophantus
A Complete Translation and Commentary
Jean Christianidis and Jeffrey Oaks

Aristarchus of Samos: *On the Sizes and Distances of the Sun and Moon*
Greek Text, Translation, Analysis, and Relevant Scholia
Christián C. Carman and Rodolfo P. Buzón

The *Spherics* of Theodosios
Nathan Sidoli and R.S.D. Thomas

For more information about this series, please visit: https://www.routledge.com/classicalstudies/series/SWAMW

Aristarchus of Samos:
On the Sizes and Distances of the Sun and Moon

Greek Text, Translation, Analysis, and Relevant Scholia

Christián C. Carman and Rodolfo P. Buzón

Routledge
Taylor & Francis Group

LONDON AND NEW YORK

First published in Spanish as "Acerca de los tamaños y las distancias del Sol y de la Luna" by Edicions de la Universitat de Barcelona in 2020.

First published in English in 2023
by Routledge
4 Park Square, Milton Park, Abingdon, Oxon OX14 4RN

and by Routledge
605 Third Avenue, New York, NY 10158

Routledge is an imprint of the Taylor & Francis Group, an informa business

British Library Cataloguing-in-Publication Data
A catalogue record for this book is available from the British Library

Library of Congress Cataloging-in-Publication Data

Names: Aristarchus, of Samos. | Aristarchus, of Samos. On the sizes and distances of the sun and moon. | Aristarchus, of Samos. On the sizes and distances of the sun and moon. English. | Aristarchus, of Samos. On the sizes and distances of the sun and moon. Greek. | Carman, Christián C., editor, translator. | Buzón, Rodolfo, 1942- editor, translator.
Title: Aristarchus of Samos: On the sizes and distances of the sun and moon : Greek text, translation, analysis, and relevant scholia / Christián C. Carman, and Rodolfo P. Buzón.
Description: Abingdon, Oxon; New York, NY: Routledge, 2023. | Series: Scientific writings from the ancient and medieval world | "First published in Spanish as "Acerca de los tamaños y las distancias del Sol y de la Luna" by Edicions de la Universitat de Barcelona in 2020." | Includes bibliographical references and index.
Identifiers: LCCN 2022052003 (print) | LCCN 2022052004 (ebook) | ISBN 9781032026732 (hardback) | ISBN 9781032026763 (paperback) | ISBN 9781003184553 (ebook)
Subjects: LCSH: Astronomy, Greek.
Classification: LCC QB21 .A713 2023 (print) | LCC QB21 (ebook) | DDC 520—dc23/eng/20230104
LC record available at https://lccn.loc.gov/2022052003
LC ebook record available at https://lccn.loc.gov/2022052004

ISBN: 978-1-032-02673-2 (hbk)
ISBN: 978-1-032-02676-3 (pbk)
ISBN: 978-1-003-18455-3 (ebk)

DOI: 10.4324/9781003184553

Typeset in Times New Roman
by Apex CoVantage, LLC

For Emi.
CCC

For Ana Victoria, Alexander, and Benjamin.
RPB

Contents

x   *Contents*

Figures

Figures in Greek text and English translation

Figures in Latin commentary and English translation

Figures in Greek scholia and English translation

Preface

On the Sizes and Distances of the Sun and Moon is the only extant work of the Greek astronomer Aristarchus of Samos. It is also the earliest record of the application of geometry to the study of the sky. In the book, Aristarchus applies geometrical reasoning to empirical values and obtains outputs that, until that time, were totally out of the reach of humankind.

In effect, Aristarchus calculates the relative distances of the Sun and the Moon from the Earth and their relative sizes. He obtains the result that the Sun is around 19 times farther away and greater in size than the Moon. In turn, the Earth is approximately three times bigger than the Moon and, hence, roughly six times smaller than the Sun. To understand the importance of these results, it is enough to recall that Epicurus asserted that "the Sun is as big as it appears" or that Anaxagoras affirmed that "the Sun is as big as the Peloponnesus is." But the most valuable contributions of the work are not the results, which are wrong, but the methods used to obtain them, which became the dominant methods for establishing the distances and sizes of the Sun and the Moon until the eighteenth century. The methods of Aristarchus's *On Sizes* were used, with small modifications, by Hipparchus, Ptolemy, the Arabic astronomers, and the vast majority of the early modern astronomers, including Copernicus and Kepler.

In 2020, the authors of the present book published a volume containing a Spanish translation of *On the Sizes and Distances of the Sun and Moon* with an introductory study. Several scholars expressed interest in having an English version because, even if Heath's classic English translation is really good, it is now over a century old. Besides, his introductory study is not exhaustive and, more important, a century out of date. Significant research has been published during the last century allowing scholars to more deeply understand *On Sizes*'s content, context, and influence. The present volume is not merely an English translation of the volume published by us in Spanish in 2020. In fact, both the text and the translation were again reviewed, taking into account all the available information (manuscripts, editions, and translations). For the introduction, we start from the text of the Spanish edition and have rewritten a significant part of Chapters 1 and 3. Some errors were found and corrected.

The volume is divided into four chapters. Chapter 1 is an introductory essay in three parts. The first one is dedicated to Aristarchus of Samos himself, reviewing the little data we have about him. In the second, we discuss the attempts to calculate the distances and sizes of the Sun and the Moon, from the beginnings of Greek science to Kepler, showing the different ways that astronomers have employed Aristarchus's method. In the last one, we review previous editions and translations of this work and describe our translation criteria.

Chapter 2 consists of the Greek text and translation. We add the Latin text and translation of the comments and notes introduced by Commandino in his edition of 1572 because we consider that after centuries they are still very useful.

In the third chapter, we analyze the content of the treatise in detail. This chapter includes a commentary that goes through the text section by section. Therefore, the notes in the translation are limited to clarifying the text itself, but not its content.

Chapter 4 contains the Greek text and translation of the scholia that we consider significant. The translation of each scholion, in turn, is preceded by a short introduction.

Finally, we include two appendices. In the first one, we describe the mathematical tools that Aristarchus uses in the treatise. In the second, we list the extant manuscripts of Aristarchus's work.

Acknowledgements

Alexander Jones gave us innumerable suggestions to improve the translation of Aristarchus's text and the scholia. Without his help, this translation would have been impossible. We are deeply grateful to him and to John Steele, editor of this series, who also generously agreed to publish this volume, for his comments on specific aspects of the translation and style advice.

To James Evans, Nathan Sidoli, and Reviel Netz, our gratitude for their willingness to respond to our queries on specific topics related to Aristarchus translation, manuscripts, and astronomy in general.

Special thanks to Beate Noack, who generously mailed us her copy of all the surviving manuscripts of the Aristarchus text; to Paolo Vian, Director of the Manuscripts Department of the Vatican Apostolic Library, for giving us access to Aristarchus's manuscripts; and to the Bodleian Library for allowing us to consult the copy of the Wallis edition that belonged to Wallis himself, as well as several copies of the editions of Commandino and Valla.

Evangelos Spandagos was kind enough to give us a copy of his book, and Ignacio Silva allowed us to access bibliography that, otherwise, we would not have been able to consult. He also helped us improve our English style. Many thanks to them and also to Xenophon Moussas and Lina Anastasiou for providing us with access to modern Greek literature, and to David Valls-Gabaud for providing us with a copy of Angelo Gioè's thesis.

The preparation of this edition has been financed by the Research Project PICT-2019-01532 of the National Agency for Scientific and Technological Promotion of the Argentine Republic. The Research Project PIP 2012–2014 IU of the National Council for Scientific and Technical Research (CONICET) and the Research Program "Philosophy and History of Science" of the National University of Quilmes partially founded Christián C. Carman's stay in Rome to work with the manuscripts in the Vatican Apostolic Library.

1   Preliminary study

Aristarchus of Samos

Ptolemy provides one of the few pieces of evidence that allows us to infer the date of Aristarchus. The Alexandrian astronomer uses an observation of the summer solstice of 280 BC made by Aristarchus.[1] Therefore, around 280 BC, Aristarchus was already an adult. In addition, we know from Aetius (I, 15.5; Diels 1879: 313) that Strato of Lampsacus, the second director of the Lyceum after Theophrastus's death, was Aristarchus's teacher. Strato headed the Lyceum around 288 or 287 BC, and he did it for 18 years. Finally, in *Sand-Reckoner* (Heath 1897: 221–222), Archimedes refers to Aristarchus's heliocentrism. Archimedes died during the Siege of Syracuse in 212 BC. Considering all this evidence, we can agree with Heath (1913: 299) that Aristarchus lived approximately between 310 and 230 BC.[2]

Aristarchus of Samos was called "Aristarchus the Mathematician," "Aristarchus of Samos," or even "Aristarchus of Samos, the Mathematician," probably to distinguish him from other known Aristarchuses,[3] as, for example, Aristarchus the Grammarian mentioned by Vitruvius (I, 1, 12; 1999: 23). It is Vitruvius who makes a beautiful compliment to our Aristarchus. He affirms that certain people pass beyond the business of architects and are turned into mathematicians when "nature has granted such wits, acuity, and good memory that they are fully skilled in geometry, astronomy, music, and related disciplines" (I, 1, 17; 1999: 24). He points out that these people are few, and among the seven that he mentions, Aristarchus of Samos ranks first, along with such figures as Eratosthenes, Apollonius, and Archimedes.

We know very little about Aristarchus's work, besides *On Sizes* and his proposal of a heliocentric doctrine. Again, according to Vitruvius, he invented a kind of sundial known as σκάφη (*scaphê*), consisting of a hemisphere inscribed with

---

1 Ptolemy mentions this observation three times in the *Almagest*. In the first two (III, 1; Toomer 1998: 137; III, 1 and Toomer 1998: 138), he attributes it not directly to Aristarchus but to his followers (ὑπὸ τῶν περὶ Ἀρίσταρχον). In the last one (III, 1; Toomer 1998: 139), he attributes it directly to Aristarchus.
2 Wall (1975) offers a detailed analysis of the extant references, proposing wider margins.
3 A list and description of other historical figures named Aristarchus can be seen in Fortia 1810: 1–196 (but this pagination begins again in the book after page 248).

DOI: 10.4324/9781003184553-1

marks for hours in which the gnomon's shadow is projected. Vitruvius also attributes to Aristarchus another sundial, including a disk on a flat surface (IX, 8, 1; 1999: 116). Furthermore, we know from Aetius (IV. 13, 8; Diels 1879: 404, 853) that he developed some topics related to optics, particularly with the nature of color. Because his teacher Strato also worked on optics, Aristarchus would have been strongly influenced by him (Heath 1913: 300). Finally, Vitruvius (1999: IX, 3) links Aristarchus with the correct explanation of the Moon's phases, and a scholion to book XX of Odyssey refers to Aristarchus, who is said to make some comments on Thales's explanation of solar eclipses (Lebedev 1990; Bowen and Goldstein 1994).

Regarding heliocentrism, in one of his most curious works, *Sand-Reckoner* (Heath 1897: 221–252), Archimedes challenges himself to calculate the number of grains of sand necessary to fulfill the entire universe. The challenge is just an excuse for introducing a new numbering system allowing numbers greater than those permitted by the Greek system. In that context, he must calculate the size of the universe. And in doing that, he mentions that Aristarchus "brought out a book consisting of some hypotheses" about the universe which are "that the fixed stars and the [S]un remain unmoved, that the Earth revolves about the [S]un in the circumference of a circle, the [S]un lying in the middle of the orbit."[4] If the stars remain unmoved, one can only explain the daily motion of the stars by the Earth's rotation. Thus, the Earth would not only revolve around the Sun but also rotate on its axis. The text is clear enough, and Archimedes is sufficiently close in time to Aristarchus that there is no doubt about its meaning and authenticity.[5]

Although the daily rotation of the Earth is only inferred from Archimedes's text, it is confirmed by a reference in Plutarch (*De facie in orbe lunae*, c. 6, 922F–923A; 1957: 55) which makes it explicit. According to Plutarch, Cleanthes suggested that Greeks ought to lay an action for impiety against Aristarchus for

> setting the home of the universe in motion because he thought to save [the] phenomena, by assuming that the sky remains at rest and the Earth revolves in an oblique circle while rotating, at the same time, on its own axis.

Indeed, we know that Cleanthes is the author of a book titled *Against Aristarchus*.[6] Also, Aetius (II, 24.8; Diels 1879: 355) tells us that Aristarchus placed the Sun in the middle of the fixed stars and set the Earth in motion around the solar circle.[7] It

---

4 "Ἀρίσταρχος δὲ ὁ Σάμιος ὑποθέσιών τινων ἐξέδωκεν γραφάς, . . . Ὑποτίθεται γὰρ τὰ μὲν ἁπλανέα τῶν ἄστρων καὶ τὸν ἄλιον μένειν ἀκίνητον, τὰν δὲ γᾶν περιφέρεσθαι περὶ τὸν ἄλιον κατὰ κύκλου περιφέρειαν, ὅς ἐστιν ἐν μέσῳ τῷ δρόμῳ κείμενος" (*Arenarius* I, 4–7, Heiberg 1913, 2: 244; Heath 1897: 221–222).
5 Only Wall (1975) doubts it.
6 Diogenes Laërtius, VII.174.
7 Other references to Aristarchus's heliocentrism are in Plutarch, *Platonic Questions*, (1006C; 2004: 63) and in Sextus Empiricus, *Adv. Math.* X.174; 2012: 113). Heath (1913: 301–310) offers an excellent summary and discussion of the texts.

is therefore indisputable that Aristarchus of Samos defended some kind of helio-centrism.[8] However, there are no traces of heliocentrism in *On Sizes*, which seems to assume geocentricism.

The calculation of the distances and sizes of the Sun and the Moon from Antiquity to the early modern period

As we will see later in detail, Aristarchus's calculation has deep roots in the history of astronomy: it will be used, practically without modification, from the third century BC until well into AD eighteenth century.[9] Aristarchus's calculation is the oldest recorded scientific attempt to obtain values for the solar and lunar distances and sizes. We know that other thinkers before him had proposed values for distances and sizes. However, the justifications for these proposals were more or less arbitrary, if they existed at all.

The values before Aristarchus

Several pre-Socratic philosophers dealt with the sizes and distances of celestial bodies. As far as we know, Anaximander was the first to speculate on this question (Simplicius, *In De Caelo*, II, 10; Heiberg 1894: 471). He asserted that the Sun and the Earth are of the same size, but that the circle in which the Sun moves is 27 times greater than the Earth. This assertion would imply that the solar orbit radius is 27 Earth radii (Aetius ii, 21.5, Diels 1879: 351).[10] As we have already mentioned, Anaxagoras held that the Sun is larger than the Peloponnese (Aetius ii, 9-21-10, Diels 1879: 351). In turn, based on his theory of the harmony of the spheres, Pythagoras obtained the distances of the planets from musical notes (Dreyer 1953: 35–52).

From Aristotle (*Meteor*. 1, 8, 345b; 1942: 61), we know that already by this time the calculation of the solar and lunar distances and sizes had aroused great interest. According to him:

> [A]stronomical researches have now shown that the size of the [S]un is greater than that of the [E]arth and that the stars are far farther away than the [S]un from the [E]arth, just as the [S]un is farther than the [M]oon from the [E]arth.

Philippus of Opus, a contemporary of Aristotle, wrote several works that did not reach us. Only the titles are extant. Still, they are eloquent enough to leave open the possibility that Aristotle was referring to him: *On the distance of the Sun and*

---

8 About the differences between Copernicus's and Aristarchus's heliocentrism, see Carman (2018a).
9 An excellent introduction to distance calculations is van Helden (1986), which we follow to orga-nize this section, particularly the last one, dedicated to modern astronomy.
10 Although a little earlier it is stated that the value is 28 (cf. Aetius ii, 20.1; Diels 1879: 348). On the other hand, Hippolytus (*Philos*. VI; Diels 1879: 560) affirms that the circle of the Sun is actually 27 times that of the Moon and not that of the Earth.

the Moon and *About the sizes of the Sun, Moon, and Earth.*[11] But Aristotle could also be referring to Eudoxus, who, like him, was a student of Plato. According to Archimedes, Eudoxus held that the Sun is nine times larger than the Moon, and having the same apparent size, it is also nine times farther away.

Along with that of Eudoxus, Archimedes mentions two other values found after Aristotle: that of his father, Phidias,[12] who proposes the ratio 12 to 1, and that of Aristarchus that we will see in the next section. We do not know how Phidias or Eudoxus obtained these values, but it is possible that a method similar to that of Aristarchus was already known at that time.[13]

Aristarchus's calculation

Aristarchus's calculation has two well-defined parts. In the first one, he obtains the ratio between the solar and lunar sizes and between their distances. In the second one, he compares these sizes with that of the Earth.

The idea behind the calculations of the first part is straightforward and robust. Aristarchus invites us to imagine the Sun, the Moon, and the Earth forming a triangle. If we knew two angles of the triangle, we could calculate the ratio of its sides, that is, the proportion between the distances.

We can obtain empirically the angle centered at the Earth by measuring the lunar elongation, that is, the angular distance from the Moon to the Sun. Aristarchus could not travel to the Moon or the Sun to measure angles there. However, since the Moon is illuminated by the Sun, the Moon's appearance, as seen from the Earth, provides information about the Sun's and the Moon's relative positions. In particular,

---

11 Neugebauer (1975: 574) doubts that they are works written by Philippus of Opus. According to Neugebauer, an author before Eudoxus or Aristarchus could hardly write those books. The list merely enumerates the typical works attributed to an astronomer. However, Dillon (2003: 180, n.4) claims that these works could have been written after Eudoxus. For information on Philippus of Opus, see Toomer (2005), also Taylor (1929), and more extensive (though focusing on his place in the Platonic school and not as an astronomer), Dillon (2003: 179–197) (but especially the earlier pages) and Tarán (1975).

12 That Phidias is Archimedes's father is a conjecture that arises from a smart proposal to correct an incomprehensible phrase in *Sand-Reckoner*'s manuscripts. They say, "Φειδία δὲ τοῦ Ἀκούπατρὸς," which would be translated "Phidias of Acupatros," but there is no record of a place called Acupatros. Blass (1883: 255) proposes to change the κ to μ. Assuming that, the text would say, "Φειδία δὲ τοῦ ἁμοῦ πατρὸς," "Phidias, our father." A scholion to a text of Gregory of Nazianzus supports this conjecture. The scholion mentions "Phidias, originally from Syracuse, astronomer, and father of Archimedes." Cf. Heath (1897, note ss on p. XV).

13 Tannery (1893) states that it was Eudoxus who used for the first time the method employed by Aristarchus and that, subsequently, Phidias also used it, both before Aristarchus. He states that the differences in the results obtained by Eudoxus, Phidias, and Aristarchus are mainly due to differences in the value used for the lunar elongation at dichotomy: while, as we shall see, Aristarchus argued that he departed from a quadrant by 3°, for Phidias it would depart by 5°, and for Eudoxus by 6°. While tenable, Tannery's analysis is highly speculative. We have too little evidence to reconstruct how Phidias and Eudoxus did their calculations, if their values actually are the results of calculations (see Heath 1913: 332).

Aristarchus knows that during a lunar dichotomy (i.e., when we see from the Earth exactly half a Moon), the angle centered at the Moon between the Sun and Earth is a right angle. Otherwise, we would not see exactly a half-moon. Therefore, at a lunar dichotomy, we have a right triangle with the right angle centered at the Moon and the angle centered at the Earth representing the lunar elongation. See Figure 1.1.

Aristarchus states that the elongation of the Moon from the Sun in the lunar dichotomy is 87°. With this datum, it would not even be necessary to make calculations. It would be enough to draw a triangle with a right angle, another of 87°, and then measure the ratio between the hypotenuse and the minor leg. In Figure 1.1, minor leg *EM* represents the Earth–Moon distance, and hypotenuse *ES* the Earth–Sun distance.

If we want to calculate it, we can easily obtain that the Sun is 19.1 times farther than the Moon, applying trigonometry.[14] But since trigonometry had not yet been developed, Aristarchus cannot get a precise value. Nevertheless, he is able to establish a maximum and minimum limit: the Sun is between 18 and 20 times farther than the Moon. We will call this method the *dichotomy method*.

Once Aristarchus obtains the ratio between the distances, he can easily get also the ratio between their sizes. According to him, the lunar and solar apparent sizes are equal. Thus, the ratio of their sizes is equal to the ratio of their distances. The

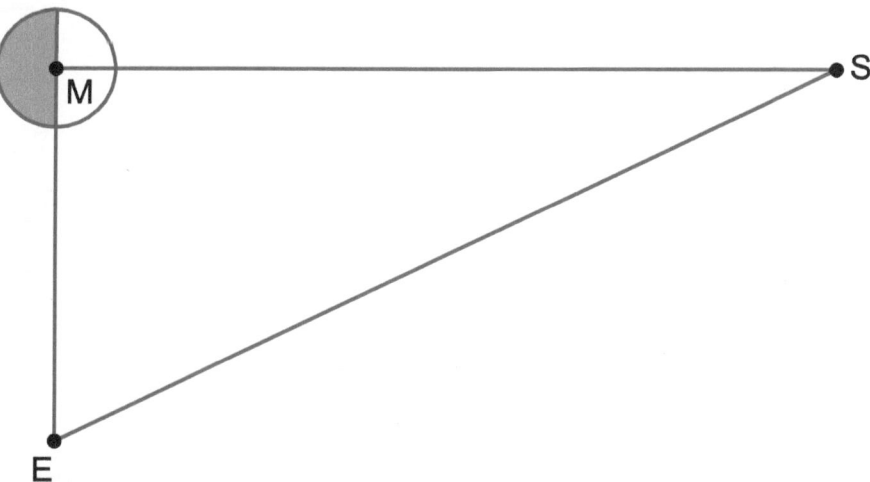

Figure 1.1 The Sun (*S*), the Earth (*E*), and the Moon (*M*) at the exact moment of lunar dichotomy.

14 See Figure 1.1. We know that the cosine of the angle with *E* as center is the adjacent side (the Earth–Moon distance) over the hypotenuse (the Earth–Sun distance). We know the angle centered at *E*. Therefore, we can obtain its cosine and the problem is solved. The cosine of 87° is 0.052. The Earth–Moon distance represents 0.052 of the Earth–Sun distance, or once inverted, the Sun is 19.1 times farther than the Moon.

Sun is between 18 and 20 times bigger than the Moon. This calculation closes the first part of the treatise.

In the second part, the great challenge is calculating the ratio between the already-known lunar and solar sizes with the Earth's size. Here the second intuition of Aristarchus comes into play.

The comparison with the Earth's size seems to be, in principle, much more complicated because as we are located on its surface, we cannot use its apparent size. However, Aristarchus knows that it is possible to determine an object's size by analyzing its shadow size. The size of an object's shadow depends on several factors: (a) the size of the object, (b) the size of the light source, (c) the distance from the object to the light source, and (d) the distance from the object to the screen on which the shadow is projected. See Figure 1.2.

If we know the shadow's size and three of these four values, we can find the remaining one. Aristarchus knows that, in a lunar eclipse, the shadow produced by the Earth illuminated by the Sun is projected on the Moon's surface. Thus, the Sun is the source of light, and the Moon, the screen. We want to determine the size of the object casting its shadow (in this case, the Earth). We already know the light source's size (the Sun) and distances from the illuminated object to the light source and the screen (the Moon). We also know the shadow's size, which, according to Aristarchus, is equal to two Moons (we will see later how he obtains that size). The Sun is 19 times[15] bigger than

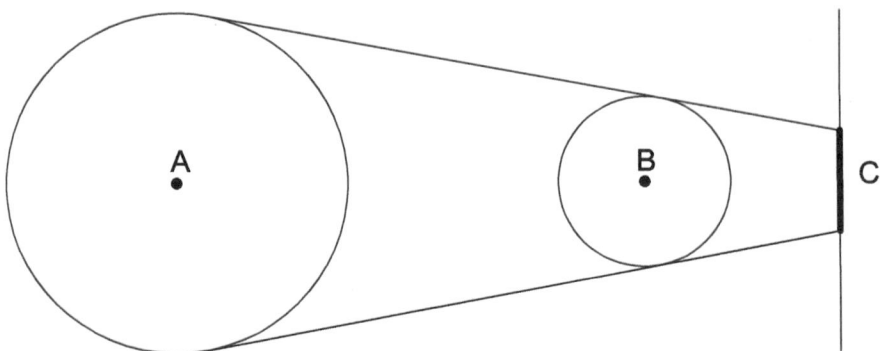

Figure 1.2 The sphere centered in *A* is the source of light that illuminates the sphere centered in *B*; the shadow of *B* is projected on the perpendicular line (which acts as a screen), producing the shadow *C*. The size of the shadow *C* depends on the distances *A–B* and *B–C* and the ratio between the sizes of *A* and *B*. Throughout the analysis, we focus on the umbra, leaving aside the penumbra, which plays no role in Aristarchus's analysis.

15 For the sake of simplicity, from now on we will assume that the ratio between the distances is plainly 19, even if what Aristarchus found is that it is between 18 and 20. We will analyze in detail each limit in Chapter 3.

the Moon, and the Earth is located 19 times farther from the Sun than it is from the Moon. We already have all the data to obtain the Earth's size based on the solar and lunar sizes and distances.

In Figure 1.3, *S* is the center of the Sun, line *SA* represents (approximately)[16] the solar radius, *M* is the center of the Moon, and line *MB* represents a lunar radius. Therefore, *SA* is 19 times larger than *MB*. *MC* is twice *MB* and represents the radius of the Earth's shadow cast on the Moon (which is equal to two Moons). Point *E* represents the center of the Earth. Line *ES* is 19 times larger than *EM* since the Sun is located 19 times farther than the Moon from the Earth. If the center of the Earth is at *E*, its radius must be *ED* to cast the shadow *MC* on the Moon. There- fore, the unknown is the value of line *ED*, representing (approximately) a terres- trial radius. Line *EG* is equal to *MC* (since *SM* and *FC* are parallel). Therefore, *EG* is equal to two Moons. We still have to find *GD* that, added to *EG*, will give *ED*. Triangles *AFC* and *DGC* are similar. Therefore, its sides are proportional. We know that *AF* measures 17 lunar radii (since *SF* measures 2 lunar radii and *AS* 19 lunar radii) and that *FC* measures 20 Moon–Earth distances (since *FG* measures 19 Moon–Earth distances and *GC* 1 Moon–Earth distance). Consequently, *AF/FC* is 17/20. Because triangles *AFC* and *DGC* are similar, *AF/FC* is equal to *DG/GC*. Therefore, *DG/GC* is 17/20 too. But we know that *GC* measures 1 Moon–Earth distance, so that *DG* will measure 17/20 = 0.85 lunar radius. The terrestrial radius, *ED*, is equal to *EG* + *DG*. That is 2.85 lunar radii. Thus, the Earth is 2.85 times

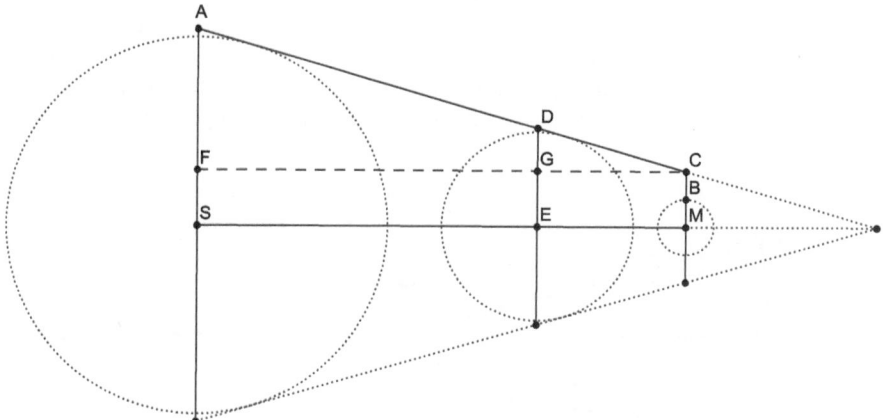

Figure 1.3 Second Aristarchus's intuition: it is possible to obtain the size of the Earth know- ing the size of its shadow projected on the Moon and the distances and sizes of the Sun and the Moon.

16 In a strict sense, line *SA* is a little bigger than a solar radius, and the same happens with line *TD* and the terrestrial radius. For calculation purposes, this is negligible. However, as we will see later, Aristarchus has not overlooked this detail. See p. 220.

bigger than the Moon and (19/2.85 =) 6.66 times smaller than the Sun. We will call this second method *the lunar eclipse method*.

Thus, two very simple intuitions constitute the heart of Aristarchus's *On Sizes*: the fact that the Sun, the Moon, and the Earth form a right triangle when the Moon is in a dichotomy will allow us to know the ratio between the distances and the ratio between the sizes of the Moon and the Sun, and the fact that, in a lunar eclipse, the shadow of the Earth produced by the Sun is projected on the Moon will allow us to know how big the Earth is in relation to the sizes of the Moon and the Sun.

To these two straightforward intuitions, Aristarchus will add an outstanding technical ability to translate them into geometrically analyzable diagrams. Aristarchus's work is even more impressive because the geometrical tools that were available to him were very rudimentary: the Pythagorean theorem and a few inequalities that relate angles and sides of triangles found in Euclid's *Elements*.

Using as input data the lunar elongation from the Sun in a dichotomy (87°) and the ratio between the apparent sizes of the Moon and the Earth's shadow when it is crossed by the Moon (two times the size of the Moon), Aristarchus obtained the lunar and solar distances and sizes.[17]

Figure 1.3 shows these ratios. Horizontal lines represent distances, and vertical lines, radii. On the one hand, *ES* is 19 times *EM*; on the other, *SA* is 19 times *MB*. But there is no defined ratio between the horizontal and vertical lines of the figure, that is, between radii and distances. Thus, this diagram does not allow us to express, for example, the distances in terrestrial radii, that is, the ratio between *ES* or *EM* and *ED*. We need some extra datum that allows us to link the radii with the distances, for example, the solar or lunar apparent radius (which are equal), angle *AES* in Figure 1.3.[18]

In Figure 1.4, *S* is the center of the Sun, *E* is the center of the Earth, and *M* the center of the Moon. Triangle *BHK* represents the cone of Earth's shadow. Line *VU* is the radius of the cone of Earth's shadow at the lunar distance. Angle α, then,

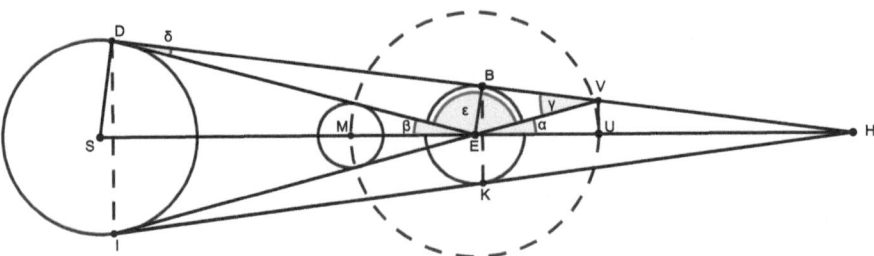

Figure 1.4 Aristarchus's lunar eclipse method.

---

17 Actually, he introduces a third empirical value, the apparent size of the Moon (2°). But as we will see later, it plays a subsidiary role. See next note and p. 194.
18 In fact, Aristarchus offers a value for the solar and lunar apparent sizes and uses it in his calculation, not to express distances in terrestrial radii, but for another reason. See pp. 202–206.

represents the apparent radius of the shadow of the Earth projected on the Moon. Angle β represents the solar and lunar apparent radii. Angle γ represents the horizontal parallax of the Moon, that is, the angular difference of the Moon seen from the center of the Earth (E) and from the point on its surface tangent to the cone (B). Finally, angle δ represents the solar horizontal parallax. The parallaxes depend on distances. Thus, knowing those angles, we can get the distances.

Since the interior angles of a triangle sum to 180°, looking at triangle DEV, we can express angle ε as:

$$\varepsilon = 180° - (\gamma + \delta)$$

But because angles α and β are on axis SEH, it is also true that both added to ε sum to 180°. Therefore, we also know that:

$$\varepsilon = 180° - (\alpha + \beta)$$

And consequently:

$$(\alpha + \beta) = (\gamma + \delta)$$

Therefore, we know that the sum of the apparent radii of the Moon (β) and the shadow (α) is equal to the sum of the lunar (γ) and solar (δ) horizontal parallaxes. Knowing only the apparent radii (α and β), we still have two unknowns, so some more data is necessary to solve the equation. In the case of Aristarchus, the additional datum is the ratio between the distances. For other authors, as we will see later, it is one of the distances expressed in terrestrial radii.

The horizontal parallaxes allow us to express the distances in terrestrial radii. Indeed, in the right triangle EBD, EB is a terrestrial radius, and ED, the hypotenuse, is similar to the Earth–Sun distance (which would strictly be ES). Thus, the inverse sine of the horizontal parallax will give us the distance expressed in terrestrial radii (ED as a function of EB).

Showing the relationship between parallaxes and apparent sizes in a lunar eclipse is the usual way of introducing Aristarchus's calculation.[19] It has the great advantage of making explicit the continuity between his calculation and that of later authors who will aim to obtain the distances expressed in terrestrial radii. Therefore, it will help us in introducing the following authors. But it is important to remember that this presentation, when strictly applied to the calculation that Aristarchus performed, is a bit unnatural and forced. Aristarchus does not pretend to obtain distances expressed in terrestrial radii. Although the lunar apparent radius plays a role, it is not the same role that it plays in this reconstruction. We will see Aristarchus's calculation in detail in Chapter 3.

---

19 See, for example, Evans (1998: 69–74).

Aristarchus's methods in Antiquity and the Middle Ages

Unlike Aristarchus, there are no extant works of Hipparchus of Nicaea (190 BC–120 BC) related to the calculation of the distances and sizes of celestial bodies. Nevertheless, we have descriptions of the calculations and values he proposed for the solar and lunar distances from Ptolemy (*Almagest* V, 11; Toomer 1998: 243–244) and Pappus of Alexandria (Toomer 1974: 126–127). These references are brief and obscure. Consequently, the reconstructions of the calculation proposed by scholars are inevitably speculative. But we can get at least some clear ideas from Ptolemy's and Pappus's texts. In the first place, Hipparchus wrote a lost book calculating the solar and lunar distances and sizes, titled *On Sizes and Distances*.[20] Hipparchus calculated the lunar distance in two steps. As a first step, he assumed that the Sun is at an infinite distance, and then employing two different methods, he calculated the maximum and minimum limits of the lunar distance. The maximum distance from the Moon would be 83 terrestrial radii (tr), and the minimum, 54.5 tr. Then, in a second step, by combining both methods, he calculated the actual distance of the Sun and the Moon rather than only their limits: the Sun would be at 490 tr, and the Moon at 67.33 tr. As far as we know, one of the two methods used by Hipparchus was introduced by him. The method is based on the observation of a solar eclipse from two different locations.[21] The other method is Aristarchus's lunar eclipse method, but Hipparchus introduces his own values for the apparent size of the Moon (0.55°) and that of the Earth's shadow (2.5 Moons).[22] Furthermore, instead of assuming a ratio between the distances, he directly takes the Sun to an infinite distance.[23]

Claudius Ptolemy[24] (ca. AD 100–AD 170) offers values for the solar and lunar distances in three different works, which, arranged chronologically, are the *Canobic Inscription*, the *Almagest,* and the *Planetary Hypotheses*.[25]

The *Canobic Inscription* (Jones 2005)[26] is included in some medieval manuscripts after the *Almagest*. It is a small work consisting mainly of a list of values with astronomical significance. According to its first lines, it is an inscription that Ptolemy himself erected on a monument in honor of the Savior God in Canopus.

20 It is likely that the name is not complete and this is the same work mentioned by Theon and Calcidius (Toomer 1974: 126, n.1) as *On the distances and sizes of the Sun and Moon*. It would have, in this case, the same title as Aristarchus's work.

21 On a simpler version of the argument proposed by Capella, see Carman (2017).

22 We know both values thanks to Ptolemy's *Almagest* (IV, 9; Toomer 1998: 205).

23 Hipparchus's calculation is analyzed by Swerdlow (1969), Toomer (1974), and Carman (2020a).

24 Boll (1894) and Fisher (1932) are the most informative biographies and the source for all the others. The biographies in Pedersen (1974: 11–13) and Neugebauer (1975: 834–836) are also useful. A good biography, more complete and up-to-date than the first two, can be found in Toomer (1975: 186–206).

25 Neugebauer (1975: 834) asserts that the *Almagest* is probably Ptolemy's earliest work, but Hamilton, Swerdlow, and Toomer (1987) have argued strongly that the *Canobic Inscription* is earlier.

26 For more details on the history of this text and its analysis, see Jones (2005: 53–55) and the bibliography referenced there.

Among many other values, it includes the solar and lunar distances, the apparent sizes of the Moon (and of the Sun, which is the same), and the apparent size of Earth's shadow. According to the inscription, the apparent diameter of the Sun and the Moon is 0.55°, that of the shadow is 1.38° (very close to 2.5 Moons), the lunar distance is 64 tr, and the solar distance is 729 tr. In the inscription, there are no traces of the calculation used for obtaining the distances. Still, the record of the apparent diameters of the Moon's and the Earth's shadow, right next to the distances, strongly suggests that he applied the lunar eclipse method to obtain the distances. And indeed, if the method is applied by introducing the apparent diameters and one of the distances, the other is obtained quite accurately (see Hamilton et al. 1987).[27] This coincidence shows that the application of the method is consistent with the set of values, but it does not reveal which of the data he used as input and which as output. As we already mentioned, besides the two apparent sizes, one needs to introduce some extra datum to apply the method. Likely, Ptolemy introduced the lunar distance obtained by another method and got the solar distance. At least, this is the procedure that he follows in the *Almagest*.

In Book V, Chapters 13 to 16 of the *Almagest* (Toomer 1998: 247–257), Ptolemy calculates the lunar and solar distances and sizes. He uses the lunar eclipse method, but unlike Hipparchus, who introduced an infinite solar distance to obtain the lunar distance, Ptolemy introduces the lunar distance to obtain the solar distance. The rationale for this change of strategy is simple: since the Moon is close enough to the Earth, Ptolemy can measure the lunar parallax. In turn, the Sun is not close enough to measure its parallax. Now, by measuring the lunar parallax, he can obtain the lunar distance. Then, Ptolemy uses the lunar eclipse method to calculate the solar distance.

However, it should be noted that Ptolemy's strategy has a significant disadvantage in the way he uses the lunar eclipse method. The method is such that the lunar distance value is very stable, reacting almost indifferently to even big changes in the solar distance. Hipparchus cleverly exploits this characteristic when he presumably uses the lunar eclipse method for obtaining the lunar distance, assuming an infinite solar distance. Assuming an infinite solar distance has almost no effect on the calculation of the lunar distance. But if, like Ptolemy, one uses the lunar distance to obtain the solar distance, the solar distance is enormously sensitive to small changes in the lunar distance. Thus, the result will not be very reliable.[28] As

---

27 The suggestion that the author of the *Canobic Inscription* used the lunar eclipse method does not appear for the first time in Hamilton et al. (1987), but some almost 20 years earlier as an appendix to Swerdlow (1969: 301–303), and even earlier in his PhD thesis defended a year earlier (Swerdlow 1968). However, there Swerdlow suggests that since the calculation takes the maximum distance of the Moon from Ptolemy and the apparent diameters of the shadow and the Moon at the mean distance, the author might not be Ptolemy who would not make such a mistake. The calculation, then, is incorporated with slight variations in Hamilton et al. (1987), of which Swerdlow is a coauthor.

28 Abers and Kennel (1975) analyze the nature of this error.

we will see, this weakness of the method will generate significant criticism during the early modernity period.

In a lunar eclipse, parallax does not interfere, meaning, that the eclipse is seen almost identically from every possible place on Earth where it is visible. This is why Ptolemy derives the parameters of his lunar model exclusively from lunar eclipses. This strategy allows him to assume that his model gives the Moon's position as measured from the center of the Earth. Thus, any difference between the value observed by him on the Earth's surface and the value calculated by the model from the center of the Earth will be due to horizontal parallax. Ptolemy calculates and observes the lunar position on a particular day and obtains a parallax implying a maximum lunar distance of 64.16 tr. Then, from Babylonian records of a couple of ancient eclipses, he gets new values for the apparent size of the Moon (0.52°) and of the Earth's shadow (2.6 Moons) when the Moon is at maximum distance. He applies the lunar eclipse method with these values and finds that the Sun is 1,210 tr from Earth. Therefore, the ratio between the solar and the lunar distance is (1,210/64;10 =) 18.86. Curiously, this value is within the limits that Aristarchus had established centuries before.[29] It is important to note that Ptolemy made his calculations using trigonometry, which emerged sometime between him and Aristarchus, probably before Hipparchus. The use of trigonometry greatly simplified the calculations, freeing Ptolemy from having to enclose the desired value between a maximum and a minimum limit, as Aristarchus did. Ptolemy solves in one page a calculation that took Aristarchus practically half a book.

The *Planetary Hypotheses* is a short book written by Ptolemy after the *Almagest*. It has a more cosmological than mathematical approach. The calculation of the distances offered in it is different from the previous ones, and it seems that Ptolemy was the first to propose it. All the methods that we have described so far are based on empirical data and geometry and can therefore be considered as belonging to mathematical astronomy. Instead, the method in *Planetary Hypotheses* uses two cosmological principles more linked to metaphysics than to mathematical astronomy. First, Ptolemy argues that nature does nothing in vain, and second, he asserts the famous Aristotelian principle of *horror vacui*, that is, that there is no vacuum. From the *Almagest* data, Ptolemy can find each planetary orbit's width, but not its distance from Earth. He also assumes a specific order of the planets according to their proximity to the Earth: Moon, Mercury, Venus, Sun, Mars, Jupiter, and Saturn. If there is no vacuum and nature does nothing in vain, the orbit of Mercury must necessarily begin where the lunar orbit ends. Otherwise, there would be an empty or at least useless space between both orbits.

Thus, applying these cosmological principles allows Ptolemy to conclude that Mercury's minimum distance must be equal to the maximum lunar distance he has

---

29 Several authors (Newton 1977: 199; van Helden 1986: 19; Dreyer 1953: 184–185; Hartner 1980: 24 and 1964: 255; Evans 1998: 73) have pointed out this coincidence. Some have even risked that Ptolemy would have forced the data to stay within Aristarchus's limits (Newton 1977: 199), but see also Carman (2008: 207, note 5) and Hartner (1980: 24).

already obtained by parallax. Knowing the width of an orbit and its minimum distance, he can get its maximum. In this way, Ptolemy knows Mercury's maximum distance, which in turn will coincide with Venus's minimum distance. Then, knowing the width of Venus's orbit, he can get its maximum distance. Now, Venus's maximum distance should coincide with the Sun's minimum distance. So Ptolemy obtains the solar distance. But Ptolemy had already calculated the solar distance by the lunar eclipse method in the *Almagest*. Interestingly, both values are very close, leaving a small gap of a few terrestrial radii between the maximum distance from Venus and the minimum from the Sun calculated in the *Almagest* (see Carman 2009). See Figure 1.5.

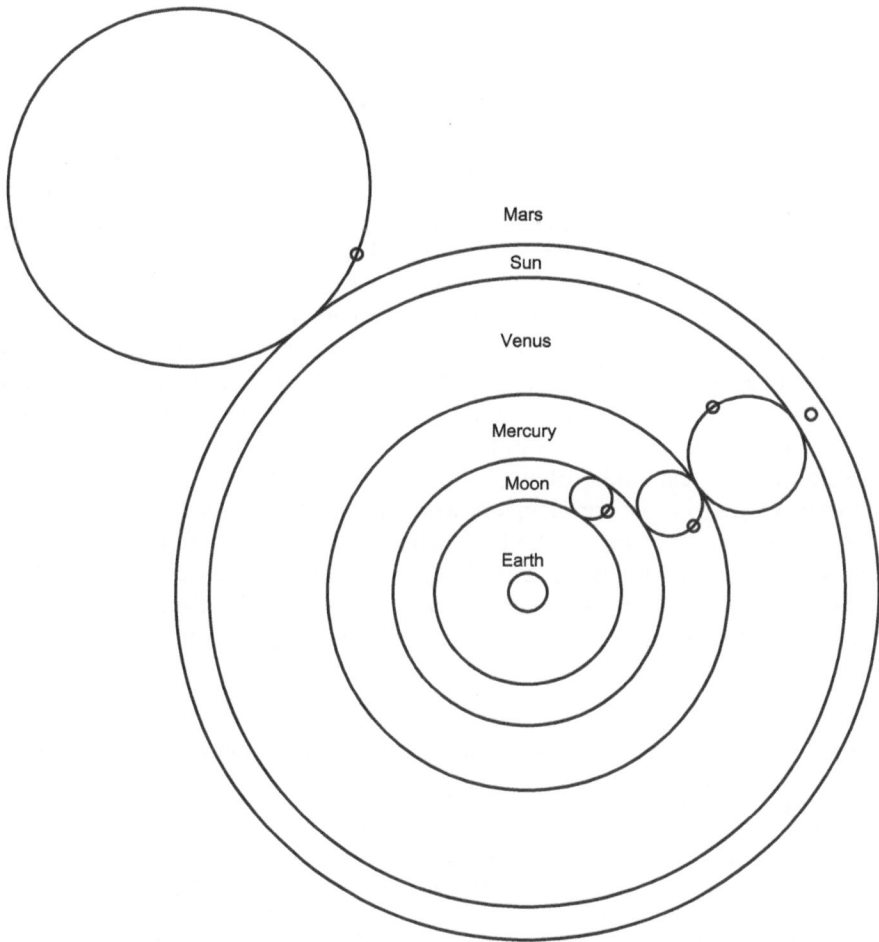

Figure 1.5 Order of the planets according to Ptolemy's Planetary Hypotheses.

But the *Planetary Hypotheses'* calculation does not stop at the solar distance. Ptolemy continues applying the method and establishes the distances of Mars, Jupiter, and Saturn. Saturn's maximum distance is equal to that of fixed stars, that is, the size of the universe itself: 19,865 tr (Goldstein 1967: 8).

We have no record of further calculations of the distances of the Sun and the Moon in Antiquity after Ptolemy. Besides Aristarchus, the two authors who calculate distances – Hipparchus and Ptolemy – use the lunar eclipse method, although with different approaches. But neither of them uses or even mentions the dichotomy method. In fact, this method was ignored for centuries, and it only re-emerged in the early modern period when astronomers began to question the precision of the lunar eclipse method. Ptolemy's second method of calculating the distances based on the contiguity of the spheres became the preferred method in Arabic astronomy and therefore dominated the calculation of distances throughout the Middle Ages.

Indeed, in Arabic astronomy, the calculation of the distances of the celestial bodies aroused a lively interest and some originality in the attempts to solve the difficulties presented by Ptolemy's method of the contiguous spheres. For example, several attempts try to solve the small gap left between Venus's maximum distance and the Sun's minimum distance. Other attempts incorporate the planet's diameters that should be added to the orbit's radii so that the calculations are more realistic, although the resulting modification of the values is negligible.[30] In general, the corrections introduced by Arabic astronomers do not modify the universe's size on a large scale, except that of Mu'ayyad al Din al-Urdi, who, in his treatise *Kitab al-Hay'ah*, establishes that the universe is approximately seven times larger than Ptolemy's. Al-Urdi notes that Venus's orbit did not fit between Mercury's and Sun's orbits, and decides to place it beyond the Sun. This switching of the Sun's and Venus's orbits causes a domino effect on the other orbit's distances and, as a consequence, enlarges the size of the universe considerably (see Goldstein and Swerdlow 1970 and Saliba 1979).

However, one Arabic astronomer deserves a separate paragraph because he used Aristarchus's lunar eclipse method: Al-Battani (858–929).[31] In his most influential work, *Kitāb al-Zīj* (Treatise on Astronomy),[32] he double-checks each of the values that Ptolemy uses in the *Almagest* for calculating the solar distance. He finally modifies only the lunar distance, to almost 60 tr. With this new value for the lunar

---

30 For the leading Arabic astronomers who have made contributions to Ptolemy's method of contiguous spheres, see Swerdlow (1968: 129–212), van Helden (1986: 28–40), and Dreyer (1953: 257–264). On how to explain some variations in the value for the volume of Venus found in some Arabic astronomers, see Carman (2008: 232, note 41).

31 For a biography of Al-Battani, see Hartner (2008). For the calculation used, see Swerdlow (1973) and the pages devoted to this subject in his doctoral thesis (Swerdlow 1968: 88–95), which constitute the basis of Swerdlow (1973).

32 The best translation is into Latin, made at the beginning of the twentieth century: Nallino (1899–1903). Hartner (2008) spares no praise when he says that it will always remain a masterpiece in the history of science.

distance, the lunar eclipse method would give an absurd solar distance, close to 340,000 tr. (Note how modifying the lunar distance by just 4 tr produces an enormous change in the solar distance. This is an excellent example of the instability of the lunar eclipse method for calculating the solar distance using the lunar distance mentioned earlier.) But Al-Battani does not apply the lunar eclipse method to obtain the solar distance. Instead, he assumes the ratio between the solar and lunar distances followed from Ptolemy's calculation (18.86) and multiplies it by his lunar distance to calculate the solar distance, obtaining a value of 1146 tr. Needless to say, this approach is absolutely inconsistent, for it assumes at the same time two different values for the lunar distance: Ptolemy's and his own.[33]

Under the influence of Arabic astronomy, popularized through Latin translations in the mid-thirteenth century, the distances obtained from the contiguous spheres method became part of the common culture. Many traces of this can be seen even in the literature (van Helden 1986: 37–40). But because this method assumes geocentricism, it was inevitably abandoned after the Copernican revolution.

The calculation of the distances and sizes in the early modern period

Undoubtedly, the *De Revolutionibus Orbium Coelestium* of Nicolás Copernicus (1473–1543) begins a revolution in astronomy and, with it, in virtually all fields of knowledge (see Kuhn 1957). Hence, it seems curious that the book itself, far from revolutionary, submissively follows not only the general structure of the *Almagest* but also many of their calculations and even their values. The case of the solar distance is paradigmatic. Janice Henderson (1991) has shown by carefully analyzing Copernicus's manuscripts that he modified the values more than once and probably faked some observations to obtain a value close to Ptolemy's and Al-Battani's.

By establishing the revolution of all the planets around the Sun, Copernican heliocentrism introduces a significant novelty in calculating the planetary distances. It is now easy to express all the planetary distances as a function of the Earth–Sun distance. However, there is no novelty in the methods used by Copernicus to obtain the Earth–Sun distance. With Ptolemy's method of contiguous spheres discarded, the methods proposed by Aristarchus gain prominence: not only that of the lunar eclipse, but also that of the dichotomy, which, after centuries of abandonment, is once again seriously considered by the astronomers of the seventeenth century.

To calculate the Earth–Sun distance, Copernicus uses the lunar eclipse method. The calculation he offers, however, is frankly disappointing. In Chapters 17, 18, and 19 of Book IV of the *De Revolutionibus*, Copernicus undertakes the task of recalculating the Earth–Sun distance. Like Ptolemy, he introduces the lunar distance in the lunar eclipse method and obtains the solar distance. He also follows Ptolemy in the way of getting the input values of the calculation. In Chapter 17, he calculates the lunar parallax and obtains the lunar distance. In Chapter 18, using a pair of lunar eclipses, he obtains the apparent sizes of the Moon and the Earth's

---

33 This distance exactly coincides with a parallax of three minutes, as van Helden (1986: 31) notes.

shadow. Finally, in Chapter 19, he deduces the Earth–Sun distance, following the Ptolemaic method step by step. In fact, Chapters 17, 18, and 19 of Book IV correspond one to one with Chapters 13, 14, and 15 of Book V of the *Almagest*. The values obtained by Copernicus are not very different from those of Ptolemy, but the observations used are much more problematic or even perhaps directly faked (see Henderson 1991: 41–46, 91). In any case, when making the calculation, Copernicus introduces other values without justification. Finally, Copernicus assumes the value of 1,142 tr for the solar distance. Much to the amusement of Aristarchus, had he been alive, since for Copernicus the mean lunar distance in dichotomy is 60.30 tr, the ratio between the solar and lunar distances is 18.93. Eighteen centuries later, Aristarchus's ratio is still valid!

Erasmus Reinhold (1511–1553) produced the Prutenic Tables based on Copernicus's new theories of planetary motion. To build the tables, Reinhold undertook a very tedious task. Aware of the significant inconsistencies that many calculations of the *De Revolutionibus* presented, he proposed redoing all of them. This task was required because it was the only way to have a trustable basis for the new tables. All his calculations are found in his commentary to *De Revolutionibus*.[34] Usually, he redoes the calculations but does not discuss the input values. However, the section devoted to calculating the solar distance does not limit itself to recalculating the results from Copernicus's calculations, but it makes new calculations using different data. It seems that Reinhold was not satisfied with the distance obtained by Copernicus. Indeed, Reinhold's manuscript shows four successive attempts to obtain the solar distance. In all of them, he applies the lunar eclipse method to the same values for the apparent size of the Moon and the Earth–Moon distance. He exclusively modifies the shadow's apparent size until obtaining a value close enough to that of Ptolemy. He finally gets a distance of 1,208 tr.

Johannes Kepler wrote a small work titled *Hipparchus or Sciametria*. He mentions it in several works and letters, but he never published it (Kepler 1620: 874). However, we can partially reconstruct its content from the manuscript that has been preserved (Frisch 1858–1870, 3, and Swerdlow 1992). In *Hipparchus*, Kepler shows his extraordinary mathematical skill by considerably simplifying the mathematics of the lunar eclipse method. Kepler finds for the first time the equation establishing that the sum of the apparent radii of the Moon and the shadow is equal to the sum of the solar and lunar diurnal parallaxes. That is:

$$(\alpha + \beta) = (\gamma + \delta)$$

---

34 The commentary was never published. The manuscript is in the Deutsche Staatsbibliothek at Berlin. The commentary is titled *Commentarius in opus revolutionum Copernici* (Books III and IV) and is on the first folios, ff. 1–63; Latin 4° 32. It was carefully analyzed by Henderson (1991), on which we based our discussion.

The explanation that we have offered a few pages earlier basically follows Kepler's (Frisch 1858–1870, 3: 520–522). In his *Astronomia Nova* (XI; Kepler 1937, 3: 225; Donahue 2015: 159), he says that, although he cannot give a precise value for the solar distance yet, in his *Hipparchus* he will prove (presumably by applying the lunar eclipse method) that it is between 700 tr and 2,000 tr (Swerdlow 1992: 25). Unfortunately, the calculation is not extant in the preserved part of the book.

On the other hand, Kepler makes a new application of the lunar eclipse method. Aristarchus introduced the ratio between the distances to obtain the size of the Earth. Hipparchus assumed an infinite solar distance to obtain the lunar distance. Ptolemy had introduced the lunar distance to obtain the solar distance. In turn, Kepler assumes an infinite solar distance and a determined lunar distance to obtain the radius of the shadow, which he needs to predict eclipses (Frisch 1858–1870, 3: 550–643).

But around 1617, Kepler began to lose confidence in the lunar eclipse method and started to explore other alternatives. It seems that after giving up on the lunar eclipse method, Kepler restored Aristarchus's dichotomy method. Indeed, in point 19 of the 1617 ephemeris, he clarifies that in the tables, he will not use the ratio of the distances that he had previously established but will assume the ratio 23/1, which corresponds to an elongation in a dichotomy of 87.5° (Frisch 1858–1870, 7: 486). In the 1619 ephemeris, Kepler manifests his hope that further observations made with telescopes would make it possible to detect the moment of dichotomy more accurately. The exact time of dichotomy is a condition for obtaining a more accurate value for the lunar elongation. But a few years later, he would be disappointed. In the 1626–1628 ephemerides, he asserts that the great irregularity of the Moon's surface makes it impossible to accurately determine the moment of dichotomy: while some parts of the line that divides the dark from the light part look concave, others look convex. He believes that even if it is useful to show that Aristarchus's elongation of 87° is not correct (Frisch 1858–1870, 7: 528), the observations are not accurate enough to determine the ratio.

By discarding the two methods of Aristarchus, Kepler had no way of measuring the Earth–Sun distance. However, this, far from discouraging him, led him to explore possible mathematical harmonies that would allow him to determine planetary distances. In 1619, Kepler published *Harmonice Mundi*. After many years of carefully studying Tycho Brahe's observations – 17 years according to Kepler himself – he was finally able to formulate his famous third law. It states that the square of a planet's orbital period is directly proportional to the cube of its distance from the Sun (Gingerich 1975). However, neither this nor any other Kepler's laws allow us to calculate the Earth–Sun distance in terrestrial radii.

Over time, the lunar eclipse method was increasingly discredited due to its instability when used to obtain the solar distance, and astronomers began to explore the dichotomy method again. However, several authors also tried to get the Earth–Sun distance from certain harmonic relations. Some astronomers, like Boulliau, still insisted on applying the lunar eclipse method; others, like Wendelin, Riccioli, and Flamsteed, applied the dichotomy method, while others, like Horrocks and Huygens, criticized it.

In his *Astronomia Kepleriana, defensa e promota* (Horrocks 1673), Jeremiah Horrocks (1618–1641) argued that neither of Aristarchus's methods are useful. At first, he had hoped that the dichotomy method might be helpful, but then he was disappointed. Indeed, Horrocks had suggested to his colleague William Craetrius that by analyzing the elongation at dichotomy, the solar parallax could be corrected. Still, in a letter from the end of April 1640, he tells him that, after analyzing the dichotomies of this month, he realized that the method would not be successful because of the irregularities of the surface of the Moon, the same problem that Kepler had found (Horrocks 1673: 333).

Horrocks dedicated the whole fifth *disputatio* of *Astronomia Kepleriana, defensa e promota* (Horrocks 1673: 102–167) to analyzing and criticizing the lunar eclipse method when applied to calculating the solar distance.[35] Much of the *disputatio* aims to show that Philippe Lansbergius (1561–1632) is wrong in boasting that he has found a trustable solar distance, using geometry and observations exclusively, and not harmonies like Kepler or mystical numbers like Tycho (Horrocks 1673: 102). Indeed, in *Uranometria* (1631), Lansbergius obtains a solar distance of 1,550.86 tr, applying the lunar eclipse method to his own observations. But Horrocks shows that it is impossible to determine the values required to use the method with enough accuracy to obtain a stable result (Horrocks 1673: 156).

The Belgian priest Gottfried Wendelin (1580–1667) based his solar distance calculations on the dichotomy method. In *Loxias seu de Obliquitate Solis Diatriba* (1626), he argues that, in June 1625, he observed that the Moon did not deviate more than 1° from quadrature at the time of the dichotomy. Consequently, the Sun was about 60 times farther than the Moon, and since the lunar distance is roughly 60 tr, the solar distance will be 3,460 tr (Wendelin 1626: 10). However, a few years later, he proposed a much greater solar distance: 14,720 tr. It is likely that he had appealed to harmonic reasons in determining this new value. However, the value would still be based on the dichotomy method, since, after careful analysis, he concluded that the difference between the dichotomy and quadrature is, at most, only 15′ (Riccioli 1651, 1: 109; van Helden 1986: 113).

Another priest, the Jesuit Giovanni Battista Riccioli (1598–1671), analysed the distances in his *Almagestun Novum*. He went over all the known values and added his own. According to him (1651, 1: 106–109), there are three methods: the direct measurement of the solar parallax (as has been done for the Moon) and the two Aristarchian methods. But the parallax method and the lunar eclipse method must be discarded. The first because the solar parallax is too small to be detected. The second because it requires such an accuracy in the input values that it is also inapplicable. The only possible method, therefore, is the dichotomy method. Riccioli's idea is to observe the Moon near a quadrature, when it is almost on the ecliptic, and to record both the moment in which one believes that it begins to be in

---

35 This work has been edited by John Wallis from a series of manuscripts from the last years of Horrocks. Wallis is the editor of the *editio princeps* of Aristarchus's *On Sizes*, as we will see later. See p. 27.

dichotomy and the moment in which one believes that it ends up being so. The exact moment of the dichotomy will be the average of the two (Riccioli 1651, 1: 108–109). He made the observations on two different occasions, on October 16, 1646, and on June 13, 1648, obtaining a difference of 30'15" and 31'34" with the quadrature. The calculated solar distances are close to 7,300 tr, the value he assumes.

But although the lunar eclipse method had lost advocates, it had not entirely disappeared. For example, Ismaël Boulliau (1605–1694), a very influential European astronomer, established in his *Astronomia Philolaica* (1645) the value of 1,485.93 tr for the maximum solar distance, using the lunar eclipse method. The input values, however, do not seem to be obtained by observations. Indeed, by analyzing eclipses, he obtains values for the lunar distance and the apparent sizes of the Moon and the shadow. Then, applying the method to those input values, Boulliau shows that they lead to an absurd result (1645: 192–193). Once he showed that one could not rely exclusively on observations, he chose without giving a justification values close to those obtained by observation, which implies a solar parallax of 2'24". Although he does not mention it, this parallax is Kepler's value in *Astronomia Nova*. As expected, he obtains a value for the solar distance of 1,459.9 tr, consistent with the solar parallax assumed (1645: 196).

According to Christiaan Huygens (1629–1695), by 1659, a sufficiently accurate method for measuring the solar distance has not yet been found. In effect, he asserts that "whether [astronomers] try to discover it using eclipses or by the dichotomies of the Moon, it can easily be shown that these efforts will be in vain" (Huygens 1888–1950, 15: 342–345). The only current methods were, still, those of Aristarchus. And this will not change until the astronomers can measure the parallax of some planet.

Giovanni Cassini (1625–1712) is often considered the first astronomer to effectively measure the parallax of a planet. Although his final result is certainly acceptable, it was likely due more to an accurate selection of data than to a correct measurement of the parallax. With his measurement methods, the margins of error were equal to or greater than the parallax he wanted to measure. Using simultaneous measurements made in France and Cayenne, Cassini obtained that the parallax of Mars, in opposition, was 25'33". As we mentioned, in a heliocentric framework, all the heliocentric distances are linked. Consequently, Cassini can obtain the Earth–Sun distance from Mars–Earth distance at opposition. The parallax of Mars obtained by Cassini implies a solar distance of 21,600 tr.

The observations took place in 1672, a year that could be considered the date of death of Aristarchus's methods. John Flamsteed (1646–1719) is also considered, with Cassini, the first astronomer to measure Mars's parallax, since he obtained very similar values. However, in May of the previous year, there were still attempts to apply the dichotomy method. In a letter to Henry Oldenburg, Flamsteed argued that the dichotomy method could still be useful if employed carefully (van Helden 1986: 135; Hall and Hall 1953: 8, 47).

Although Cassini and Flamsteed had agreed on the result, the scientific community had not immediately reached a consensus. For several decades, new measurements and calculations gave different values. All of them, however, implied always a solar parallax bellow 15″ and, consequently, a solar distance greater than 13,750 tr, that is, ten times greater than the values accepted from Ptolemy to Kepler. Therefore, the dimensions of the solar system had definitively broken with tradition. Van Helden (1986: 155) rightly points out that it must have been very shocking to eighteenth-century astronomers who, having been trained in the dimensions of Ptolemy, had discovered in their maturity that the universe is ten times greater in size.

Edmund Halley (1656–1742), who assisted Flamsteed in many observations, was one of the strongest opponents of Cassini's value, insisting that, although Mars in opposition is much closer than the Sun, the observation of this phenomenon implies inevitable mistakes that vitiate the results. The best opportunity to definitively measure a parallax, Halley pointed out, was reserved for the astronomers of the following century, when Venus passes over the solar disk, which would only occur on May 26, 1761 (Halley 1679: 4). Thus, Halley predicted not only the return of a comet but also the end of a very long effort of humanity. He had predicted that the comet bearing his name would return in 1758. When this happened, his fame grew even more, and many astronomers were observing Venus in transit a few years later. Indeed, the expeditions to measure the transits of Venus in 1761 and 1769 brought the definitive end to the effort initiated by Aristarchus some 20 centuries earlier.

History of the text and introduction to the translation

In this section, we first review the history of Aristarchus's text, briefly describing each edition, and then offer the characteristics of our translation.

History of the text and description of the editions

Aristarchus's *On Sizes* is one of the few pre-Ptolemaic books on astronomy that have come down to us complete (or almost complete). In Antiquity and the Middle Ages, copying a text required a great effort in time and resources. Thus, copyists stopped copying books when they were no longer interesting and, consequently, they usually got lost. Too often, competing with the work of a great author, those of previous writers were not read any longer, with the sad consequence of their irreparable loss. The paradigmatic case is that of pre-Socratic philosophers, eclipsed by the genius of Plato and Aristotle. In general, the little we know about them comes from what Plato and Aristotle transmitted to us in their works. Something similar happened in the field of astronomy with Claudius Ptolemy. His great work of astronomy, the *Almagest*, organized and surpassed all previous astronomy in such a brilliant way that earlier works became superfluous. Like in the case of the pre-Socratics, much of what we know about pre-Ptolemaic astronomers comes from Ptolemy's remarks or commentaries on Ptolemy's works. The paradigmatic case is Hipparchus. Although an important and influential astronomer, he pales before Ptolemy, and consequently, almost nothing of his astronomical works have survived.

Fortunately, in astronomy, there has been an exception. In Late Antiquity, the astronomical curricula included a series of nine or ten short treatises used to prepare the study of the *Almagest*. Consequently, this set of books continued to be copied even after the *Almagest*. The collection was known as "The Little Astronomy," probably in opposition to "The Great Astronomy," as the *Almagest* was known. The collection includes Autolycus of Pitane's *On the moving sphere* and *On risings and settings*; Euclid's *Optics* and *Phenomena*; Theodosius of Bithynia's *The Sphaerics, On habitations,* and *On days and nights*; Hypsicles's *On Ascensions*; probably also the Menelaus's *The Sphaeric*; and Aristarchus's *On Sizes and Distances of the Sun and Moon*.

Noack 1992 is the only study on the Greek manuscripts of *On Sizes*. The author identified and analyzed 30 manuscripts, going from the ninth to the seventeenth century. The oldest manuscript, presumably written in the ninth century, is known as *ms. Vaticanus Graecus 204* (Vat.Gr.204, for us, ms. A).[36] Along with other scientific works, the copy includes the treatises of "The Little Astronomy." The codex also contains several scholia. The manuscript is excellent and seems to be the source of all the others. So superior is it to the others that Heath (1913) decided to prepare his critical edition based exclusively on it. One single hand wrote the manuscript, and another drew all the diagrams. The diagrams, drawn in red, are clear, neat, and precise. We have reproduced them in our translation of the treatise and scholia.

Noack (1992: 45–47) enumerates three opportunities during the Middle Ages when the book could have been translated into Latin or perhaps was actually translated, but the translation had not survived. Gerard of Cremona (1114–1187) was a very fruitful translator at the Toledo School of translators. Many of his translations are preserved, but not that of Aristarchus's *On Sizes*. However, in the ninth place of a complete list of his translations included in a Parisian manuscript, we find the *On Sizes* title. Accordingly, Noack suggests that there are reasons to think that he, Gerard, translated it. Second, *On sizes* could have been translated in the Sicilian translation campaign of the twelfth century. In fact, during that campaign, some scientific works were translated directly from *ms. A*, which we also know contained Aristarchus's treatise. The third and last chance was in the thirteenth century. William of Moerbeke (1215–1286), the most prolific translator of Greek works in the Middle Ages, was chaplain and confessor of Popes Urban IV and Clement IV during their entire papacy. It is known that among the manuscripts that Charles I of Anjou gave to Clement IV, on the occasion of the victory of the Battle of Benevento in 1266, was *ms. A*. William translated many scientific works. There are no reasons to think that he did not do it with Aristarchus's.

* * *

The oldest preserved Latin translation of *On Sizes* is not a manuscript but a book printed in Venice in 1498 by Simonem Papiensem Bevilaquam. The translator is Giorgio Valla (1447–1500), and the volume collects some of his translations,

---

36 See Appendix 2 at p. 297 for a list of the manuscripts.

including works by Galen, Euclid, Aristotle, and the edition of some of his own works.

It is titled Giorgio Valla Placentino Interprete. Hoc in volumine hec continetur: Nicephori Logica, Georgij Valle libellus de argumentis, Euclidis quartusdecimus elementorum, Hypsiclis interpretatio eiusdem libri euclidis, Proclus de astrolabo, Aristarchi Samij de magnitudinibus et distantijs solis et lune, Timeus de mundo, Cleonidis musica, Eusebii pamplhili de quibysdam theologicis ambiguitatibus, Cleomedes de mundo, Athenagore philosophi de resurrectione, Aristotelis de celo, Aristotelis magna ethica, Aristotelis ars poetica, Rhazes de pestilentia, Galenus de in equali distemperantia, Galenus de bono corporis habitu, Galenus de confirmatione corporis humani, Galenus de presagitura, Galenus de presagio, Galni introdutorium, galenus de succidaneis, Alexander aphroditeus de causis febrium, Pselus de victu humano.[37]

The translation is dedicated to the Venetian nobleman Giovanni Baduaro. Valla points out in the proem that, despite the most interesting mathematical arguments put forward by excellent scholars, *longe omnium pulcherrima de solis lunaeque magnitudine Samii Aristarchi traditio est* (Aristarchus of Samos's treatise *On the size of the Sun and the Moon* is by far the production most beautiful of all), combining science with compelling subjects.

Giorgio Valla was born on January 1, 1447, in Piacenza, the son of Andrea Valla and Cornelia Corvini. He began his education in his hometown. At the age of 15, he moved to Milan, where he studied Greek and where, for a time, he was tutor to Francisco Sforza's son. Later, he went to Pavia, where he deepened his knowledge of Greek and studied mathematics. It was then that he met Leonardo da Vinci. In 1485, he was in Venice, working on his main work, *De expetendis et fugiendis rebus*, published *post mortem* by Giovanni Pietro Valla Cademosto, Valla's adopted son, in the workshop of the humanist Aldus Manutius in 1501. It is an excellent encyclopedia in which, among other topics, it explains various planetary models. Doctor, astronomer, Latin and Greek philologist, and musicologist, it is in this work where Valla's broad culture is revealed. With his lectures on Greek thinkers and his translations by authors such as Aristotle, he contributed to arousing the interest of his contemporaries in ancient science. Valla died in Venice on January 23, 1500, and was buried in the church of *Santa María della Carità*.

The innumerable reprints of Valla's works tell us of its popularity, but his translations were not exempt from adverse judgments. For example, Robert Balfour, author also of a Latin version and comment to *De motu circulari corporum*

---

37 Heath (1913: 321, n.3) mentions that the 1498 edition was a reprint of a publication ten years earlier and refers to the *Bibliotheca Graeca* by Fabricius (1795, IV:19). In the entry of *Bibliotheca Graeca*, the date of publication is correctly indicated as 1498. The text includes a footnote with the content and the imprint of the text. There, both editions are cited: *Venet. per Anton. de Strata. 1488, – per Simonem Papiensem dictum Beuilaquam, 1498*. The only publication edited by Antonio of Chremona at 1488 that can be taken into account is the one that brings together three Latin translations, of which only the first one is due to Valla: *Alexandri aphrodisei p[ro]blemata per Georgiu[m] ualla[m] in latinu[m] co[n]uersa. Aristotelis problemata p[er] Theodoru[m] gaza[m]. Plutarchi problemata per Iohanne[m] petru[m] lucensem impressa Venetiis per Antonium de strata Cremonensem. [1488]*. Already, Rosen (1995: 142, note 14) denounces this confusion.

caelestium by Cleomedes (already published by Carolus Valgulius, Brescia 1497, and by Giorgio Valla, Venice 1498), praises Valla's labor in his Preface (CTC 1960–2003: 9) for having been the first to make it accessible to the Latin world. However, regarding Valla's translations, Balfour says that many times Valla starts from incorrect assumptions, which lead him to a wrong translation. Later scholars began to recognize the positive aspects of Valla's work, so Gardenal (1981: 44–53) thinks that his translations are not free of imperfections, but that they make it clear that Valla's intention to replicate the original text constitutes a scientific objective. Gardenal asserts: "Despite some indecision and uncertainties, in his translations [he] displays a scientific aim that is particularly noticeable in his search for a meticulous closeness to the original text" (p. 52). Similar is the position of Piero Tassinari, for whom the characteristics of Valla's translations are the exact fidelity to the Greek text, which may make it difficult to understand, and its implication in the creation of a clear scientific language (Tassinari 1994: 90).

According to Noack (1992: 49), Valla used *ms. L, Vat. Barb. Gr. 186* for his translation of *On Sizes*. One of its merits is to be both the first translation of *On Sizes* and the
earliest printed edition. Another one is to have made a fundamental text of Greek
astronomy available to a public that did not have the knowledge of the Greek language
to read it (or at least not enough) and that, perhaps, did not even know of its
existence.

However, it is a very inconsistent translation. In it, there can be found confusions
of the Greek letters referring to points, lines, or circles with numbers (which in
Greek are also expressed with letters) and vice versa, and several missing texts.
However, the responsibility is divided equally between the translator and the manuscript used, which also contains many errors. Moreover, Simone Bevilaqua's workshop seems not to have been completely reliable. He edited works of different
subjects:

> [A]ll of high culture and some not insignificant, it must nevertheless be agreed
> that, typographically, they never went beyond the type of books for current
> sale: without pomp and, above all, without adequate correction, so much so
> that many texts are ugly by too many errors deriving from hasty and badly
> revised composition.
>
> (Cioni 1998)

A single example will be more than enough. See Figure 1.6. In a paragraph of
proposition 15, the number 10,125 appears five times. In Greek, 10,125 is written
Μ̅ρκε. Each of those five times is translated differently by Valla: the first time as
10,000; the second he translates 10000.yrke., interpreting that ρκε (125) is not part
of the number but refers to letters; the third time, he directly omits the sentence in
which the number appears; in the fourth, he translates it correctly, as 10,125; and
in the fifth, as 10,425. In addition, in the same paragraph, the number 979 (ϡοθ)
appears three times. Valla omitted it once, and he translated it by incorrect
and different numbers twice. However, as we noted, part of the source of the error
is in the manuscript. In Figure 1.6, we enclose in rectangles the number 10,125
(Μ̅ρκε). There are only four squares, because the one overlooked by Valla is also

omitted in his manuscript. As you can see, the three times that he translates it incorrectly, there is between M̱ and ρκε an ü that does not make sense. Probably, this confused the translator. The only time the number is correct in the manuscript, it is also correct in the translation. In this case, therefore, the source of the error is mainly in the manuscript. But with the number 979, the situation is quite different. Again, when Valla omits it, it is also omitted in the manuscript, but the two times that it appears, it is well written, and yet Valla translates it wrong. We must recognize that, in the manuscript, the spelling of the sampi (ϡ) – an archaic Greek letter that later became a symbol for 900 – which follows the also-archaic form (Ⲧ), is very similar to a tau (τ). This similarity could explain the second error, in which, presumably confusing the sampi with a tau, he translates 379 (τοθ). But it is not an excuse for the first error, in which he translates 579 because the letter for 500 is φ. In Figure 1.6, we enclosed in dotted rectangles the two times that ϡοΘ appears, and we encircled two tau to show their resemblance to the sampi.

The translation style (practically a word-by-word translation) was already out of fashion in his time (Noack 1992: 53–58), although, as we pointed out, his intention may have been to give a translation as close as possible to the original language. The great number of errors – of shared responsibility – suggests that it was difficult for anyone to understand Aristarchus's work from this translation. Finally, the figures, in many cases, are practically incomprehensible, although, admittedly, the primary source of the error here is the manuscript. Figure 1.7 reproduces the diagram corresponding to proposition 7, both the manuscript's version (left) and Valla's version (right). It is patent that Valla's draftsman did his best to clarify the diagram. But at least one mistake must be attributed to Valla's draftsman. In Valla's diagram, line *ab* representing the Earth–Sun distance is longer than the circle representing the solar orbit centered at *b*. Accordingly, it makes the quadrilateral *bafe* a rectangle and not a square, which is necessary for the proof.

* * *

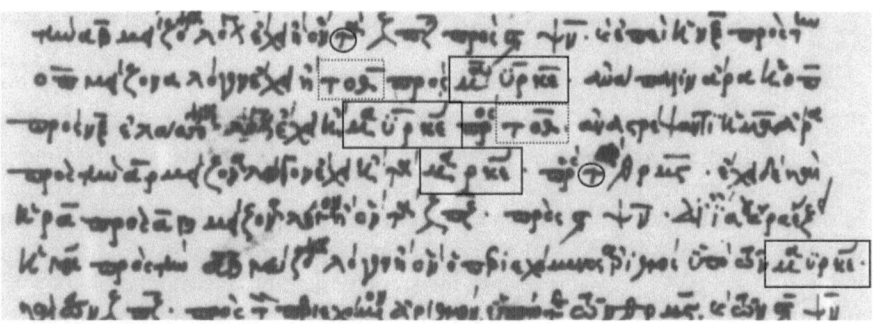

Figure 1.6 Detail of ms. L, probably used by Valla.

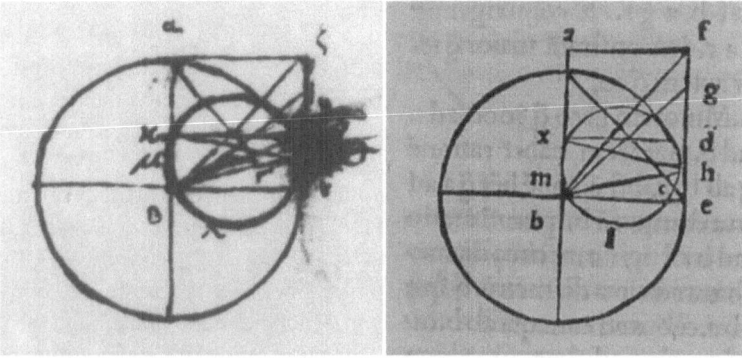

Figure 1.7 Diagram of proposition 7 in ms. L (folio 20.a.) and in Valla's edition.

Radically different is the translation of Federico Commandino (1509–1575), who was a physician and mathematician born in Urbino, Italy. His disciple, Bernardino Baldi, wrote a biography that is still now the best source for knowing his life (Baldi 1714). Commandino studied Latin and Greek with Giacomo Torelli da Fano and mathematics with Gian Pietro de' Grassi, who took him to Rome to serve for a few years as the private secretary of Pope Clement VII. After the pope's death, Federico Commandino went to Padua, where he studied philosophy and medicine. He eventually returned to Urbino, where he came back to translate and became the general practitioner of Guidobaldo, Duke of that city. In 1549, he returned once again to Rome, where the study of manuscripts became increasingly relevant to humanists. He stayed there for about 16 years and befriended the Jesuit Cristobal Clavius (Baldi 1714: 534). From Rome, he came back to his hometown, in which he died ten years later. In his biography, Baldi attempts a curious apology for the reasons for his teacher's death. The rumor was that he died from having "surrendered to the venereal pleasures at that [advanced] age." Still, the real reason, Baldi clarifies, is what Commandino told him when his illness had begun: that so many translations did not leave him time to do exercise. And as proof of this, Baldi adds that, in the frequent deliria that the illness provoked to him in his last days, when he found "out of himself, as if dreaming, he reasoned confusedly about things related to this profession [the intellectual task]." Close to the end of the biography, Baldi asserts that Commandino

> was a man of such a great kindness in practical things and of such a great science in intellectual matters that, had he not been inclined to feminine pleasures, not even the very Momus[38] would have found something to reproach him.
>
> (Baldi 1714: 536)

38 It refers to the god Momus, who, in Greek mythology, personified sarcasm, irony, unfair and malicious criticism.

Therefore, he does not deny that his master surrenders to feminine pleasures but that this surrender was the cause of his death.

The edition of Commandino, published in 1572, is titled Aristarchi de magnitudinibus et distantiis solis et lunae liber cum Pappi Alexandrini explicantionibus quibusdam a Federico Commandino Urbinate in latinum conversus et commentariis illustratus. Cum Priuilegio Pont. Max. In annos X. PISAVRI, Apud Camillum Francischinum. MDLXXII.

It is not so clear which manuscript Commandino used, and he does not mention it in his work. John Wallis, who published the *editio princeps*, erroneously assumes that they both used the same manuscript (*ms. W, Oxford Savilianus 10*). But this manuscript should be excluded as the source since Henry Savile, the copyist of this manuscript, undertook the trip to Italy in which he found and copied the manuscript as late as 1578, some six years after the publication of Commandino's book, and Commandino never left Italy. After a detailed analysis, Noack (1992: 61) concludes that the manuscript used by Commandino is *ms. U, Ambrosianus A 101 Suppl.* – which does not contain diagrams – and probably also *ms. 3, Ambrosianus C 263 inf.* From the latter, Savile would have copied his own manuscript, which Wallis finally uses. The two Ambrosian manuscripts and many others were part of the library of Gian Vicenzo Pinelli. And there are reasons to think that Commandino belonged to the circle of scholars who used that library.[39]

Unlike Valla's, the translation of the "admirable and indefatigable translator" – as Heath (1913: 321) calls him – is impeccable. His mastery of Latin and Greek is indisputable, but so is his understanding of the subject. His Latin is beautiful and clear, and he deviates from Greek only long enough to obtain a successful Latin translation. Of course, one can find some terminological errors or mistakes (Noack 1992: 64), but his translation helped us in numerous complicated or ambiguous passages. His notes and comments, with constant references to the sources of the theorems that Aristarchus uses and also the detailed demonstration of theorems that Aristarchus supposes, are extremely useful, even for a current reader. That is why we decided to include the Latin text and translation of his comments in this edition. His figures are also clear and accurate. They show that Commandino reflected on the diagrams before drawing them. Heath's diagrams are inspired in those of Commandino and through Heath's edition, the Commandino's diagrams have been the source of the figures in modern editions.[40]

Like many of the time, his edition has the friendly characteristic of reprinting the same figure on all the pages in which it is referred, avoiding the annoying task of turning the page to consult the figure again. Thus, for example, the full-page figure of proposition 13 is reproduced seven times. Commandino does not mention Valla's translation. Perhaps he did not know it. A facsimile reproduction of the Commandino's edition is found in Massa (2007: 77–157).

* * *

---

39 Cfr. Ciocci (2022) for new research on the manuscripts used by Commandino.
40 Massa (2007) directly reproduces the figures of Commandino; the ancient Greek part of Stamatis (1980) is a facsimile of Heath (1913), and Spandagou (2001) reproduces Heath's figures.

The English mathematician John Brehaut Wallis (1616–1703) is responsible of the *editio princeps*. He was born in Ashford, Kent, and studied at Emmanuel College in Cambridge, where he defended an argument in favor of the circulation of the blood (Autobiography of Wallis (1697) in Scriba 1970: 29).

He was ordained a Presbyterian priest but stood out as an important cryptographer of the time. To prevent his expertise from being used against his homeland, as late as 1697, he still rejected an invitation to teach cryptography in Hannover and to write a treatise on cryptography. The request, by letter, had been made by Gottfried Leibniz himself, who was a younger generation than Wallis and a great admirer of his mathematical work. At that time, cryptography was in its infancy, but Wallis took it to a higher stage. The following passage of his autobiography narrates the first time he decrypted an encrypted message:

> About the beginning of our Civil Wars, in the year 1642 <or 1643> a Chaplain of Sr. William Waller shewed me one evening an intercepted Letter written in Cipher. He shewed it me as a curiosity on{e} evening just as we were sitting down to supper at the Lady Vere's, (and it was indeed the first thing I had ever seen written in Cipher.) He asked me (between jest and earnest) if I could make any thing of it. And was surprised when I sayd (upon the first view) perhaps I might.
>
> 'Twas about ten a clock when we rose from supper, and I withdrew to my chamber to consider of it. By the number of different characters in it (not above 22 or 23) I judged it could be no more than a new Alphabet, and, before I went to bed, I found it out, which was my first attempt upon Deciphering.
>
> (autobiography of Wallis (1697) in Scriba 1970: 37)

As a mathematician, he made significant contributions to the development of infinitesimal calculus and was the one who introduced the symbol ∞ to represent the notion of infinity. In 1649, he obtained the Savilian professorship at Oxford, which he held for more than 50 years, until his death. The edition of Aristarchus's work belongs to this time. Wallis was not only a great mathematician and prodigious cryptographer but also wrote on logic, philosophy, and English grammar. Towards the end of his life, at the request of his friend Thomas Smith (1368–1710), he wrote a short autobiography used as a source in these paragraphs.

His edition of Aristarchus's work was published at Oxford in 1688 and is titled Aristarchi Samii De Magnitudinibus et Distantiis Solis et Lunae, Liber. Nunc primum Graeche editus cum Federici Commandini versiones Latina, notisque; illius et Editoris. Pappi Alexandrini Secundi Libri Matematicae Collectionis, Fragmentum, Hactenus Desideratum. E. Codice MS. Edidit, latinum fecit, Notisque illustravit. Johannes Wallis, S.T.D. Geometriae Professor Savilianus; et Regalis Sociatetis Londini, Sodalis. Oxoniae, et Theatro Sheldoniano, 1688. Luego, fue reeditado en la edición que reúne los trabajos matemáticos de Wallis, Johannis Wallis Opera Mathematica, 1693–1699, vol. 3, pp. 565–594.

In the Preface, Wallis asserts that for his translation, he used two manuscripts, the manuscript *Oxford Savilianus 10* (*ms. W*) and a copy of this manuscript made by Edward Bernard. In addition to the Greek text, Wallis's edition incorporates the

Latin translation of Commandino and even uses it in some passages to amend the Greek text. Along with some of his own notes, Wallis also includes all of Commandino's. The Greek edition of Wallis is careful and accurate. In several cases, he even had the understanding to intuit not only that a text was missing from his manuscript but also to reconstruct it in a practically identical way to how it is found in other manuscripts (see note 204 of the translation, on p. 161, and Fortia 1810: 234).

* * *

The third Latin edition comes from Agricol-Joseph-François-Xavier-Pierre-Esprit-Simon-Paul-Antoine, conde de Fortia d'Urban (1756–1843), a French mathematician and writer who was born in Avignon. He was named Knight of the Legion of Honor in 1811, Corresponding Member of the Belgian Academy in 1828, and Free Member of the *l'Académie des Inscriptions et belles-lettres* in 1830. He died in Paris. In addition to the *Histoire d'Aristarque de Samos*, he published numerous works on the most diverse topics: history, religion, customs, etc.

The volume, published in 1810, is titled Historie d'Aristarque de Samos, suivie de la traduction de son ouvrage sur les distances du Soleil et de la Lune, de l'histoire de ceux qui ont porté le nom d' Aristarque avant Aristarque de Samos, et le commencement de celle des Philosophes qui ont paru avant ce méme Aristarque. Par M. de F****. Paris, Chez Madame Veuve Duminil-Leuseur, 1810.

Undoubtedly, it is an odd volume. The first thing that stands out is that the author's name is not complete, just a few initials followed by asterisks: M. de F **** (presumably for *Monsieur de Fortia*). Without any Preface, the bilingual Greek Latin edition begins on the first page. The bilingual text concludes on page 87; in the following, there is a note explaining which manuscripts the author used in the edition. It then contains the first edition of almost all scholia, along with a Latin translation running up to page 199, and then a large set of critical notes on the work (up to p. 236) and the scholia (up to p. 248). Next begins a section titled "History of Aristarchus of Samos, in which the complete translation of the works of him that have arrived until us will be found." However, the title is misleading, for the translation of Aristarchus's work already took place in the previous section. In this section, nothing says about the history of Aristarchus de Samos. It contains two chapters. The first develops the biography of all the known *Aristarchuses* that preceded the Samian (13 in total). The second contains description of the works and ideas of "the philosophers who have been useful to Aristarchus of Samos." Still, the authors and topics developed are so broad that they are difficult to link with Aristarchus, such as Indian, Chinese, Babylonian, Egyptian philosophy, and even Egyptian alchemy and the history of god Hermes-Thot.

Let us say something about the Greek edition and the Latin translation. First of all, we should notice that the volume does not include diagrams, although it constantly refers to them, not only in Aristarchus's text, but also in the notes. The diagrams are also missing in the edition of the scholia. The Greek edition presents numerous differences with Wallis's explained in detail in the notes following the translation. It is difficult to judge whether, as the author suggests, his edition surpasses that of Wallis. Nevertheless, his principal value lies in using another set of manuscripts and pointing out certain text variations. The Latin translation of Aristarchus's reproduces that of Commandino, with few modifications.

In addition to Wallis's edition, Fortia consulted the Parisian manuscripts *Par.Gr. 2342, 2363, 2364, 2366, 2386, 2472, and 2488 (mss. B, R, Y, 5, Z, Q, and C)* together with a Vatican manuscript that, at that time, was in the library of Paris (Fortia 1810: 68). Heath (1913: 327) assumes that the Vatican manuscript used by Fortia is the *Vat. Gr. 204* (ms. A), the same one he used. This manuscript has the *Bibliotheque Nationale*'s seal. It had been brought together with other manuscripts in 1808 to Paris by order of Napoleon Bonaparte and returned to the Vatican after the Congress of Vienna in 1814. Fortia affirms that the Vatican manuscript is no better than the ones extant in French libraries. Heath (1913: 327), probably caused in part by his own patriotism, argues that Fortia's judgment is motivated primarily by his patriotism. Noack (1992: 34, n. 122) agrees with Heath (1913: 326, n. 2) in pointing out a mistake made by Fortia. According to Heath, in Fortia's critical notes, there are references to readings from the *Paris Gr. 2483* manuscript, although he does not include it in the list of manuscripts used by him. This manuscript does not contain Aristarchus's treatise; the reference should be to the manuscript *Paris Gr. 2472 (ms. Q)*, used by Fortia. Noack, then, refers to an error made by Heath himself (1913: 325), who includes among the manuscripts that transmit the treatise the *codex Marc. Gr. 301* (Venice), a codex that does not contain it.

* * *

Thirteen years later, Fortia publishes a French translation of Aristarchus's *On Sizes* (Fortia 1823). The volume is titled Traité d'Aristarque de Samos sur les grandeurs et les distances du Soleil et de la Lune, traduit en francais pour la premiere fois, par M. le Comte de Fortia d'Urban. Paris, Firmin Didot Peré et fils, Libraires, 1823. The very first page contains a note where he states:

> Those who have the Greek text with a Latin translation and notes, published in Paris, in 1810, without the author's consent, will be able to add this translation to it, which will complete the work, and whose figures are absolutely necessary to read the text.

Later, in the Preface, he explains the curious origin of the publication of his bilingual edition:

> The text of the work of Aristarchus of Samos, which I had revised from eight manuscripts in the King's library, and which had been printed in France where it had not been published until then, with absolutely unpublished scholia, having been put up for sale without my permission, has appeared in an almost ridiculous way. There are references to figures that I had drawn on every page, but that annoying circumstances have made disappear during my stay in Italy. I will try to supplement it with the publication of this translation that will be

accompanied by new figures in which I have had the Greek letters added for those who want to join this translation to the text.

(Fortia 1823: 1)

We agree with Heath (1913: 324) when he affirms that Fortia's translation is "meritorious and useful." The French translation is excellent and contains few errors. However, we should note that Fortia has a sometimes "too free" style, although never incorrect. This edition also includes a French translation of the scholia. Again, this is undoubtedly a significant contribution of Fortia.

At the end of the book, there are three plates containing figures that, as he says, have the Latin letters but also the Greek ones so that they can be used together with his 1810 edition.

* * *

In 1854 was published a German translation by Anton Nokk (1797–1869), a German politician and high school teacher at the Gymnasium of Bruchsal. In 1838, he became Director of that Gymnasium, and from 1848 on, he held the same position at the Freiburg Lyceum. Nokk devoted almost all his life to teaching, mainly mathematics. In 1857, he was honored with the *honoris causa* doctorate in philosophy at the University of Freiburg due to his contribution to the knowledge of Greek mathematics. Nokk was also elected deputy of the Lower House of the General States of Baden in Karlsruhe and made Knight of the Order of the Lion of Zähringer (*Zähringer Löwen*), an order founded in 1812 by Charles II, Grand Duke of Baden.

The book, titled Aristarchos über die Grössen und Entfernungen der Sonne und des Mondes. Uebersetzt und erläutert von A. Nokk. Als Beilage zu dem Freiburger Lyceums-Programme von A. Nokk, is part of a supplement to the programs of the Freiburg Lyceum of 1854.

The book does not include any Introduction or Prologue, but it does have notes at the end where Nokk explains which source he consulted for doing the translation. He worked mainly with Wallis's edition but also had access to Fortia's 1810 edition. We can presume that Nokk was not aware of Fortia's French translation of 1823, for he says (Nokk 1854: 31) not to be sure who the author of Fortia's 1810 edition was. In the notes, he promises a Greek edition of the text that was never published (Nokk 1854: 30). The translation is correct, although it has a very free style, particularly with the translation of mathematical relations. This feature undoubtedly makes reading easier for a modern reader. However, when translating classical works, we prefer to keep closer to the original text. For instance, what we translate as (end of proposition 13):

Therefore, the more so has line YA to line AP a greater ratio than the [ratio] that 89 has to 90. And [the same holds for] the double [of them]. Therefore, the diameter of the Sun has to line ΠP a ratio greater than the [ratio] that 89 has to 90. And it has also been proved that line ΞN has to the diameter of the Sun a ratio greater than the [ratio] that 22 has to 225. Therefore, by equality of terms, the more so has line ΞN to line ΠP a greater ratio than the [ratio] that the product of 22 and 89 has to the product of 90 and 225, that is, than the [ratio] that 1958 has to 20250; and to the halves [of these numbers], that is, to the [ratio] that 979 has to 10125.

Nokk translates it:

un um so mehr YA : AR > 89 : 90;
folglich 2YA : 2AR > 89 : 90,
od. Sonnendurchmesser: PR > 89 : 90.
Es war aber XN : Sonnendurchmesser > 22 : 225;
woraus folgt XN : PR > 89 x 22 : 90 x 225,
d.i.   > 1958 : 20250,
oder   > 979 : 10125.

He did his translation mainly to help Lyceum students appreciate Aristarchus's reasoning. With this goal in mind, he has done a great job.

* * *

Two years after the publication of Nokk's translation, a new Greek edition was also published in Germany, by Johann Ernest Nizze (1788–1872), a German mathematician and educator.

Nizze began working as an assistant teacher at the Friedrich-Wilhelms-Gymnasium in Berlin, obtained his doctorate in 1812 at the Faculty of Philosophy of the University of Erlangen, and was immediately appointed Deputy Headmaster of the Prenslau Gymnasium. He interrupted his teaching career for two years while he fought in the infantry of the Lützowsches Freikorps. Once peace was reestablished, Nizze returned to his teaching career, reaching his highest degree when, in 1832, he succeeded Karl Kirchner as Rector of the Sundischen Gymnasium in Stralsund, a position he held for 33 years. From this time is the publication of the work of Aristarchus (Krössler 2008; Häckermann 1886: 744–745).

The volume is titled Ἀριστάρχου Σαμίου βιβλίον περὶ μεγεθῶν καὶ ἀποστημάτων ἡλίου καὶ σελήνης, mit kritischen Berichtigungen von E. Nizze. Mit swei Figuren – Tafeln. Druck der Koeniglichen Regierungs – Buchdruckerei. Stralsund, 1856. After a brief introduction in which he presents the previous editions (including the then-new edition of Nokk and the two Fortia editions), he offers the Greek text with a series of notes. He prepared the Greek edition without consulting any manuscript at all. The text is based exclusively on the editions of Fortia and Wallis. Thus, it has no particular value. We agree with Heath (1913: 324) when he states that the text is unreliable since it has not been prepared with enough care.

* * *

But without a doubt, the best modern edition is the one that Thomas Heath published a little over a century ago, at Oxford University Press.

Sir Thomas Little Heath (1861–1940) was born in Lincolnshire, England. He studied at Clifton College and then went to Trinity College, Cambridge, where he graduated in mathematics and classics in 1883. Heath excelled in both fields in which he worked, the civil service and the study of Greek mathematics. He belonged to the first from 1884 until 1926, when he retired. He was first Assistant Secretary to the Treasury, then Permanent Secretary. In 1919, he continued serving at the National Debt Office. For his results, Heath was honored as Knight of different orders.

In addition to his administrative career, he stood out as a historian of ancient science, particularly Greek mathematics. He wrote the articles on *Pappus* and *Porism* for the *Encyclopaedia Britannica* when he had not yet graduated. He translated countless Greek scientific works into English, widely spreading their knowledge in the Anglo-Saxon world. His translations include the works of Archimedes, Euclid, and Apollonius, among others. Heath also wrote histories of Greek mathematics (1921, 1931) and Greek astronomy (1932), which are still reference works. He was elected to the *Royal Society* in 1913 and was awarded *honoris causa* doctorates at Oxford (1913) and Dublin (1920). He knew how to join his administrative and academic career, with a great love for music and mountaineering (Cfr. Heath 2002: xxi)

His translation and new critical edition are part of a volume dedicated to Aristarchus, titled Aristarchus of Samos. The Ancient Copernicus. A History of Greek Astronomy to Aristarchus together with Aristarchus's Treatise on the Sizes and Distances of the Sun and Moon. A new Greek Text with Translation and Notes by Sir Thomas Heath. Oxford: Oxford University Press, 1913.

According to Heath, the translation was prompted by the request of Helbert Hall Turner, a former colleague of Heath at *Clifton College* who became *Savilian Professor* of astronomy at Oxford. Turner's particular interest was to find out why Aristarchus had proposed such a patently exaggerated value for the apparent size of the Moon. The name of the work is not innocent. A few years earlier, the highly regarded historian of science Giovanni V. Schiaparelli (1898) had published an article in which he attempted to show that the first scientist to hold heliocentrism was Heraclides and not Aristarchus. In his volume, Heath defended Aristarchus's merits. The first part of the work is a long and careful history of Greek astronomy up to Aristarchus. The second part includes the translation, a new Greek edition, an introductory study, and notes.

As we have already mentioned, for his Greek edition, Heath based primarily on *ms. A*, the oldest surviving manuscript, which he had access to through photographs. He also used Wallis's and Fortia's editions. The latter allowed him indirect access to the Parisian manuscripts used by Fortia and, when necessary, includes references to them. Heath's Greek edition is simply excellent, and the manuscript used is also so good that practically nothing of relevance is lost by being based on a single manuscript. It remains, after a century, the Greek text of reference. His translation is also very successful, and without a doubt, it is the best, after that of Commandino. For it, he consulted all the then-existing translations.

In translations of other books made by Heath, in which he intended to make ancient science accessible to the twentieth-century reader, he chose a much freer style, translating Greek expressions into mathematical symbols. The emblematic case of this style is his translation of the works of Archimedes (Heath 1897). In the case of *On Sizes*, however, the translation is respectful of the original text.

* * *

In 1941, the Loeb Classical Library published two volumes with a selection of Greek mathematics texts: Selections Illustrating the History of Greek Mathematics with an English Translation by Ivor Thomas in two volumes. The volume that interests us is the second, titled From Aristarchus to Pappus. Cambridge: Harvard University Press, 1941.

Ivor (Bulmer)-Thomas (1905–1993), British journalist, politician, and writer, was for eight years Member of Parliament. He was awarded a scholarship to St. John's College in Oxford, where he studied mathematics, classics, and theology. He then moved to Magdalen College, where he became Demy Senior in Theology. Ivor Thomas was a great athlete and represented Oxford in several of the classic races against Cambridge between 1925 and 1927; he even represented Wales and had a chance to compete for the British team at the 1928 Olympics in Amsterdam. In 1952, he added his wife's last name, "Bulmer," to his name. He was a committed Christian and a lover of ancient churches. In 1957, he founded *Friends of Friendless Churches*, an organization dedicated to recovering churches in poor condition, sometimes buying them by members' funding (Saunders 1993).

From the time as Demy Senior of Magdalen College is the partial translation of *On Sizes*. It consists of the Greek edition of Heath and a new translation by Ivor Thomas of some fragments of Aristarchus's work. The section dedicated to Aristarchus is the first one and occupies pages 8 to 15. It begins with the bilingual edition of two ancient references to Aristarchus, that of Aëtius and of Archimedes's famous *Sand Reckoner*. It follows the translation of Aristarchus's introduction before the first proposition, and then the complete proposition 7, in which he calculates the limits of the ratios between the Earth–Sun and Earth–Moon distances. It concludes with two short fragments of propositions 13 and 15. The edition, although careful, does not add anything to Heath's.

* * *

In 1970, Aguilar Press published in Madrid a volume titled Científicos Griegos. Recopilación, estudio preliminar, preámbulos y notas por Francisco Vera.[41] Francisco Vera Fernández de Córdoba (1888–1967) was born in Alconchel, Spain, and died in Buenos Aires, Argentina. After the bachillerato in Badajoz, he studied mathematics in Salamanca and Madrid. It seems that, like Wallis, he was also a cryptographer and had to run away from Spain for having written the cryptographic code of the army loyal to the Republic (Cobos Bueno and Pecellin Lan-

41 We thank Daniel Alonso Carrara for making us aware of this work and providing us with a copy.

carro 1997: 510) in 1939. He went in exile first to France, then he moved to the Dominican Republic and Colombia, and finally, he settled in Argentina in 1944. There he taught courses and conferences at various institutions, including the University of Buenos Aires. His most important work is the five-volume *Historia de la Cultura Científica* (Vera 1956–1969).

The founder of the publishing house, Manuel Aguilar, asked Francisco Vera to prepare a volume bringing together the main Greek scientific works. Vera's last published work consists of two big volumes, in which "twenty-three authors and the works and fragments that best reveal his ideas" (Vera 1970, I: 7) are offered to the Spanish reader. The first volume includes fragments from Plato, Aristotle, Eudemus of Rhodes, Theophrastus, Euclid, and Aristarchus. The section dedicated to Aristarchus is found between pages 983 and 992. Unfortunately, only very short fragments of *On Sizes* were translated, practically only the titles of the propositions. It has a brief introduction that contains some interesting data along with some basic errors. For example, he suggests that the limits obtained by Aristarchus for the ratio between the Earth–Sun and Earth–Moon distance, that is, 18 and 20, are due to two different presumed measures of lunar elongation in a dichotomy, one that would have been of 86° 49′, and the other of 87° 8′. However, a quick reading of proposition 7 reveals that Aristarchus only uses 87°. The difference between upper and lower limits is due to the calculations performed without trigonometry, not to two different input data. But towards the end, he makes a valuable bibliographic reference, reviewing the most significant editions of Aristarchus's work. His translation includes Aristarchus's introduction, the statement of propositions 8, 10–12, 15–18, and a complete translation of Pappus's commentary. The translation is correct, and although the proportion translated is almost insignificant, it has the merit of being the first translation into Spanish.

* * *

In 1980, Evangelos S. Stamatis published in Athens a translation to modern Greek of Aristarchus's *On Sizes*. Ευάγγελος Σ. Σταμάτης (1897–1990) was a historian of ancient Greek mathematics. His name is well-known because he was in charge of the correction and re-edition of Archimedes's *Opera Omnia* in Teubner's prestigious edition (1972), originally made by the unmatched editor of classical scientific works, J. L. Heiberg (1880). In 1923, Stamatis graduated in physics from the University of Athens. However, during the last two years of his studies, he enlisted in the Army and participated in the Asia Minor campaign. In 1931, he obtained a scholarship to study at the University of Berlin. In addition to his work as an academic and translator (he translated the *Elements* of Euclid, the complete works of Archimedes, and the *Conics* of Apollonius), his tireless effort to make the knowledge of ancient science more popular in Greece deserves to be highlighted. He was a high school teacher for more than 35 years. Stamatis published numerous articles and popular books. He was the first Greek to become a member of the *Académie Internationale d'Historie des Sciences* in 1966 (Christianidis and Kastanis 1992).

The volume is titled Αριστάρχου Σαμίου περὶ μεγεθῶν καὶ ἀποστημάτων ἡλίου καὶ σελήνης. 2300 τὴ ἐπέτειος τῆς γεννήσεώς του 320 π.Χ. –1980. Αθηεναι 1980.

The book begins with a short introduction. The core of the volume is a bilingual ancient Greek–modern Greek edition. For the ancient Greek text, Stamatis's edition includes a facsimile of Heath's pages. Although we have consulted this edition, our modern Greek knowledge is not good enough to judge the translation.

* * *

Even if it is not a translation, we cannot omit here the monumental work of Beate Noack, Aristarch von Samos. Untersuchungen zur Überlieferungsgeschichte der Schrift περὶ μεγεθῶν καὶ ἀποστημάτων ἡλίου καὶ σελήνς Serta Graeca: Beitrage zur Erforschung griechischer Texte, Band I; Dr. Ludwig Reichert Verlag, Wiesbaden, 1992.

Beate Noack (1961–) showed interest in the study of manuscripts since her studies at the Universities of Heidelberg and Freie Universität, Berlin (classical philology and ancient history), between 1980 and 1990, and then at the University degli Studi di Roma "La Sapienza" (codicology). Later, Noack went to Rome (Biblioteca Apostolica Vaticana) and Venice (Centro Tedesco di Studi Veneziani), where she researched on Greek manuscripts from the Middle Ages. Her doctoral thesis (Freie Universität Berlin, 1990) dealt with the history of the transmission of the text of Aristarchus's *On Sizes*, whose treatise she had been working on since 1984 (published in Noack 1992). At first, she worked as an assistant at the University of Hamburg, and since 1993, she has taught inter alia Greek and Latin and devoted herself to the research of late ancient Greek at the University of Tübingen. Now she is retired.

The book consists of an expanded and improved version of her doctoral directed by Prof. D. Harlfinger, who had suggested the topic to her a few years before. It is the only serious and complete study of the manuscripts of Aristarchus's work. She managed to identify 30 Greek manuscripts having the work of Aristarchus. With them, she made the collation of the text and the scholia. She also identified the manuscripts used by Valla and Commandino. Her work is impeccable and constitutes an outstanding contribution to the study of Aristarchus's *On Sizes*. It includes a philological analysis of the scholia, including the identification, classification, edition, German translation, and brief commentary of more than 40 scholia not found in Fortia's edition. Her work is a must for anyone who wants to prepare a new edition of the Greek text. As she says, her work is a preliminary study for a critical edition that is still waiting to be done.

* * *

In 2001, a new modern Greek translation by Ευάγγελος Σπανδάγος appeared. Evangelos Spandagos (1940–) graduated in mathematics from the University of Athens; he spent 28 years in high school education and authored more than 100 books on the history of mathematics. Spandagos is the director of his own publishing house, in which most of his books are published. In June 1997, he was awarded in Paris by the World Group for the Study of Ancient Cultures for his work related to ancient Greek science's lost works. Since 1999, he directs radio and television programs on science in ancient Greece.

The book, titled περὶ μεγεθῶν καὶ ἀποστημάτων ἡλίου καὶ σελήνης τοῦ Ἀριστάρχου τοῦ Σαμίου. Εἰσαγωγή, Ἀρχαῖο κείμενο. Μετάφραση, Ἐπεξηγήσεις, Σχόλια Ἱστορικά Στοιχεία. Αἴθρα Ἀθήνα, 2001, has three chapters. The first one is divided into two parts. In the first part, he develops an introduction to the life and work of Aristarchus. The second is divided into several sections. After a few general words, he (1) discusses the hypotheses and (2) reviews the main results of *On Sizes* by comparing them with those of other ancient astronomers. Third, he (3) explains and translates Pappus's commentary. After that, the author (4) reviews the other editions of Aristarchus's work, ending with Stamatis's. He (5) examines the mathematical techniques that Aristarchus uses and explains ancient Greek numbering. He then (6) lists some 34 types of instruments used by ancient Greek astronomers (gnomes, diopters, etc.). He (7) concludes the chapter with a short glossary in which he translates the expressions of Aristarchus into modern Greek. The second chapter contains the edition of the ancient Greek text, which is Heath's, but without critical apparatus and diagrams. The third chapter consists of his translation, where the diagrams and explanatory notes do appear. We are also not in a position to judge the quality of his translation.

* * *

At the Congress of the Spanish Society for the History of Sciences and Techniques that took place in Cádiz in September 2005, Cándido Martín, from the University of Cádiz, proposed María Rosa Massa Esteve to translate Aristarchus's work into Spanish. Two years later, the University of Cádiz Press published the first complete translation into Spanish of *On Sizes*. The book is titled Aristarco de Samos sobre los tamaños y las distancias del Sol y la Luna, texto latino de Federico Commandino. Introducción, traducción y notas de María Rosa Massa Esteve. Cádiz: Universidad de Cádiz, 2007.

María Rosa Massa Esteve (1954–), former secretary of the *European Society for the History of Science* from 2012 to 2014 and a high school math teacher for many years, is professor at the *Universitat Politècnica de Catalunya*. She is known for being one of the greatest authorities on the work of Pietro Mengoli (1626– 1686) – see, for example, Massa Esteve (1999, 2006). Aiming to select historical texts that facilitate mathematics learning, she directs a research project on the history of trigonometry, the topic that interests her particularly. She has also been a high school math teacher for many years (Lusa Monforte 2008).

The book has a brief but informative Introduction describing the historical context in which the work appears, and overviewing Aristarchus's life and work. The Introduction concludes by revising some of the previous editions and explaining his own translation characteristics. Massa Esteve translated not from the Greek text but from the Latin translation of Commandino. However, her colleague

Joaquín Ritoré has revised the translation from Heath's Greek edition and English translation. The translation, although with some minor errors, is generally adequate, elegant, and accurate. In several passages, however, there is too much dependence on Heath's English text and some inconsistencies, probably the result of having translated from one language and corrected from another. Although it is a translation of a translation, the version of Commandino that she correctly translates is so good that the result constitutes a reliable and intelligible text. The figures are facsimiles of the Commandino diagrams. The work includes a translation of Pappus's commentary – translated from the French edition of Paul ver Eecke (1933)– and a complete facsimile of the copy of the Commandino's edition kept in the Royal Institute and the San Fernando Navy Observatory.

* * *

In 2007, Angelo Gioè defended his PhD, Aristarque de Samos. Sur les dimensions et les distances du Soleil et de la Lune. Édition critique, présentation, traduction et notes, at the University of Paris IV–Sorbonne. The volume offers a general Introduction in which he summarizes the life and work of Aristarchus, his intellectual context, his connection with the heliocentric tradition, and a mathematical introduction to the content of the treatise. The author also presents transmission of *On Sizes* and gives an account of his own principles of editing. But without a doubt, the most valuable aspects of his contribution lie in the first Italian translation, in a new French translation, and in the first step towards a critical edition of the text. In general, the French translation is correct. The style is rather free, but not as free as Fortia's. Although a new translation into French after almost 200 years is welcome, frequent errors (especially the confusion of letters that designate points and the omission of some passages) make it difficult to understand. Something similar happens in the Italian translation. His critical apparatus consists of pointing out the differences that Noak detected between the oldest surviving manuscripts of the five families that make up the transmission. According to Gioè (p. 97), he checked with the manuscripts all the variants introduced.

* * *

Finally, in 2020, our Spanish edition has been published. It is titled Aristarco de Samos. Acerca de los tamaños y las distancias del Sol y de la Luna. Estudio preliminar, revisión del texto griego y traducción al castellano de Christián C. Carman y Rodolfo P. Buzón. Barcelona. Univesitat de Barcelona Editions. 2020. As we already mentioned, the present volume is based on that, but introducing significant modifications, extensions, and corrections.

In summary, five editions of the original Greek text have been published so far (Wallis 1688; Fortia d'Urban 1810; Nizze 1856; Heath 1913 and Gioè 2007), three Latin translations (Valla 1498; Commandino 1572 and Fortia d'Urban 1810), two translations into French (Fortia 1823 and Gioè 2007), one into German (Nokk 1854), one into English (Heath 1913), one into Italian (Gioè 2007), two into modern Greek (Stamatis 1980 and Spandagos 2001), a translation from Latin to Spanish (Massa Esteve 2007), one complete translation into Spanish (Carman and Buzón 2020), and two partial translations, one into English (Thomas 1941) and the other into Spanish (Vera 1970).

Our translation

In preparing our translation, we have evaluated the most important variants found in the 30 manuscripts identified by Noack (1992: 64), all the Greek editions, and all the translations into Latin and modern languages mentioned in the previous section. We also considered the autograph fragments of Commandino's translation preserved in the Central Library of the Università degli Studi di Urbino (contenitore 121, cartella 6, fogli 400–405).

Our initial intention was to make a new edition of the Greek text based on all the available testimonies: manuscripts, editions, and Noack's preliminary study. Still, in the course of preparation, we have realized that the differences with the Greek text of Heath, undoubtedly the best to date, would be so insignificant that it was not worth the effort. The few differences and some errors in his edition do not justify a new edition. Therefore, we have decided to reproduce Heath's text and, when necessary, point out in a footnote the differences that, nevertheless, are always insignificant from a conceptual point of view. We referred to the manuscripts in order to check a reading or find the source of some differences in the translations of Heath, Valla, Commandino, or Fortia, the only ones that used manuscripts as sources for their work. When the first draft of our translation was finished, we compared it with all the other ones. Commandino's, Fortia d'Urban's, and Heath's translations have been constructive in that process. In footnotes/endnotes, we have pointed out the most relevant differences with the existing translations. In the process of getting our translation into English, the advice of Alexander Jones was invaluable. He revised and corrected every sentence of the English translation of Aristarchus's text and the scholia.

Otto Neugebauer (1975: 751–755) and Reviel Netz (1999) have insisted that in Greek mathematical works, such as those of Archimedes, Apollonius, and also Aristarchus, diagrams play a fundamental role that has usually been neglected in modern editions.[42] The diagrams cannot be reconstructed only by reading the text, because they contain information that is not found in it (Netz 1999: 19) or that, if in the text, is ambiguous. The reader has to refer to the diagram to disambiguate that information. For example, when letters refer to points or lines that appear in the text without having been previously introduced, as points E and Λ of the diagram of proposition 6.

In addition, as preserved in manuscripts, diagrams are very different from the ones we would construct today. As we have argued recently (Carman 2018b, 2020b), this is probably due to minor imperceptible but accumulative errors during transmission. However, in any case, the manuscript's diagrams must be interpreted

---

42 Cfr. Saito and Sidoli (2012). See also notes 3 and 4 in Sidoli (2007: 526).

to render them intelligible to the modern reader. Thus, together with the Greek text, we decided to include the diagrams as they are preserved in the oldest extant manuscript (ms. A). In the English translation, we have completely recreated the diagrams, elaborating them directly from the diagrams in the manuscripts and the text. Therefore, we offer a translation of the diagrams as well.

A scientific discipline requires a technical vocabulary. During Aristarchus's lifetime, Greek geometry was still developing. Nevertheless, probably thanks to Euclid – only a generation older than Aristarchus – there was already a specific technical vocabulary governing mathematics. A technical vocabulary is characterized by using the same word for the same concept and avoiding synonyms. Thus, although Greek had several terms to refer to a line, center, angle, sphere, and others, in these mathematical works, there is a tendency to always use the same single word or expression for each concept. Technical vocabularies are not only made up of technical terms. As can be seen in the work of Aristarchus, ordinary words are employed in a technical sense. For example, ἐφάπτω means *to touch*. But in mathematical context, it means *to be tangent*, that is, *to touch only at one point*. The technical vocabulary includes not only words but also constructions that always keep the same structure and meaning. Following a practice established in Homeric studies, Netz (1999: 127) calls these structures *formulas*, understanding them not as mathematical equations but as a (relatively) rigid way of using a group of words. Aristarchus's work is plenty of these formulas. In our translation, we tried to respect this characteristic of technical vocabulary. Accordingly, even if it may be a bit repetitive to a modern reader, we have translated the same expression in the same way, as long as the good use of English allows it. In no case have we looked for synonyms to embellish the translation, since we consider it to be a scientific work.

Let us briefly review the main formulas and the reasons for our translation.

First, when referring to geometric objects, Greek authors use an article (sometimes also with a preposition) followed by the letter or letters that refer to the object. Thus, the neuter article with a letter, such as τὸ B, means *the [point] B*. A similar structure is used for line, but here the article is feminine and usually (but not always) accompanied by two letters. Thus, ἡ AB means *the [line] AB*. In both cases, the article is introduced to indicate the nature of the geometrical object (point or line) and not that the noun is known to the reader. Therefore, we decided to omit the article and translate τὸ B simply by *point B* and ἡ AB by *line AB*. Because the article determines the geometrical object, we decided to introduce *line* and *point* without square brackets. Sometimes, Aristarchus omits the article before the letter or letters, and the context allows the reader to interpret the nature of the geometrical object. In those cases, we introduced point or line into square brackets.

The formula Ἐπεζεύχθω ἡ AB constitutes an interesting exception to this rule. Literally, it should be translated by *join the [line] AB*. The meaning is clearly *join points A and B to make line AB*. However, the subject, ἡ AB, refers to a line, not to points, as can be seen both in the feminine gender and in the singular number of the article.

This inconsistency in the formula is probably due to a displacement of the subject from points to a line, a displacement that is not odd in the context of Greek geometry since it works with lines, not with points.[43] To translate it, perhaps we could have looked for a verb like *build or draw the line AB*; in that case, we would have respected the subject, but not the meaning of the verb. Another option would have been to ignore the feminine article that refers to line and translate it by *join points A and B*, but it would not be satisfactory either because we would have to change the number of the verb and ignore what the article refers to. A more extended and more explicit expression could have been sought, like *join points A and B to make line AB*, but the frequency of this formula does not make it recommendable. Regardless, we have decided, in this case, to translate it simply by *join AB*, keeping the ambiguity. However, the reader should remember that it means *join points A and B to make line AB*. In some cases, Aristarchus uses the plural when he asks to join several pairs of points to form several different lines: ἐπεζεύχθωσαν αἱ ΚΒ, ΒΘ, ΚΑ, ΑΘ, ΒΔ, which means *join points K and B to make line KB, points B and Θ to make line BΘ*, and so on. We translated it by *join KB, BΘ, KA, AΘ, and BΔ*.

The masculine article followed by three letters is used to name a circle. For example, ὁ ΑΒΓ means *the [circle] ABΓ*. For referring to the diameter of a circle, the word διάμετρος is explicitly used. But for the radius, a formula is introduced: ἡ ἐκ τοῦ κέντρου τοῦ ΑΒΓ κύκλου. This formula would mean something like *the [line that starts from] the center of circle ABΓ*. We translate it by *the radius of circle ABΓ*. Another formula is used to identify a circle performing a definite action: the masculine article followed by a participle. It refers to the circle that performs the action designated by such participle. For example, we find very often the expression ὁ διορίζων ἐν τῇ σελήνῃ τό τε σκιερὸν καὶ τὸ λαμπρόν that we translate by *the [circle] delimiting the shaded and bright parts on the Moon*.

Greek mathematicians use περιφέρεια both to refer to a circumference and a part of it, that is, to an arc. In Aristarchus's work, the 40 times the word appears, it always refers to an arc. Thus, although Heath translates περιφέρεια by *circumference* and Massa Esteve as *circumferencia*, translating Commandino's *circumferentia*, we have decided to translate it by *arc*, since it is not common usage in English to use the word *circumference* to refer to a portion of it.

There are two possibilities to refer to an angle. If it belongs to a specific triangle that is already known from the context, ἡ πρὸς τῷ Α, we translate it by *the angle at point A*. Nevertheless, when Aristarchus wants to mention an angle without reference to a particular triangle, he uses ἡ ὑπὸ τῶν ΑΒΓ, which should be translated by *the [angle] [formed by the points] A, B and Γ*. To simplify the translation, we translate it by *angle ABΓ*. This second option is much more frequent in Aristarchus's treatise.

Some formulas refer to specific relationships between geometric objects. Aristarchus frequently expresses *perpendicular* in two different ways. The word κάθετος is reserved exclusively for lines perpendicular to each other. Instead, when referring to a line perpendicular to a plane, he uses the formula πρὸς ὀρθάς, which

43 We thank Reviel Netz for this suggestion in personal communication.

means *at right angles to*. However, this last expression is also used exceptionally for two lines. We have decided to keep the difference translating κάθετος by *perpendicular* and the formula πρὸς ὀρθὰς by *at right angles to*. Something similar happens with the expressions for parallel. If it refers to two parallel lines, Aristarchus always uses the word παράλληλος. Nevertheless, to say that one circle is parallel to another, he employs the expression ὁ παρὰ τὸν . . . κύκλος. For example, the formula ὁ παρὰ τὸν διορίζοντα μέγιστος κύκλος means the great circle parallel to the [circle] delimiting. In our translation, since these formulas are so frequent, we have decided to omit the use of square brackets and translate it by *the great circle parallel to the delimiting one*.

To say that a line is tangent to a circle, Aristarchus uses the verb ἐφάπτω, in the middle aorist. The verb means to touch, but in a mathematical context, it means *to touch at a point*, that is, *to be tangent*. In fact, the Latin word that translates ἐφάπτω is *tango* (touch), from which comes *tangent*. As in this context, the verb clearly has its technical meaning, we translate it by *be tangent to*.

Second, the numbers are generally expressed in the Greek number system, in which the letters of the alphabet plus three ancient symbols represent the numbers. However, Aristarchus sometimes uses words, especially when dealing with fractional numbers.

Thus, *18 times* it appears either as ιη (18) or as ὀκτωκαιδεκαπλάσιον (eighteen). We have reflected this difference in our translation. The treatise also has specific arithmetic operators, such as the square of a number or line. In those cases, Aristarchus uses the formula τὸ ἀπὸ (τῆς) ΑΒ. If the τῆς is present, it is making explicit that it is the square of a line, and thus we will translate it by *the square of line AB*; if it is not, we translate it by *the square of AB*. When referring to the cube, he uses ὁ ἀπὸ τῆς Α κύβος, which we will translate *the cube of line A*. This is an example where the line is designated by a single letter. For referring to the product between two numbers, there is a common expression: ὁ περιεχόμενος ἀριθμὸς ὑπὸ τῶν Ν καὶ τῶν Μ (*the number product of N and M*). However, more frequently, Aristarchus employs another, not-so-common one: ὁ συνηγμένος ἔκ τε τῶν κβ καὶ πθ, which we translate by *the product of 22 and 89* (Berggren and Sidoli 2007: 246).

Third, there are distinctive expressions for the different operations with quotients. The meaning of these is developed in Appendix 1. Here, we only explicate how we translate them. ἐναλλάξ is translated by *by alternation*, ἀνάπαλιν by *by inversion*, συνθέντι by *by composition*, διελόντι by *by separation*, ἀναστρέαντι by *by conversion*, and δι 'ἴσου by *by equality of terms*. To assert that one quotient is equal to another, that is, ΑΚ/ΚΒ = ΑΓ/ΒΖ, Aristarchus very frequently uses the expression ὡς ἡ ΑΚ πρὸς τὴν ΚΒ, οὕτως ἡ ΑΓ πρὸς τὴν ΒΖ, which we translate by *as line AK is to line KB, so is line AΓ to line BZ*. In a few cases, οὕτως is omitted. In those cases, we put *so* into square brackets. For asserting inequalities between quotients, like ΑΒ/ΒΓ > ΔΕ/ΗΘ or ΑΒ/ΒΓ < ΔΕ/ΗΘ, Aristarchus uses the formula ἡ ΑΒ πρὸς τὴν ΒΓ μείζονα / ἐλάσσονα λόγον ἔχει ἢ ἡ ΔΕ πρὸς τὴν ΗΘ. We systematically translated by *line AB has to line BΓ a greater/smaller ratio than line ΔΕ [has] to line ΗΘ*. This formula offers some little differences that are not worth explaining here but are reflected in our translation.

The logical connectors have very precise formulae too. Some logical connectors indicate conclusions, and some others premises. We indicate in what follows the rules that we have followed in our edition. To indicate a conclusion, the most frequent is ἄρα; we have translated it by *therefore*. Also frequent is δή. Its meaning is similar to that of ἄρα, but with a clearly different use. Sometimes, δή introduces a conclusion that is accompanied by an epistemic judgment: *it is evident that* (φανερὸν δὴ ὅτι), *I say that* (λέγω δὴ ὅτι), *it will be proved that* (δειχθήσεται δή). Sometimes it is used in the description of the construction of the diagrams. In general, to indicate the sections a drawn plane will make in a figure. For example: [τὸ ἐπίπεδον] ποιήσει δὴ τομάς (*[the plane] will make sections*). Nevertheless, as Alexander Jones told us in a personal communication, ancient Greek particles are more like herbs than spices, giving a subtle flavor, whereas the common English counterparts often draw too much attention to themselves. Often it is best to simply not represent δή at all in the translation; it has something of the same effect as a sentence in English that follows another without any explicit connection. Consequently, we decided not to translate it.

Several particles are used to express premises. If the premise precedes the conclusion, it is indicated either by καὶ ἐπεὶ or by ἐπεὶ γάρ. We translate καὶ ἐπεὶ by *and since*. However, in the case of ἐπεὶ γάρ, the meaning of γάρ is so weak that we decided to translate the whole expression directly by *since*. Often, ἐπεὶ γάρ introduces a premise, and the conclusion follows immediately, preceded by ἄρα, giving rise to the expression "since *p*, therefore *q*." In these cases, we translate ἄρα by *therefore*. We also translated γάρ by *therefore* when it is not preceded by ἐπεὶ and indicates the following conclusion. Nevertheless, most of the time, γάρ is linked to an imperative used to draw the diagrams. In those cases, again, its meaning is so weak that we do not translate it, as in Ἔστω γὰρ (*Let X be*), Ἐπεζεύχθω γὰρ (*Let X be joined*), Ἤχθω γὰρ (*Let X be drawn*). Occasionally, γάρ may appear indicating premises, and we translate it by *since*. For example, in proposition 3, we read καὶ ἔστιν ἐλάσσων ἡ ΒΓ τῆς ΒΔ· κέντρον γάρ ἐστι τὸ Α τοῦ ΓΔ κύκλου (*And line ΒΓ is smaller than line ΒΔ, since the center of circle ΓΔ is point A*). ὥστε and ὥστε καὶ are used very frequently to indicate a premise following a conclusion. In those cases, we translate it by *so that* and *so that also*, respectively.

Finally, Aristarchus uses a series of formulas to indicate the construction of the diagrams. In proposition 1, he asks to build spheres. In all the others, he will assume that they are already made, and begin by identifying their centers. To do this, he will use a formula similar to Ἔστω γὰρ ἡ μὲν ἡμετέρα ὄψις πρὸς τῷ Α, ἡλίου δὲ κέντρον τὸ Β, σελήνης δὲ κέντρον τὸ Γ, which we translate by *Let our eye be at point A, let the center of the Sun be point B, and let the center of the Moon be point Γ*. Usually, these spheres are contained by a cone. After the identification of the centers of the spheres, he asks that a plane be extended, passing through the center of the spheres: ἐκβεβλήσθω διὰ τῆς ΑΓΒ ἐπίπεδον. We translate it by *let a plane be produced through line ΑΓΒ*. The plane makes it possible to transform a figure that was initially three-dimensional, with spheres and cones, into a two-dimensional one, with circles and triangles. He usually adds, therefore, ποιήσει δὴ τομάς, ἐν μὲν ταῖς σφαίραις μεγίστους κύκλους, ἐν δὲ τῷ κώνῳ εὐθείας, which we translate by *it will make sections on the spheres [that are] great circles, and on the cones, straight lines*. And then, he completes the construction with a phrase that serves to identify the circles and triangles attributing the corresponding letters: ποιείτω οὖν ἐν μὲν ταῖς σφαίραις μεγίστους κύκλους τοὺς ΖΗ, ΚΛΘ, ἐν δὲ τῷ κώνῳ εὐθείας τὰς ΑΖΘ, ΑΗΚ. We translate it by *so let [the plane] make on the spheres great circles ΖΗ and ΚΛΘ, and on the cone, the straight lines ΑΖΘ and ΑΗΚ*. This structure, with slight variations, is repeated in nearly all the propositions.

As mentioned, we have chosen to be as faithful as possible to the incipient technical vocabulary governing Aristarchus's *On Sizes*. Reading the repeated formulas will give the reader the feeling that it is a technical work. We have preferred to be faithful to the text and, in any case, clarify in a note the meaning of a specific expression that might not be entirely clear to a modern reader. In the few cases we found it convenient to deviate from a faithful translation, we have clarified it in a footnote, justifying our choice.

To this new English translation of Aristarchus's *On Sizes*, we added the English translation of the scholia that we consider relevant and the English translation from the Latin of Commandino's comments, since we believe that even today they can be of great utility for the modern scholar. For the translation of the scholia and Commandino's Latin text, we have followed the same criteria as for Aristarchus's text already explained. In the translation of Aristarchus's text, we have introduced, between square brackets, the number of the scholium preceded by *Sch* that refers to that text. For example, "[Sch. 81(51)]" means that the scholion 81(51) refers to this paragraph. Commandino introduces his comments in two ways. Most of the time, he introduces endnotes, identified by capital Latin letters. In the translation of *On Sizes*, we have added the letters corresponding to Commandino's notes between square brackets and the note's text in a column to the side. Occasionally, he adds a small reference in the margin of the text without index letters. In most cases, they are short references to propositions of Euclid's *Elements* or cross-references to other parts of *On Sizes*. We marked them with [*] in the text and placed the translation in the margin. These brief references often appear in the text of Commandino's endnotes themselves. In those cases, we have added the reference between curly braces at the right place in our translation of Commandino's note. Commandino's work includes Aristarchus's diagrams and some of his own that he introduces in the notes. In all cases, the label letters used by Commandino belong to the Latin alphabet. We have preserved the Latin letters in the Latin text. Still, when the letters refer to Aristarchus's diagrams, which we have preserved in Greek, we have replaced them in the English translation with the corresponding Greek letters.

The English translation of *On Sizes* and the scholia are accompanied by the corresponding Greek text, and the translation of Commandino's notes, by his Latin text.

2 Greek text and translation

ΑΡΙΣΤΑΡΧΟΥ ΠΕΡΙ ΜΕΓΕΘΩΝ ΚΑΙ
ΑΠΟΣΤΗΜΑΤΩΝ ΗΛΙΟΥ ΚΑΙ ΣΕΛΗΝΗΣ

<ΥΠΟΘΕΣΕΙΣ>
α΄. Τὴν σελήνην παρὰ τοῦ ἡλίου τὸ φῶς λαμβάνειν.
 β΄. Τὴν γῆν σημείου τε καὶ κέντρου λόγον ἔχειν
πρὸς τὴν τῆς σελήνης σφαῖραν.

DOI: 10.4324/9781003184553-2

Aristarchus, *On the Sizes and Distances of the Sun and Moon*[1]

<Hypotheses>[2]

1. That the Moon gets its light from the Sun.
2. That the Earth has a ratio of a point and center with respect to the sphere of the Moon.[3,4]

---

1 In the notes, we use the following abbreviations: V: Valla (1498), C: Commandino (1572), W: Wallis (1688), FG: Comte´de Fortia (1810), F: Comte de Fortia (1823), No: Nokk (1854), Ni: Nizze (1856), H: Heath (1913), Th: Thomas (1941), M: Massa Esteve (2007), G: Gioé (2007), S: Spandagos (2001). In the translation, we have indicated the text discussed by the translated scholia introducing [*Sch*], followed by scholion number.

2 ΥΠΟΘΕΣΕΙΣ does not appear in the manuscripts. W:569 introduces ΘΕΣΕΙΣ and translates it as *Positiones* (following the translation of C:1b). Ni:5 follows W and introduces ΘΕΣΕΙΣ. H:352 adds ΥΠΟΘΕΣΕΙΣ because Aristarchus refers to them explicitly using that term (see, for example, pp. 48–49) and because Pappus also uses it in his commentary. Th:4 follows H. In his translation, No:3 introduces: "Voraussetzungen." S also introduces ΥΠΟΘΕΣΕΙΣ in his Greek edition and translation. F, FG y V follow the manuscript and do not introduce any title.

3 "The sphere of the Moon" does not refer to the sphere of the Moon itself, but to the sphere that carries the Moon, that is, to the sphere in which the Moon moves. The meaning is so obvious that most modern translators decided to incorporate it into the translation (H:353, F:5, M:1), in contrast with the Latin translators that keep the literal meaning (V:66, C:1b). We prefer to offer a literal translation. The meaning of "the sphere of the Moon" is evident in proposition 3. It is enough to see the figure, where point *A* represents at the same time the center of the Earth and our eye, which can only be true if the Earth has a ratio of a point and center with respect to the sphere in which the Moon moves, represented by arc *ΓΔ*.

4 G:112 translates: *comme un point au centre de la sphère* (like a point in the center of the sphere . . .).

γ΄. Ὅταν ἡ σελήνη διχότομος ἡμῖν φαίνηται, νεύειν εἰς τὴν ἡμετέραν ὄψιν τὸν διορίζοντα τό τε σκιερὸν καὶ τὸ λαμπρὸν τῆς σελήνης μέγιστον κύκλον.

δ΄. Ὅταν ἡ σελήνη διχότομος ἡμῖν φαίνηται, τότε αὐτὴν ἀπέχειν τοῦ ἡλίου ἔλασσον τεταρτημορίου τῷ τοῦ τεταρτημορίου τριακοστῷ.

ε΄. Τὸ τῆς σκιᾶς πλάτος σεληνῶν εἶναι δύο.

ς΄. Τὴν σελήνην ὑποτείνειν ὑπὸ πεντεκαιδέκατον μέρος ζῳδίου.

3. That, when the Moon appears to us bisected, the great circle[5] delimiting the shaded and the bright parts of the Moon points towards our eye.[6]

4. That, when the Moon appears to us bisected, it is then distant from the Sun by one quadrant minus one-thirtieth part of a quadrant.[7]

5. That the width of the shadow [of the Earth] is two moons.

6. That the Moon subtends one-fifteenth part of a sign of the zodiac. [Sch. 3 (5)].

5 Ni:5 omits μέγιστον because Aristarchus proves later that it is not a great circle (see note a). Berggren and Sidoli (2007: 215, note 8) assert that it should be omitted, even if Greek manuscripts, Pappus's commentary, and the Arabic editions attest its existence. We prefer a translation reflecting the text of the manuscripts. It is true that it is not a great circle. Indeed, proposition 5 asserts that the circle that verges to our eye is not the great circle but one parallel to it, dividing the dark and bright portions of the Moon. However, proposition 4 proves that its difference with respect to a great circle is imperceptible to the human eye. So it seems that hypothesis 3 is still confusing what propositions 4 and 5 will distinguish. Assuming that, it makes perfect sense that hypothesis 3 asserts that the dividing circle is also maximum before the following propositions show that the dividing circle cannot be the maximum. Consequently, proposition 5 will clarify what hypothesis 3 roughly says. Similarly, hypothesis 5 asserts that "the width of the shadow [of the Earth] is two moons" when this is not accurate. Proposition 12 later shows that the width of the shadow is equal to two circles of visibility of the Moon, which are smaller than two Moons.

6 In the title of proposition 5, (pp. 88–89) Aristarchus makes explicit what this means when he asserts ὁ παρὰ τὸν διορίζοντα ἐν τῇ σελήνῃ τό τε σκιερὸν καὶ τὸ λαμπρὸν νεύει εἰς τὴν ἡμετέραν ὄψιν, τουτέστιν, ὁ παρὰ τὸν διορίζοντα μέγιστος κύκλος καὶ ἡ ἡμετέρα ὄψις ἐν ἑνί εἰσιν ἐπιπέδῳ (the great circle parallel to the one delimiting the shaded and bright parts on the Moon points towards our eye, that is, the great circle parallel to the delimiting one and our eye are in one plane). Cf. H:453, note 2. In the Latin translations, C:1b and V:66 use "vergere," which means the same as the Greek term νεύω.

7 That is, it is smaller than 90 degrees by 1/30 part of 90 degrees, that is, 87 degrees. M:35 mistakenly translates: . . . *entonces dista del Sol menos de una treintava parte de un cuadrante* (consequently, it is far from the Sun less than one-thirtieth part of a quadrant), but this means smaller than 3°, which is absurd.

Ἐπιλογίζεται οὖν τὸ τοῦ ἡλίου ἀπόστημα ἀπὸ τῆς γῆς τοῦ τῆς σελήνης ἀποστήματος μεῖζον μὲν ἢ ὀκτωκαιδεκαπλάσιον, ἔλασσον δὲ ἢ εἰκοσαπλάσιον, διὰ τῆς περὶ τὴν διχοτομίαν ὑποθέσεως· τὸν αὐτὸν δὲ λόγον ἔχειν τὴν τοῦ ἡλίου διάμετρον πρὸς τὴν τῆς σελήνης διάμετρον· τὴν δὲ τοῦ ἡλίου διάμετρον πρὸς τὴν τῆς γῆς διάμετρον μείζονα μὲν λόγον ἔχειν ἢ ὃν τὰ ιθ πρὸς γ, ἐλάσσονα δὲ ἢ ὃν μγ πρὸς ς, διὰ τοῦ εὑρεθέντος περὶ τὰ ἀποστήματα λόγου, τῆς <τε> περὶ τὴν σκιὰν ὑποθέσεως, καὶ τοῦ τὴν σελήνην ὑπὸ πεντεκαιδέκατον μέρος ζῳδίου ὑποτείνειν.

It is concluded, then, that the distance of the Sun from the Earth is greater than eighteen[8] times, but smaller than twenty times the distance of the Moon[9] [from the Earth], from the hypothesis related to the halving [of the Moon];[10] that[11] the diameter of the Sun has the same ratio to the diameter of the Moon[12] [as the ratio between the distances, and] that the diameter of the Sun has to the diameter of the Earth a ratio greater than the [ratio] that 19 has to 3, but smaller than [the ratio that] 43 has to 6,[13] from the ratio found[14] concerning the distances,[15] from the hypothesis about the shadow,[16] and from [the fact that] the Moon subtends one-fifteenth part of a sign of the zodiac [Sch. 5 (4)].[17]

8 V:66 translates *octupla* (eight times).
9 This refers to proposition 7.
10 Pappus's commentary quotes this paragraph verbatim. In it, the expression διὰ τῆς περὶ τὴν διχοτομίαν ὑποθέσεως is located at the end of the following sentence, immediately after διάμετρον, and preceded by τοῦτο δὲ. Consequently, according to Pappus's version, the hypothesis related to the dichotomy of the Moon is used to prove not only the ratio between the distances but also the ratio between the sizes. Ni:5 and F:6 follow Pappus's version (even if Ni omits τοῦτο δὲ and in the Greek edition and Latin translation of F the text has not been modified (FG:4,5)).
11 V:66 omits τὴν τοῦ ἡλίου διάμετρον πρὸς τὴν τῆς σελήνης διάμετρον· τὴν δὲ τοῦ ἡλίου διάμετρον πρὸς τὴν τῆς γῆς διάμετρον μείζονα μὲν λόγον ἔχειν.
12 This refers to proposition 9. M:36 translates *Que el diámetro del Sol está en la misma razón que el diámetro de la Luna* (that the diameter of the Sun is in the same ratio as the diameter of the Moon). This is a mistake, because it would imply that the Sun and the Moon are equal in size. Even if this is true when referring to the apparent size, it is clear that here Aristarchus is not comparing apparent sizes.
13 This refers to proposition 15.
14 C:1b omits εὑρεθέντος and, consequently, also do it M:36 and W:570 in his Latin translation. F. omits it in his Latin (FG:4) and French translation (F:6).
15 This refers to proposition 7.
16 This refers to hypothesis 5.
17 This refers to hypothesis 6.

α'. Δύο σφαίρας ἴσας μὲν ὁ αὐτὸς κύλινδρος περιλαμβάνει, ἀνίσους δὲ ὁ αὐτὸς κῶνος τὴν κορυφὴν ἔχων πρὸς τῇ ἐλάσσονι σφαίρᾳ· καὶ ἡ διὰ τῶν κέντρων αὐτῶν ἀγομένη εὐθεῖα ὀρθή ἐστιν πρὸς ἑκάτερον τῶν κύκλων, καθ' ὧν ἐφάπτεται ἡ τοῦ κυλίνδρου ἢ ἡ τοῦ κώνου ἐπιφάνεια τῶν σφαιρῶν.

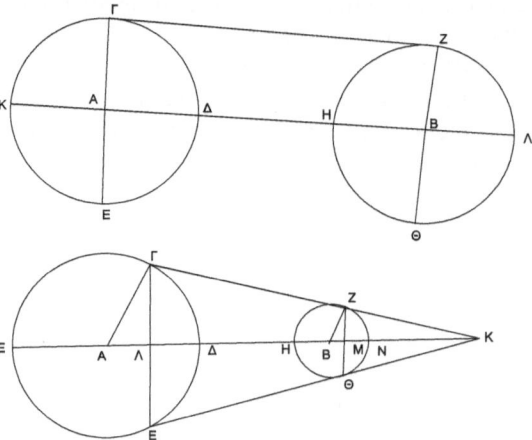

1. **One and the same cylinder envelops two equal
    spheres, and one and the same cone [envelops
    two] unequal [spheres, the cone] having its
    vertex in the same direction as the smaller
    sphere; and the straight line drawn through
    their centers[18] is at right angles to each of the
    circles in which the surface of the cylinder or
    of the cone is tangent to the spheres.[19]**

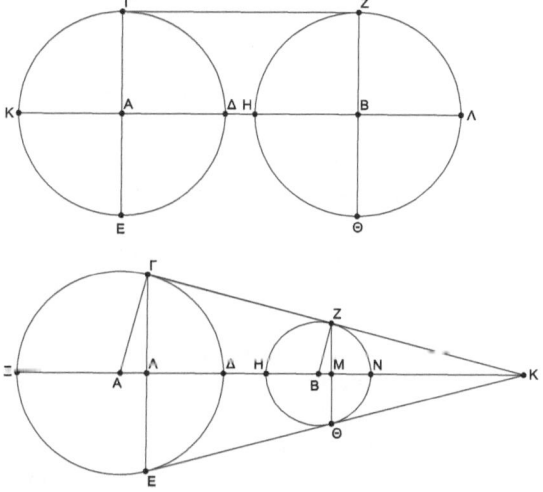

Figure 2.1.1 Diagram of proposition 1: Diagram A (above)
and diagram B (below).

18 This refers to the center of the spheres.
19 The numbering of propositions varies in the different manuscripts. In some of them (for example,
    ms. A), proposition 1 is split into two. We prefer to follow the division of W., which all authors,
    except F. and V., adopted. In any case, as Berggreen and Sidoli (2007: 218) recall, most likely, the
    numbering of the propositions is not Aristarchus's but a later intercalation.

A. Ex prima[m][1] propositione sphaericorum Theodosii.

Ἔστωσαν ἴσαι σφαῖραι, ὧν κέντρα ἔστω τὰ Α, Β σημεῖα, καὶ ἐπιζευχθεῖσα ἡ ΑΒ ἐκβεβλήσθω, καὶ ἐκβεβλήσθω διὰ τῆς[2] ΑΒ ἐπίπεδον· ποιήσει δὴ τομὰς ἐν ταῖς σφαίραις μεγίστους κύκλους. ποιείτω οὖν τοὺς ΓΔΕ, ΖΗΘ κύκλους, καὶ ἤχθωσαν ἀπὸ τῶν Α, Β τῇ ΑΒ πρὸς ὀρθὰς αἱ ΓΑΕ, ΖΒΘ, καὶ ἐπεζεύχθω ἡ ΓΖ.

καὶ ἐπεὶ αἱ ΓΑ, ΖΒ ἴσαι τε καὶ παράλληλοί εἰσιν, καὶ αἱ ΓΖ, ΑΒ ἄρα ἴσαι τε καὶ παράλληλοί εἰσιν. παραλληλόγραμμον ἄρα ἐστὶν τὸ ΓΖΑΒ, καὶ αἱ πρὸς τοῖς Γ, Ζ γωνίαι

B. Ex 34 primi Eucl. paralelogrammorum enim locorum anguli, qui ex opposito aequales sunt et sunt recti qui ad A B anguli. Ergo et qui ad C F recti erunt.

ὀρθαὶ ἔσονται· ὥστε ἡ ΓΖ τῶν ΓΔΕ, ΖΗΘ κύκλων ἐφάπτεται.

ἐὰν δὴ μενούσης τῆς ΑΒ τὸ ΑΖ παραλληλόγραμμον καὶ τὰ ΚΓΔ, ΗΖΛ ἡμικύκλια περιενεχθέντα εἰς τὸ αὐτὸ πάλιν ἀποκατασταθῇ ὅθεν ἤρξατο φέρεσθαι, τὰ μὲν ΚΓΔ, ΗΖΛ ἡμικύκλια

C. Ex 16 tertii libri elementorum.

ἐνεχθήσεται κατὰ τῶν σφαιρῶν, τὸ δὲ ΑΖ παραλληλόγραμμον γεννήσει κύλινδρον, οὗ βάσεις ἔσονται οἱ περὶ διαμέτρους τὰς ΓΕ, ΖΘ κύκλοι, ὀρθοὶ ὄντες πρὸς τὴν ΑΒ, διὰ τὸ ἐν πάσῃ μετακινήσει διαμένειν τὰς ΓΕ, ΘΖ

D. Ex 21 diffinitione undecimi libri elementorum.

ὀρθὰς τῇ ΑΒ. καὶ φανερὸν ὅτι ἡ ἐπιφάνεια αὐτοῦ ἐφάπτεται τῶν σφαιρῶν, ἐπειδὴ ἡ ΓΖ κατὰ πᾶσαν μετακίνησιν ἐφάπτεται τῶν ΚΓΔ, ΗΖΛ ἡμικυκλίων.

Ἔστωσαν δὴ αἱ σφαῖραι πάλιν, ὧν κέντρα ἔστω τὰ Α, Β, ἄνισοι, καὶ μείζων ἧς κέντρον τὸ Α· λέγω ὅτι τὰς σφαίρας ὁ αὐτὸς κῶνος περιλαμβάνει τὴν κορυφὴν ἔχων πρὸς τῇ ἐλάσσονι σφαίρᾳ.

---

1 C:4 *ex primam*, but in the manuscript of C, *ex prima*.
2 All the manuscripts: τῆς. H:354: τοῦ.

Let there be equal spheres, and let their centers be points A and B, and let AB be joined, and let it be produced,[20] and let a plane be produced through line AB. It will make sections[21] on the spheres that are great circles [A]. So let [the plane] make the circles ΓΔE and ZHΘ, and let lines ΓAE and ZBΘ be drawn from points A and B at right angles to line AB, and let ΓZ be joined.

And since lines ΓA and ZB[22] are equal and parallel, therefore lines ΓZ and AB are also equal and parallel.[23] Therefore, ΓZAB is a parallelogram, and the angles at points Γ and Z will be right [B], so that line ΓZ is tangent to the circles ΓΔE and ZHΘ [C]. If line AB remains fixed, and parallelogram AZ and semicircles KΓΔ and HZΛ are rotated and restored to the same [place] whence they started to travel, [then,] semicircles KΓΔ and HZΛ will travel along [the surface of] the spheres,[24] and parallelogram AZ will generate a cylinder [D], whose bases will be the circles [drawn] around diameters ΓE and ZΘ, at right angles to line AB, because within the entire movement,[25] lines ΓE and ΘZ remain at right angles to line AB. And it is evident that its surface[26] is tangent to the spheres since, during the entire movement, line ΓZ is tangent to semicircles KΓΔ and HZΛ.

Let there be the spheres again, and let their centers be points A and B, [and let the spheres be] unequal, and let the greater be the one whose center is point A. I say that one and the same cone envelops the spheres, [the cone] having its vertex in the same direction as the smaller sphere.

A. From the first proposition of the *Sphaerics* of Theodosius.

B. From [proposition] 34 of the first [book of the *Elements*] of Euclid, for the angles of the parallelogrammical areas that are opposite, are equal, and angles at A and B are right. Consequently, also [angles] at Γ and Z will be right.

C. From [proposition] 16 of the third book of the *Elements* [of Euclid].

D. From definition 21 of the eleventh book of the *Elements* [of Euclid].

---

20 G:116 omits: ἐκβεβλήσθω. Netz (2004) asserts that, in mathematical works, the text must be interpreted based on the figures. The figure of the proposition shows that line AB must be produced in both directions. Similarly, No:4 translates *man verbinde AB, verlängere die Verbindungslinie (nach beinden Seiten bis K, L)*, interpreting ἐκβεβλήσθω as producing line AB in both directions to K and L. He seems to be inspired in F:6–7: *dont les centres soient A et B, et soit prolongée la ligne AB qui joint ces centres (jusqu'à ce qu'elle rencontre la surface de chaque sphère aux points K, L).*

21 V:67 *faciant.*

22 In the French version, G:116 puts CF and AB instead of CA and FB; in the Italian one (G:197), it is correct.

23 FG:7 does not translate καὶ αἱ ΓZ, AB ἄρα ἴσαι τε καὶ παράλληλοί εἰσιν in the Latin version, but he does translate it in the French translation (F:7) and includes it in his Greek edition (FG:6). In FG:202, he makes explicit that it was a misprint.

24 For the meaning of κατὰ τῶν σφαιρῶν, see note 1 in H:357.

25 Aristarchus's description of a cylinder's construction closely follows definition 21 of the eleventh book of Euclid's *Elements*, referred by Commandino in comment D.

26 "Its" refers to the cylinder.

E. Ex prima sphaericorum
Theodosii.

F. Illud autem punctum hoc
modo inveniemus. Ducatur
seorsum ea, quae ex centro
circuli maioris CDE sitque
AD: et ex ipsa AD abscindatur
AO aequalis ei, quae ex centro
minoris circuli: fiatque ut DO
ad OA, ita AB ad aliam, quae
sit BK. Erit enim componendo,
ut DA ad AO, hoc est ut quae
ex centro circuli maioris ad
eam quae ex centro minoris,
ita AK ad KB.

A   O D   B ───────────── K

Ἐπεζεύχθω ἡ ΑΒ, καὶ ἐκβεβλήσθω διὰ τῆς ΑΒ
ἐπίπεδον· ποιήσει δὴ τομὰς ἐν ταῖς σφαίραις
κύκλους. ποιείτω τοὺς ΓΔΕ, ΖΗΘ·

μείζων ἄρα ὁ ΓΔΕ κύκλος τοῦ ΗΖΘ κύκλου·
ὥστε καὶ ἡ ἐκ τοῦ κέντρου τοῦ ΓΔΕ κύκλου μείζων
ἐστὶ τῆς ἐκ τοῦ κέντρου τοῦ ΖΗΘ κύκλου. δυνατὸν
δή ἐστι λαβεῖν τι σημεῖον, ὡς τὸ Κ, ἵν' ᾖ, ὡς ἡ ἐκ
τοῦ κέντρου τοῦ ΓΔΕ κύκλου πρὸς τὴν ἐκ τοῦ
κέντρου τοῦ ΖΗΘ κύκλου, οὕτως ἡ ΑΚ πρὸς τὴν
ΚΒ. ἔστω οὖν εἰλημμένον τὸ Κ σημεῖον, καὶ ἤχθω
ἡ ΚΖ ἐφαπτομένη τοῦ ΖΗΘ κύκλου, καὶ ἐπεζεύχθω
ἡ ΖΒ, καὶ διὰ τοῦ Α τῇ ΒΖ παράλληλος ἤχθω ἡ ΑΓ,
καὶ ἐπεζεύχθω ἡ ΓΖ.

Let AB be joined, and let a plane be produced through line AB. It will make sections[27] on the spheres that are circles [E]. Let [the plane] make circles ΓΔE and ZHΘ.

Therefore, circle ΓΔE is greater than circle HZΘ, and so also the radius of circle ΓΔE is greater than the radius of circle ZHΘ. It is possible to take a point, like K [F], in order that, as the radius of circle ΓΔE is to the radius of circle ZHΘ, so will line AK be to line KB. So, let the point K be taken, and let line KZ be drawn, and let it be tangent to circle ZHΘ, and let ZB be joined, and, from[28] point A, let line ΑΓ be drawn parallel to line BZ, and let ΓZ be joined.

E. From the first [proposition] of the *Sphaerics* of Theodosius.

F. But we will find that point in this way. Let a radius of the greater circle CDE be drawn separately and let it be AD. And, from the same AD, let AO, equal to one radius of the smaller circle, be separated, and let it be that, as DO is to OA, so [is] AB to another [line], let it be BK. For, by composition, it will be that, as DA is to AO, that is, as the radius of the greater circle is to the radius of the smaller circle, so will AK be to KB.

Figure 2.2.1 Diagram of Commandino's note F to proposition 1.

---

27 V:67: *faciet sectionem*, according to all manuscripts, in which appear τομὴν. In his Greek edition, W:573 changes it to the plural τομὰς, and he makes it explicit in note (c). H:356, C:3r, No:5, Ni:6 y Σ:71,99 follow him; M:39 translates "cortará" (will cut), avoiding deciding to introduce the singular or plural. FG:9 puts τομὴν in singular, in the Greek text, but he translates it in the plural in the Latin (FG:8) and French translations (F:8).

28 διὰ + genitive should be translated as "through," not "from," but in this case, "from" seems better because the line starts at point A. The same happens in the text referred to in note 92.

G. Hoc est si a puncto C ad
K ducatur recta linea,
transibit ea per F. Quod
nos demonstravimus in
commentariis in deciman
propositionem libri
Archimedis de iis, quae in
aqua vehuntur, lemmate primo.

[*Lemma I: Sit recta linea ab,
quam secent duae lineae inter
sese aequidistantes ac, de, ita
ut quam proportionem habet
ab ad bd, eandem habeat ac
ad de. Dico lineam, quae cb
puncta coniungit, etiam per
ipsum e transire.*

Si enim fieri potest, non
transeat per e, sed uel supra
uel infra. Transeat primum
infra, ut per f. Erunt triangula
abc, dbf inter se similia.
Quare ut ab ad bd, ita ac
ad df. Sed ut ab ad bd, ita
erat ac ad de, ergo df ipsi de
aequalis erit, videlicet pars
toti, quod est absurdum. Idem
absurdum sequetur, si linea
cb supra e punctum transire
ponatur. Quare cb etiam per
e necessario transibit. Quod
oportebat demonstrare.]

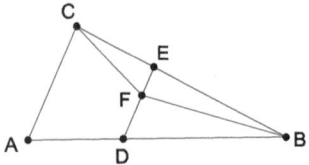

H. Ex 18 tertii elementorum.

K. Ex 29 primi elementorum.

L. Ex 17 tertii elementorum.

καὶ ἐπεί ἐστιν, ὡς ἡ ΑΚ πρὸς τὴν ΚΒ, ἡ ΑΔ πρὸς
τὴν ΒΝ, ἴση δὲ ἡ μὲν ΑΔ τῇ ΑΓ, ἡ δὲ ΒΝ τῇ ΒΖ,
ἔστιν ἄρα, ὡς ἡ ΑΚ πρὸς τὴν ΚΒ, ἡ ΑΓ πρὸς τὴν
ΒΖ. καὶ ἔστιν παράλληλος ἡ ΑΓ τῇ ΒΖ· εὐθεῖα ἄρα
ἐστὶν ἡ ΓΖΚ. καὶ ἔστιν ὀρθὴ ἡ ὑπὸ τῶν ΚΖΒ· ὀρθὴ
ἄρα καὶ ἡ ὑπὸ τῶν ΚΓΑ. ἐφάπτεται ἄρα ἡ ΚΓ τοῦ
ΓΔΕ κύκλου.

And since, as line AK is to line KB, line AΔ is to line BN; and line AΔ is equal to line AΓ, and line BN is equal to line BZ; therefore, as line AK is to line KB, line AΓ is to line BZ. And line AΓ is parallel to line BZ. Therefore, line ΓZK is straight [G]. [Sch. 13 (8)] And angle KZB is right [H]. Therefore, angle KΓA is also right [K]. Therefore, line KΓ is tangent to circle ΓΔE [L].[29]

G. That is, if a straight line is drawn from point C to K, it will go through F, what we demonstrated in the comments to the tenth proposition of Archimedes's book *On Floating Bodies*, first lemma.

[From Commandino, *Archimedis, De Iis . . .*, p. 31b]

Lemma 1: Let straight line AB be, which is intersected by two equidistant lines, AC and DE, so that AC has to DE the same ratio that AB has to BD. I say that the line joining points C and B also goes through E itself.

For, if it were possible, do not pass through E, but above or below. Let it go first below, as through point F. Triangles ABC and DBF will be similar to each other. Therefore, as AB is to BD, so [is] AC to DF. But as AB is to BD, so was AC to DE. Consequently, DF will be equal to DE itself; that is, the part [will be] equal to the whole, which is absurd. The same absurdity will follow if line CB is assumed to go above point E. Therefore, CB also will necessarily go through E, which had to be demonstrated.

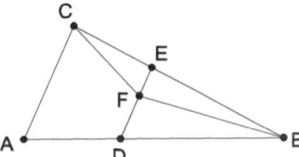

Figure 2.2.2 Diagram of Lemma 1 mentioned by Commandino in note G to proposition 1.

H. From [proposition] 18 of the third [book] of the *Elements* [of Euclid].

K. From [proposition] 29 of the first [book] of the *Elements* [of Euclid].

L. From [proposition] 17 of the third [book] of the *Elements* [of Euclid].

29 F:9 omits ἐφάπτεται ἄρα ἡ ΚΓ τοῦ ΓΔΕ κύκλου in his French translation, but not in his Greek edition and Latin translation (FG:10–11).

M. Ex 18 deffinitione undecimi
libri elementorum.

ἤχθωσαν δὴ αἱ ΓΛ, ΖΜ ἐπὶ τὴν ΑΒ κάθετοι. ἐὰν δὴ μενούσης τῆς ΚΞ τά τε ΞΓΔ, ΗΖΝ ἡμικύκλια καὶ τὰ ΚΓΛ, ΚΖΜ τρίγωνα περιενεχθέντα εἰς τὸ αὐτὸ πάλιν ἀποκατασταθῇ ὅθεν ἤρξατο φέρεσθαι, τὰ μὲν ΞΓΔ, ΗΖΝ ἡμικύκλια ἐνεχθήσεται κατὰ τῶν σφαιρῶν, τὸ δὲ ΚΓΛ τρίγωνον καὶ τὸ ΚΖΜ γεννήσει κώνους, ὧν βάσεις εἰσὶν οἱ περὶ διαμέτρους τὰς ΓΕ, ΖΘ κύκλοι, ὀρθοὶ ὄντες πρὸς τὸν ΚΛ ἄξονα· κέντρα δὲ αὐτῶν τὰ Λ, Μ· καὶ ὁ κῶνος τῶν σφαιρῶν ἐφάψεται κατὰ τὴν ἐπιφάνειαν, ἐπειδὴ καὶ ἡ ΚΖΓ ἐφάπτεται τῶν ΞΓΔ, ΗΖΝ ἡμικυκλίων κατὰ πᾶσαν μετακίνησιν.

Let lines ΓΛ and ZM be drawn perpendicular to line AB[30] [Sch. 14 (9)]. If line KΞ remains fixed, and semicircles ΞΓΔ and HZN and triangles KΓΛ and KZM are rotated and restored to the same [place] whence they started to travel, [then,] semicircles ΞΓΔ and HZN will travel along [the surface of] the spheres,[31] and triangle KΓΛ and [triangle] KZM will generate cones [M], whose bases are the circles [drawn] around diameters ΓE and ZΘ being at right angles to axis KΛ, and their centers [are] points Λ and M. And the cone[32] will be tangent to the spheres on its surface since line KZΓ is also tangent to semicircles ΞΓΔ and HZN[33] during the entire movement.

M. From definition 18 of the eleventh book of the *Elements* [of Euclid].

---

30 C:4a: *AM* (in the manuscript of C, *M* is written over a crossed-out letter). W:573 and M:39 do not follow C in this change; W makes it explicit in a note. Since points A, B, and M are in a straight line, if a line is perpendicular to AB, it is also perpendicular to AM, so the change from B to M does not imply a change in meaning.

31 Ni:6 omits τῶν. He makes explicit the reason in note (e): *Die Ausgaben haben τῶν σφαιρῶν ganz gegen die sonstige art des Aristarch.* However, on one side, κατὰ σφαιρῶν is never used in Aristarchus's text, and on the other side, κατὰ τῶν σφαιρῶν is used once in the same context when, a few paragraphs earlier, he refers to two equal spheres, and there Ni does not correct it.

32 C:4b translates: *coni . . . contingent* (the cones . . . will touch). W:573 and M:40 do not follow C in this change; W makes it explicit in a note.

33 FG:13 puts *NZH* following the three manuscripts he used (mss. A, B, Y). However, in his Latin (FG:12) and French (F:9) translations, he puts *HZN*. NI:7 puts *NZH*.

β' Ἐὰν σφαῖρα ὑπὸ μείζονος ἑαυτῆς σφαίρας
φωτίζηται,   μεῖζον   ἡμισφαιρίου
φωτισθήσεται.

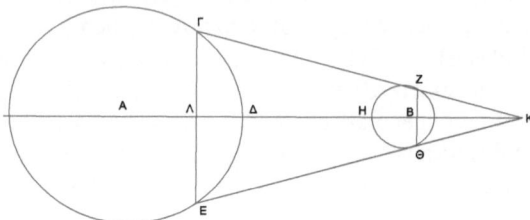

2. If a sphere is illuminated by a sphere greater than it, more than a hemisphere will be illuminated.

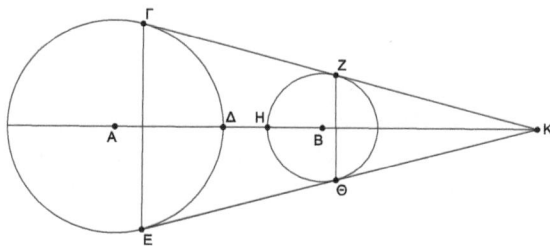

Figure 2.1.2 Diagram of proposition 2.

Σφαῖρα γάρ, ἧς κέντρον τὸ Β, ὑπὸ μείζονος ἑαυτῆς σφαίρας φωτιζέσθω, ἧς κέντρον τὸ Α· λέγω ὅτι τὸ φωτιζόμενον μέρος τῆς σφαίρας, ἧς κέντρον τὸ Β, μεῖζόν ἐστιν ἡμισφαιρίου.

Ἐπεὶ γὰρ δύο ἀνίσους σφαίρας ὁ αὐτὸς κῶνος περιλαμβάνει τὴν κορυφὴν ἔχων πρὸς τῇ ἐλάσσονι σφαίρᾳ, ἔστω ὁ περιλαμβάνων τὰς σφαίρας κῶνος, καὶ ἐκβεβλήσθω διὰ τοῦ ἄξονος ἐπίπεδον· ποιήσει δὴ τομὰς ἐν μὲν ταῖς σφαίραις κύκλους, ἐν δὲ τῷ κώνῳ τρίγωνον. ποιείτω οὖν ἐν μὲν ταῖς σφαίραις κύκλους τοὺς ΓΔΕ, ΖΗΘ, ἐν δὲ τῷ κώνῳ τρίγωνον τὸ ΓΕΚ.

φανερὸν δὴ ὅτι τὸ κατὰ τὴν ΖΗΘ περιφέρειαν τμῆμα τῆς σφαίρας, οὗ βάσις ἐστὶν ὁ περὶ διάμετρον τὴν ΖΘ κύκλος, φωτιζόμενον μέρος ἐστὶν ὑπὸ τοῦ τμήματος τοῦ κατὰ τὴν ΓΔΕ περιφέρειαν, οὗ βάσις ἐστὶν ὁ περὶ διάμετρον τὴν ΓΕ κύκλος, ὀρθὸς ὢν πρὸς τὴν ΑΒ εὐθεῖαν· καὶ γὰρ ἡ ΖΗΘ περιφέρεια φωτίζεται ὑπὸ τῆς ΓΔΕ περιφερείας· ἔσχαται γὰρ ἀκτῖνές εἰσιν αἱ ΓΖ, ΕΘ· καὶ ἔστιν ἐν τῷ ΖΗΘ τμήματι τὸ κέντρον τῆς σφαίρας τὸ Β· ὥστε τὸ φωτιζόμενον μέρος τῆς σφαίρας μεῖζόν ἐστιν ἡμισφαιρίου.

A. Ex I. sphaericorum Theodosii. Ut superius dictum est.

B. Ex 3 propositione primi libri conicorum Apollinii.

Let a sphere, whose center is point B, be illuminated by a sphere greater than it, whose center is point A. I say that the illuminated part of the sphere, whose center is B, is greater than a hemisphere.[34]

Since the same cone envelops two unequal spheres, [with the cone] having its vertex in the same direction as the smaller sphere, let there be the cone containing the spheres, and let a plane be produced through the axis. It will make sections on the spheres that are circles [A], and on the cone, a triangle [B]. So let [the plane] make circles ΓΔΕ and ΖΗΘ on the spheres and triangle ΓΕΚ on the cone.

It is evident that the segment of the sphere along arc ΖΗΘ, whose base is the circle [drawn] around diameter ΖΘ, is the part illuminated by the segment [of the sphere] along arc ΓΔΕ, whose base is the circle [drawn] around diameter ΓΕ, being at right angles to straight line AB (for arc ΖΗΘ is illuminated by arc ΓΔΕ – since lines ΓΖ and ΕΘ are extreme rays). And the center of the sphere, point B, is in segment ΖΗΘ.[35] So that, the illuminated part of the sphere is greater than a hemisphere.

A. From [proposition] 1 of the *Sphaerics* of Theodosius, as said above.

B. From proposition 3 of the first book of the *Conics* of Apollonius.

34 F:10 omits λέγω ὅτι τὸ φωτιζόμενον μέρος τῆς σφαίρας, ἧς κέντρον τὸ B, μεῖζόν ἐστιν ἡμισφαιρίου, but his Greek edition and his Latin translation include it (FG:12–13).
35 C:5a: *in proportione* (in proportion), but in his manuscript, Commandino wrote *in portione*. Ni W:574 ni M:43 follow him in this mistake.

γ΄. Ἐν τῇ σελήνῃ ἐλάχιστος κύκλος διορίζει
τό τε σκιερὸν καὶ τὸ λαμπρόν, ὅταν ὁ
περιλαμβάνων κῶνος τόν τε ἥλιον καὶ τὴν
σελήνην τὴν κορυφὴν ἔχῃ πρὸς τῇ ἡμετέρᾳ
ὄψει.

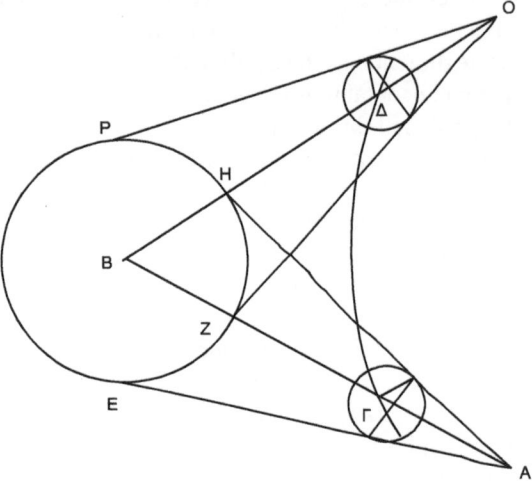

3. **On the Moon, the smallest circle**[36]**delimits the shaded and the bright parts, when the cone enveloping the Sun and the Moon has its vertex towards our eye.**

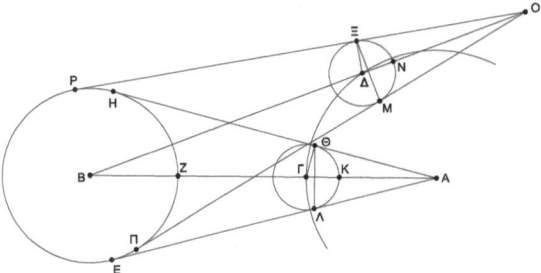

Figure 2.1.3 Diagram of proposition 3.

36 The proposition states that the circle that delimits the shaded and bright parts of the Moon – which varies in size depending on its distance from the Sun – reaches the smallest possible size every time the cone enveloping the Sun and the Moon has its vertex directed towards our eye.

A. Faciet enim triangula. Ex
   3 primi libri conicorum
   Apollonii.

B. Ex positione secunda huius.
   Ponitur enim terram puncti,
   ac centri habere rationem ad
   sphaeram lunae.

Ἔστω γὰρ ἡ μὲν ἡμετέρα ὄψις πρὸς τῷ Α, ἡλίου δὲ κέντρον τὸ Β, σελήνης δὲ κέντρον, ὅταν μὲν ὁ περιλαμβάνων κῶνος τόν τε ἥλιον καὶ τὴν σελήνην τὴν κορυφὴν ἔχῃ πρὸς τῇ ἡμετέρᾳ ὄψει, τὸ Γ, ὅταν δὲ μή, τὸ Δ· φανερὸν δὴ ὅτι τὰ Α, Γ, Β ἐπ' εὐθείας ἐστίν. ἐκβεβλήσθω διὰ τῆς ΑΒ καὶ τοῦ Δ σημείου ἐπίπεδον· ποιήσει δὴ τομάς, ἐν μὲν ταῖς σφαίραις κύκλους, ἐν δὲ τοῖς κώνοις εὐθείας. ποιείτω δὲ καὶ ἐν τῇ σφαίρᾳ, καθ' ἧς φέρεται τὸ κέντρον τῆς σελήνης, κύκλον τὸν ΓΔ· τὸ Α ἄρα κέντρον ἐστὶν αὐτοῦ· τοῦτο γὰρ ὑπόκειται· ἐν δὲ τῷ ἡλίῳ τὸν ΕΖΡ κύκλον, ἐν δὲ τῇ σελήνῃ, ὅταν μὲν ὁ περιλαμβάνων κῶνος τόν τε ἥλιον καὶ τὴν σελήνην τὴν κορυφὴν ἔχῃ πρὸς τῇ ἡμετέρᾳ ὄψει, κύκλον τὸν ΚΘΛ, ὅταν δὲ μή, τὸν ΜΝΞ, ἐν δὲ τοῖς κώνοις εὐθείας τὰς ΕΑ, ΑΗ, ΠΟ, ΟΡ, ἄξονας δὲ τοὺς ΑΒ, ΒΟ.

Let our eye be at point A; let the center of the Sun be point B; and let the center of the Moon be point Γ when the cone enveloping the Sun and the Moon has its vertex towards our eye, and when not, [let it be] point Δ. It is evident that points A, Γ, and B are in a straight line. Let a plane be produced through line AB and point Δ. It will make sections on the spheres [that are] circles, and on the cones, straight lines.[37] [A]. And let [the plane] make also on the sphere along which the center of the Moon travels, circle ΓΔ. Therefore, point A is its center – since this is hypothesized [B][3]–. [Let the plane make] on the Sun, circle EZP, and on the Moon, when the cone enveloping the Sun and the Moon has its vertex towards our eye, circle KΘΛ, and when not, [circle] MNΞ; and in the cones,[38] the straight lines EA, AH, ΠO, and OP; and the axes AB and BO.

A. For it will make triangles. From [proposition] 3 of the first book of the *Conics* of Apollonius.

B. From the second hypothesis of this [treatise]. For it is assumed that the Earth has a ratio of point and center with respect to the sphere of the Moon.

---

37 F:11 translates *triangles*. It is true that when a plane cuts a cone, it forms a triangle. But it seems that Aristarchus mentioned lines and not triangles on purpose because the calculation will not use the bases of the cones (lines HΠ and PE), which do not appear traced in the figure or used in reasoning.

38 M:42 translates *cono* (cone).

C. Ex 7. quinti elementorum
eadem ad aequales eandem
habet proportionem.

D. Iungatur enim CH et per B
ipsi CH parallela ducatur
BG. Erit triangulum ABG
simile triangulo ACH. Quare
ut GB ad BA, ita HC ad CA
ex 4. sexti: et permutando
ut GB ad HC quae sunt ex
centro circulorum EFG HKL,
ita BA ad AC. Et similiter
demonstrabitur, ut quae ex
centro circuli EFG ad eam,
quae ex centro circuli MNX,
ita esse BO ad OD.

E. Ex 11 quinti elementorum

F. Ex 8 tertii elementorum

G. Ubi hoc lemma sit, nondum
comperi, sed tamen illud idem
in 24 propositione perspectivae
Euclidis demonstratur.
Quoniam enim AC minor
est, quam OD, oculo posito
in A minus de corpore lunae
cernetur, quam eo posito in O.
ergo iunctis HL, MX, erit HL
minor ipsa MX.

καὶ ἐπεί ἐστιν, ὡς ἡ ἐκ τοῦ κέντρου τοῦ ΕΖΗ κύκλου πρὸς τὴν ἐκ τοῦ κέντρου τοῦ ΘΚΛ, οὕτως ἡ ἐκ τοῦ κέντρου τοῦ ΕΖΗ κύκλου πρὸς τὴν ἐκ τοῦ κέντρου τοῦ ΜΝΞ· ἀλλ᾽ ὡς ἡ ἐκ τοῦ κέντρου τοῦ ΕΖΗ κύκλου πρὸς τὴν ἐκ τοῦ κέντρου τοῦ ΘΛΚ κύκλου, οὕτως ἡ ΒΑ πρὸς τὴν ΑΓ· ὡς δὲ ἡ ἐκ τοῦ κέντρου τοῦ ΕΖΗ κύκλου πρὸς τὴν ἐκ τοῦ κέντρου τοῦ ΜΝΞ κύκλου, οὕτως ἐστὶν ἡ ΒΟ πρὸς τὴν ΟΔ· καὶ ὡς ἄρα ἡ ΒΑ πρὸς τὴν ΑΓ, οὕτως ἡ ΒΟ πρὸς τὴν ΟΔ. καὶ διελόντι, ὡς ἡ ΒΓ πρὸς τὴν ΓΑ, οὕτως ἡ ΒΔ πρὸς τὴν ΔΟ, καὶ ἐναλλάξ, ὡς ἡ ΒΓ πρὸς τὴν ΒΔ, οὕτως ἡ ΓΑ πρὸς τὴν ΔΟ. καὶ ἔστιν ἐλάσσων ἡ ΒΓ τῆς ΒΔ· κέντρον γάρ ἐστι τὸ Α τοῦ ΓΔ κύκλου· ἐλάσσων ἄρα καὶ ἡ ΑΓ τῆς ΔΟ. καὶ ἔστιν ἴσος ὁ ΘΚΛ κύκλος τῷ ΜΝΞ κύκλῳ· ἐλάσσων ἄρα ἐστὶν καὶ ἡ ΘΛ τῆς ΜΞ [, διὰ τὸ λῆμμα] · ὥστε καὶ ὁ περὶ διάμετρον τὴν ΘΛ κύκλος γραφόμενος, ὀρθὸς ὢν πρὸς τὴν ΑΒ, ἐλάσσων ἐστὶν τοῦ περὶ διάμετρον τὴν ΜΞ κύκλου γραφομένου, ὀρθοῦ πρὸς τὴν ΒΟ.

And since, as the radius of circle EZH is to the radius of circle ΘΚΛ, so is the radius of circle EZH to the radius of circle ΜΝΞ [C]. But, as the radius of circle EZH is to the radius of circle ΘΛΚ,[39] so is line BA to line ΑΓ [D] [Sch. 19 (11)]. And, as the radius of circle EZH is to the radius of circle ΜΝΞ, so is line BO to line ΟΔ; and therefore, as line BA is to line ΑΓ, so is line BO to line ΟΔ[40] [E]. And, by separation,[41] as line ΒΓ is to line ΓΑ, so is line ΒΔ to line ΔΟ and, by alternation, as line ΒΓ is to line ΒΔ, so is line ΓΑ to line ΔΟ. And line ΒΓ is smaller than line ΒΔ, since the center of circle ΓΔ is point A. Therefore, also line ΑΓ is smaller than line ΔΟ [F]. And circle ΘΚΛ is equal to circle ΜΝΞ. Therefore, line ΘΛ is also smaller than line ΜΞ [by the lemma [G] [Sch. 21 (13); 23 (19)],[42] so that also the circle drawn around diameter ΘΛ, being at right angles to line AB, is smaller than the circle drawn around diameter ΜΞ, at right angles to line BO.

C. From [proposition] 7 of the fifth [book] of the *Elements* [of Euclid], that the same [magnitude] has the same proportion with equal [magnitudes].

D. For let ΓΘ be joined and through B let BH be drawn parallel to ΓΘ itself. Triangle ABH will be similar to triangle ΑΓΘ. Therefore, as HB is to BA, so [is] ΘΓ to ΓΑ – [It follows from the definition] 4 of the sixth [book of the *Elements* of Euclid]. And, by permutation, since as HB is to ΘΓ, which are the radii of the circles EZH and ΘΚΛ, so [is] BA to ΑΓ. And similarly, it will be demonstrated that, as the radius of the circle EZH is to the radius of the circle ΜΝΞ, so is BO to ΟΔ.

E. From [proposition] 11 of the fifth [book] of the *Elements* [of Euclid].

F. From [proposition] 8 of the third [book] of the *Elements* [of Euclid].

G. I have not found out yet where this lemma is, but that very same thing is nevertheless demonstrated in proposition 24 of Euclid's *Optics*. For, since ΑΓ is smaller than ΟΔ, if the eye is placed in A, it will be perceived [a portion] of the body of the Moon smaller than if it is placed in O. Consequently, joined ΘΛ and ΜΞ, ΘΛ will be smaller than ΜΞ itself.

---

39 G:126: HKL.

40 G:199: BD, the French translation is correct.

41 G:199: componendo (in the French translation, it is correct; G:126: separando).

42 The lemma does not appear below in the treatise. Therefore, it is likely that it is an interpolation in the text, as suggested by H:363.

ἀλλ' ὁ μὲν περὶ διάμετρον τὴν ΘΛ κύκλος
γραφόμενος, ὀρθὸς ὢν πρὸς τὴν ΑΒ, ὁ διορίζων
ἐστὶν ἐν τῇ σελήνῃ τό τε σκιερὸν καὶ τὸ λαμπρόν,
ὅταν ὁ περιλαμβάνων κῶνος τόν τε ἥλιον καὶ τὴν
σελήνην τὴν κορυφὴν ἔχῃ πρὸς τῇ ἡμετέρᾳ ὄψει· ὁ
δὲ περὶ διάμετρον τὴν ΜΞ κύκλος, ὀρθὸς ὢν πρὸς
τὴν ΒΟ, ὁ διορίζων ἐστὶν ἐν τῇ σελήνῃ τό τε σκιερὸν
καὶ τὸ λαμπρόν, ὅταν ὁ περιλαμβάνων κῶνος τόν τε
ἥλιον καὶ τὴν σελήνην μὴ ἔχῃ τὴν κορυφὴν πρὸς τῇ
ἡμετέρᾳ ὄψει· ὥστε ἐλάσσων κύκλος διορίζει ἐν τῇ
σελήνῃ τό τε σκιερὸν καὶ τὸ λαμπρόν, ὅταν ὁ
περιλαμβάνων κῶνος τόν τε ἥλιον καὶ τὴν σελήνην
τὴν κορυφὴν ἔχῃ πρὸς τῇ ἡμετέρᾳ ὄψει.

But, on one hand, the circle drawn around diameter ΘΛ, being at right angles to line AB, is the one delimiting the shaded and bright parts on the Moon, when the cone enveloping the Sun and the Moon has its vertex towards our eye. And, on the other, the circle [drawn] around diameter MΞ, being at right angles to line BO, is the one delimiting the shaded and bright parts on the Moon, when the cone enveloping the Sun and the Moon has not its vertex towards our eye. So that the smaller circle delimits the shaded and bright parts on the Moon when the cone enveloping the Sun and the Moon has its vertex towards our eye.

δ' Ὁ διορίζων κύκλος ἐν τῇ σελήνῃ τό τε σκιερὸν καὶ τὸ λαμπρὸν ἀδιάφορός ἐστι τῷ ἐν τῇ σελήνῃ μεγίστῳ κύκλῳ πρὸς αἴσθησιν.

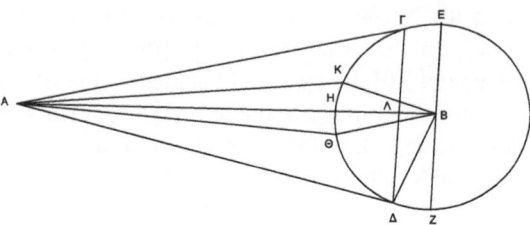

4. The circle delimiting the shaded and the bright parts on the Moon is perceptibly indistinguishable from the great circle on the Moon.[43]

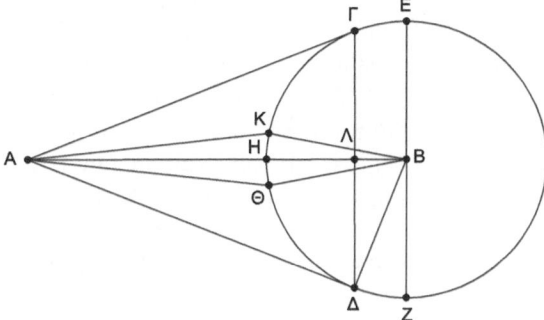

Figure 2.1.4 Diagram of proposition 4.

---

43 The letter *Λ* does not appear in the diagram of any of the Renaissance, modern, or contemporary editions, except in S:104. It is true that point Λ is not mentioned in the body of the text and seems to play no role in the argument. However, it does appear in the manuscripts. Instead, the diagrams of Commandino and Wallis show the chord of arc HΘ that is not in the manuscripts, except in ms. R, which certainly was not consulted by Commandino. This chord is not mentioned in the text and does not play any role in the argument.

Ἔστω γὰρ ἡ μὲν ἡμετέρα ὄψις πρὸς τῷ Α, σελήνης δὲ κέντρον τὸ Β, καὶ ἐπεζεύχθω ἡ ΑΒ, καὶ ἐκβεβλήσθω διὰ τῆς ΑΒ ἐπίπεδον· ποιήσει δὴ τομὴν ἐν τῇ σφαίρᾳ μέγιστον κύκλον. ποιείτω τὸν ΕΓΔΖ, ἐν δὲ τῷ κώνῳ εὐθείας τὰς ΑΓ, ΑΔ, ΔΓ· ὁ ἄρα περὶ διάμετρον τὴν ΓΔ, πρὸς ὀρθὰς ὢν τῇ ΑΒ, ὁ διορίζων ἐστὶν ἐν τῇ σελήνῃ τό τε σκιερὸν καὶ τὸ λαμπρόν. λέγω δὴ ὅτι ἀδιάφορός ἐστι τῷ μεγίστῳ πρὸς τὴν αἴσθησιν.

Ἤχθω γὰρ διὰ τοῦ Β τῇ ΓΔ παράλληλος ἡ ΕΖ, καὶ κείσθω τῆς ΔΖ ἡμίσεια ἑκατέρα τῶν ΗΚ, ΗΘ, καὶ ἐπεζεύχθωσαν αἱ ΚΒ, ΒΘ, ΚΑ, ΑΘ, ΒΔ.

καὶ ἐπεὶ ὑπόκειται ἡ σελήνη ὑπὸ ιε΄ μέρος ζῳδίου ὑποτείνουσα, ἡ ἄρα ὑπὸ ΓΑΔ γωνία βέβηκεν ἐπὶ ιε΄ μέρος ζῳδίου. τὸ δὲ ιε΄ τοῦ ζῳδίου τοῦ τῶν ζῳδίων ὅλου κύκλου ἐστὶν ρπ΄, ὥστε ἡ ὑπὸ τῶν ΓΑΔ γωνία βέβηκεν ἐπὶ ρπ΄ ὅλου τοῦ κύκλου· τεσσάρων ἄρα ὀρθῶν ἐστιν ἡ <ὑπὸ> ΓΑΔ ρπ΄.

Let our eye be at point A, and the center of the Moon be point B; and let AB be joined, and a plane be produced through line AB. It will make a section on the sphere [that is] a great circle. Let [the plane] make [the circle] EΓΔZ and, on the cone, straight lines AΓ, AΔ and ΔΓ. Therefore, [the circle drawn] around diameter ΓΔ, being at right angles to line AB, is the one delimiting the shaded and bright parts on the Moon. I say that it is perceptibly indistinguishable from the maximum [circle].

Let line EZ be drawn through point B, parallel to line ΓΔ, and let each of [arcs] HK and HΘ be half of [arc] ΔZ, and let KB, BΘ, KA, AΘ, and BΔ be joined.

And since it is hypothesized that the Moon subtends 1/15 part of a sign of the zodiac, therefore angle ΓAΔ extends over[44] 1/15 part of a sign of the zodiac. And 1/15 [part of a sign of] the zodiac is 1/180 [part] of the whole circle of the zodiac [signs] so that angle ΓAΔ stands on a 1/180 [part] of a whole circle. Therefore, angle ΓAΔ is 1/180 [part] of four right angles.

---

44 G:130 translates βέβηκεν ἐπὶ as *est à peu près* (it is approximately) and as *fait à peu près* (made approximately). In his Italian version, he translates both by *è circa* (it is approximately) (G:200, 201). Angle ΓAΔ, which represents in the diagram the apparent size of the Moon, cannot be *approximately* but *exactly* the value that the sixth hypothesis attributes to the apparent size of the Moon. Therefore, ἐπὶ must be understood as a proper construction of the verb (βαίνω + ἐπὶ = occupy) and not forming a complement with ιε΄ μέρος ζῳδίου that could be translated as *approximately 1/15 part of a zodiac sign*.

διὰ δὴ τοῦτο ἡ ὑπὸ ΓΑΔ γωνία ἐστὶν με΄ ὀρθῆς· καὶ
ἔστιν αὐτῆς ἡμίσεια ἡ ὑπὸ ΒΑΔ γωνία· ἡ ἄρα ὑπὸ
τῶν ΒΑΔ ἡμισείας ὀρθῆς ἐστι <με΄> μέρος.

Because of that, angle ΓΑΔ is 1/45 [part] of a right angle, and angle ΒΑΔ is half of it.[45] Therefore, angle ΒΑΔ is 1/45[46] part of half a right angle.

45 *This* refers to ΓΑΔ. G:132, in French, and G:201, in Italian, omit καὶ ἔστιν αὐτῆς ἡμίσεια ἡ ὑπὸ ΒΑΔ γωνία.

46 The text, as it is preserved in the manuscripts, offers two difficulties: (a) the existence of a lectio μιᾶς instead of ἡμισείας and (b) the omission of the number of parts angle ΒΑΔ should be divided into. Thus, there would be two possible translations: 1/45 (με′) parts of half (ἡμισείας) of a right angle or 1/90 (ϟ′) parts of a (μιᾶς) right angle. Regarding the lectio μιᾶς, although all the manuscripts have ἡμισείας – which led the editors to add με′ – in the oldest, ms. A, it can be clearly seen that in the place where one should expect ἡμισείας, a word has been erased and μιᾶς added above it. The erased word was probably ἡμισείας, but it is not possible to know it with certainty from the lines that are preserved, even after revision of the original. FG:22 does not add με′ in his Greek edition or translate it in Latin and replaces ἡμισείας by μιᾶς (ἡ ἄρα ὑπὸ τῶν ΒΑΔ μιᾶς ὀρθῆς ἐστι μέρος). Therefore, it states that it is a part of a right angle, without clarifying how many parts the right angle must be divided into. In FG:203–204, he justifies his choice. F:14 translates *ainsi cet angle BAD est d'un degré* (thus, this angle BAD is one degree). Hence, he interprets μιᾶς μέρος as referring to degrees, in an anachronistic interpretation (on the introduction of degrees in Greek astronomy, see p. 183). To solve the problem of the omission of the number of parts of angle ΒΑΔ, it should be noticed that in ms. Y, the ϟϛ of the last μέρος was erased, for it to be read με′, thus retaining the number (1/45) but having to suppose μέρος. A scholion of ms. F intendeds to correct the text. Between ἐστι and μέρος, it is read τὸ αὐτὸ, so it would mean ἡ ἄρα ὑπὸ τῶν ΒΑΔ ἡμισείας ὀρθῆς ἐστι τὸ αὐτὸ μέρος (therefore, angle ΒΑΔ is the same part of half of a right angle), and *the same* (τὸ αὐτὸ) would refer to 1/45, mentioned in the previous sentence, in which it is mentioned 1/45 parts of a right angle and not half a right angle. There are not enough elements to solve the question with absolute certainty. On the one hand, in proposition 12, Aristarchus seems to refer to this text remembering that angle ΒΑΓ (which is equal to angle ΒΑΔ) is 1/90 part of a right angle (ἡ δὲ ὑπὸ τῶν ΒΑΓ ὀρθῆς ϟ′ μέρος), without using μιᾶς. On the other hand, scholion 29 explicitly states that "angle ΔΑΒ will be a part of the 45 parts that half a right angle has" (ἤτοι ἡ ὑπὸ ΔΑΒ ἑνὸς οὖσα τεσσαρακοστόπεμπτον ἔσται τῆς ἡμισείας ὀρθῆς). Finally, μιᾶς is not to be found in Aristarchus's treatise, except in the presumed correction of the ἡμισείας. We prefer to keep H's proposal, even though ms. A seems to have preferred the μιᾶς. Actually, in the following sentence, Aristarchus explicitly speaks of 1/45 of a right angle. Furthermore, the fact that the beginning of the next word (μέρος) is the same as the presumed omitted one (με′) could explain the omission. The correction of ἡμισείας by μιᾶς could be a hypercorrection, that is, a reader who, warned of the absence of με′, has considered it appropriate to correct the sentence from the text of proposition 12.

A. Describatur seorsum
triangulum ADB et ab ipsa DA
abscindatur DL aequalis DB,
et BL iungatur. Erit trianguli
BLD anguli DBL, DLB inter
se aequales {5 primi.}. Et
cum angulus ad D sit rectus,
uterque ipsorum recti dimidius
erit {32 primi.}. Itaque duo
triangula rectangula sunt
ADB, LBD, quorum anguli
ad D recti, trianguli vero
ABD latus BD est commune
triangulo LDB, et latus AB
maius latere LB. Ergo ex iis,
quae nos demonstravimus
in commentariis in librum
Archimedis de numero arenae,
angulus BLD ad angulum
BAD maiorem quidem
proportionem habet, quam BA
latus ad latus BL, minorem
vero, quam latus AD ad latus
DL. Quare convertendo ex 26
quinti elementorum, quam nos
addidimus ex Papo, angulus
BAD ad angulum BLD, hoc
est ad dimidium recti maiorem
proportionem habet quam latus
DL, hoc est BD ipsi aeq[u]ale,
ad latus DA. {7. quinti.}

καὶ ἐπεὶ ὀρθή ἐστιν ἡ ὑπὸ τῶν ΑΔΒ, ἡ ἄρα ὑπὸ τῶν
ΒΑΔ γωνία πρὸς ἥμισυ ὀρθῆς μείζονα λόγον ἔχει
ἥπερ ἡ ΒΔ πρὸς τὴν ΔΑ, ὥστε ἡ ΒΔ τῆς ΔΑ
ἐλάσσων ἐστὶν ἢ με´ μέρος, ὥστε καὶ ἡ ΒΗ τῆς ΒΑ
πολλῷ ἐλάσσων ἐστὶν ἢ με´ μέρος.

And since angle AΔB is right, therefore angle BAΔ has to half of a right angle a greater ratio than line BΔ⁴⁷ [has] to line ΔA⁴⁸ [A], so that line BΔ is smaller than 1/45 part of line ΔA [B]. So that also the more so is line BH smaller than 1/45 part of line BA [C].

A. Let triangle ADB be described separately and, from AD itself let DL, equal to DB, be separated, and let BL be joined. Angles DBL and DLB of triangle BLD will be equal to each other {[*Elem.*] I, 5.}. And, since angle at D is right, each one of the [remaining] two will be half a right angle {[*Elem.*] I, 32.}. In the same manner, ADB and LBD are two right triangles, whose angles at D are right angles, but side BD of triangle ABD is common to triangle LDB, and side AB is greater than side LB. Consequently, from what we demonstrated in the Comments on Archimedes's book *Sand Reckoner*,⁴⁹ angle BLD has to angle BAD a ratio undoubtedly greater than the [ratio] that side BA has to side BL, but smaller than the one that side AD has to side DL. Therefore, by inversion, from proposition 26 of the fifth [book] of the *Elements* [of Euclid], that we added from Pappus,⁵⁰ angle BAD has to angle BLD, that is, to half a right angle, a ratio greater that the [ratio] that side DL, that is equal to BD itself, has to side DA {[*Elem.*] V, 7.}.

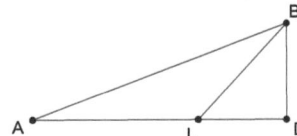

Figure 2.2.3 Diagram of Commandino's note A to proposition 4.

---

47 G:132 in French and G:201 in Italian: BA.
48 G:132 in French: BD (in Italian is correct: DA).
49 Commandino, *Archimedes, Opera nonnulla*, p. 62–63.
50 Commandino's translation of the *Elements* (1572: 69) adds some propositions to the last one, which is XXV. The XXVI says: "Si prima ad secundam maiorem habeat proportionem, quam tertia ad quartam; et convertendo secunda ad primam minorem proportionem habebit, quam quarta ad tertiam" (If the first has a greater ratio to the second than the third has to the fourth, and it is converted, the second will have a smaller ratio to the first than the fourth has to the third).

B. Sit enim, ut angulus BAD ad
    dimidium recti, ita quaepiam
    recta linea, in qua M ad ipsam
    DA, erit M quadragesima quinta
    pars ipsius DA, et habebit ad
    DA maiorem proportionem,
    quam BD ad DA. {10. quinti.}
    Ergo BD minor est, quam M;
    ac propterea minor, quam pars
    quadragesima quinta ipsius DA.

C. Est enim BG aequalis ipsi BD,
    et BA maior quam AD, cum
    maiori angulo subtendatur.

B. For, as angle BAΔ is to half a right angle, so let a certain line, M, be to ΔA itself; M[51] will be one forty-fifth part of ΔA itself and will have to ΔA a ratio greater than the [ratio] that BΔ has to ΔA {[*Elem.*] V, 10.}. Consequently, BΔ is smaller than M, and for this reason, smaller than one forty-fifth part of ΔA itself.

C. For BH is equal to BΔ itself, and BA is greater than AΔ, because it subtends a greater angle.

---

51 In the diagram of proposition 4, C:9a added line M to the original diagram. It is obviously an addition of Commandino, referring to this note. The line is in M:44, as she incorporates the figures of Commandino, and in W:594, which copies Commandino's notes.

D. Nam BH est aequalis ipsi BG; HA vero maior, quam GA, ex 8 tertii elementorum.

E. Describatur circa triangulum ABH circulus AHB, habebit recta linea AH ad rectam HB minorem proportionem, quam circumferentia AH ad HB circumferentia, ex demonstratis a Ptolemeo in principio magnae constructionis. Ut autem circumferentia AH ad circumferentia HB, ita angulus ABH ad BAH angulum {VIt. sexti.}. Recta igitur linea AH ad rectam HB minorem habet proportionem, quam angulus ABH ad angulum BAH {11. quinti.}.

διελόντι ἡ ΒΗ τῆς ΗΑ ἐλάσσων ἐστὶν ἢ μδ΄ μέρος, ὥστε καὶ ἡ ΒΘ τῆς ΑΘ πολλῷ ἐλάσσων ἐστὶν ἢ μδ΄ μέρος. καὶ ἔχει ἡ ΒΘ πρὸς τὴν ΘΑ μείζονα λόγον ἤπερ ἡ ὑπὸ τῶν ΒΑΘ πρὸς τὴν ὑπὸ τῶν ΑΒΘ· ἡ ἄρα ὑπὸ τῶν ΒΑΘ τῆς ὑπὸ τῶν ΑΒΘ ἐλάσσων ἐστὶν ἢ μδ΄ μέρος. καὶ ἔστιν τῆς μὲν ὑπὸ τῶν ΒΑΘ διπλῆ ἡ ὑπὸ τῶν ΚΑΘ, τῆς δὲ ὑπὸ τῶν ΑΒΘ διπλῆ ἡ ὑπὸ τῶν ΚΒΘ· ἐλάσσων ἄρα ἐστὶν καὶ ἡ ὑπὸ τῶν ΚΑΘ τῆς ὑπὸ τῶν ΚΒΘ ἢ τεσσαρακοστοτέταρτον μέρος. ἀλλὰ ἡ ὑπὸ τῶν ΚΒΘ ἴση ἐστὶν τῇ ὑπὸ τῶν ΔΒΖ, τουτέστιν τῇ ὑπὸ τῶν ΓΔΒ, τουτέστιν τῇ ὑπὸ τῶν ΒΑΔ· ἡ ἄρα ὑπὸ τῶν ΚΑΘ τῆς ὑπὸ τῶν ΒΑΔ ἐλάσσων ἐστὶν ἢ μδ΄ μέρος. ἡ δὲ ὑπὸ τῶν ΒΑΔ <ἡμισείας> ὀρθῆς ἐστιν <με΄> μέρος, ὥστε ἡ ὑπὸ τῶν ΚΑΘ ὀρθῆς ἐστιν ἐλάσσων ἢ ͵γϡξ΄.

By separation, line BH is smaller than 1/44 part of line HA. So that also the more so is line BΘ[52] smaller than 1/44 part of line AΘ [D]. And line BΘ has to line ΘA a greater ratio than angle BAΘ [has] to angle ABΘ [E] [Sch. 32 (16)]. Therefore, angle BAΘ is smaller than 1/44 part of angle ABΘ [F]. And angle KAΘ is the double of angle BAΘ, and angle KBΘ, the double of angle ABΘ.[53] Therefore, angle KAΘ is also smaller than one forty-fourth part of angle KBΘ[54] [G]. But angle KBΘ is equal to angle ΔBZ [H], that is, to angle ΓΔB [K], that is, to angle BAΔ [L]. Therefore, angle KAΘ is smaller than 1/44 part of angle BAΔ. And angle BAΔ is 1/45 part of half a right angle,[55] so that angle KAΘ is smaller than 1/3960 [part] of a right angle.

D. For BK is equal to BH itself, but AK is greater than HA, from [proposition] 8 of the third [book of the] *Elements* [of Euclid].

E. Let circle AHB be described around triangle ABH; straight line AH will have to straight line HB a ratio smaller than the [ratio] that arc AH has to arc HB, as it was demonstrated by Ptolemy in the beginning of the *Great Construction [Almagest]*.[56] But, as arc AH is to arc HB, so [is] angle ABH to angle BAH {[*Elem.*] VI, last [33].}. Accordingly, straight line AH has to straight line HB a ratio smaller than the [ratio] that angle ABH has to angle BAH {[*Elem.*] V, 11.}.

---

52 G:134, in French: BN; G:201, in Italian: BM, but it should be BH.

53 FG:24 put *BAH* in his Latin translation; neither his Greek edition (FG:25) nor his French translation (F.15) changes the order of the letters.

54 G:136, in French: KBN; G:201, in Italian: KBM.

55 This passage presents difficulties similar to the ones discussed in note 46. The Greek text of the manuscripts reads, ἡ δὲ ὑπὸ τῶν ΒΑΔ ὀρθῆς ἐστιν μέρος, which is translated "and angle ΒΑΔ is a part of a right angle." H:368 adds ἡμισείας (half) and με′ (1/45), and the text turns out ἡ δὲ ὑπὸ τῶν ΒΑΔ <ἡμισείας> ὀρθῆς ἐστιν <με′> μέρος (And angle ΒΑΔ is 1/45 part of half a right angle). Heath probably based his correction on the existence of a very similar passage a few lines earlier, in which only 1/45 has to be supposed (discussed in note 47). However, this is not the only solution adopted. The minimalist translation is that of Fortia. In his Greek edition, he does not add anything: FG:25, 12: ἡ δὲ ὑπὸ τῶν ΒΑΔ ὀρθῆς ἐστιν μέρος and translates, freely, F:15, 9–10: *or j'ai prouvé que l'angle BAD est d'un degré*, consistent with the translation of the other passage. His Latin translation is *at angulus BAD recti est pars* (F:24, 12–13). In FG:205, he explains why he adds neither ἡμισείας nor με′. V clarifies that it is a part if the right angle is divided into 90 parts. Thus, his translation says *q sub bad. recit e 90 ps* (V:64, 34–35, *angle BAD is a part of 90 of a right angle*). W:577, in turn, makes the same addition to the Greek text as Heath: ἡ δὲ ὑπὸ τῶν ΒΑΔ <ἡμισείας> ὀρθῆς ἐστιν <με′> μέρος <τουτέστι τῆς ὀρθῆς ϟ′ μέρος>. And at the end: *1/90 of a right angle*. He explains in note I that he adds that because Commandino has done so: "*Com. Ita enim ponitur*." Ni:9 follows the Greek of W, although he leaves only the last addition into brackets, incorporating the ἡμισείας and the με′ into the text. No:10 is similar to Ni text. C:8a–8b translates into Latin a text similar to that of W: *at angulus BAD est quadragesima quinta pars dimidii recti, hoc est unius recti pars nonagesima*. M:46 departs here from C and translates with H: *pero el ángulo BAD es una cuadragésima quinta parte de la (mitad) de un ángulo recto (but angle BAD is one forty-fifth part of [half of] a right angle)*, although it is not clear why "half of" is between parentheses and not "forty-fifth"). We agree that something is missing in the original text that should specify how many parts one or half right angle has been divided into. There are then two possibilities: (1) or it is stated that it has been divided into 45 parts, but then it must be assumed that a ἡμισείας has also been omitted before ὀρθῆς, since by context we know that it can be 1/45 parts of half a right angle, (2) or it is assumed that it is divided into 90 parts of a right angle, in which case we would have to add the ϟ′ in front of μέρος. Both alternatives are plausible, but we favor H's solution that adds both terms to make it similar to an expression that appeared a few lines earlier.

Quare convertendo ex 26 quinti,
recta linea BH ad rectam HA
maiorem proportionem habebit,
quam angulus BAH ad ABH
angulum.

F. Immo vero multo minor.

G. Ex 15 quinti elementorum.

H. Ita enim ponitur.

K. Ex 29 primi elementorum.

L. Ex 8 sexti elementorum.

56 In *Alm.* I, 10 (H: 43–44, Toomer (1998: 54–55).

Therefore, by inversion, from [proposition] 26 of the fifth [book of the *Elements* of Euclid],[57] straight line BH will have to straight line HA a ratio greater than the [ratio] that angle BAH has to angle ABH.

Figure 2.2.4 Diagram of Commandino's note E to proposition 4.

F. But rather, much smaller.

G. From [proposition] 15 of the fifth [book of the] *Elements* [of Euclid].

H. For it was assumed so.

K. From [proposition] 29 of the first [book of the] *Elements* [of Euclid].

L. From [proposition] 8 of the sixth [book of the] *Elements* [of Euclid].

M. Pappus in eodem loco.

τὸ δὲ ὑπὸ τηλικαύτης γωνίας ὁρώμενον μέγεθος ἀνεπαίσθητόν ἐστιν τῇ ἡμετέρᾳ ὄψει· καὶ ἔστιν ἴση ἡ ΚΘ περιφέρεια τῇ ΔΖ περιφερείᾳ· ἔτι ἄρα μᾶλλον ἡ ΔΖ περιφέρεια ἀνεπαίσθητός ἐστι τῇ ἡμετέρᾳ ὄψει. ἐὰν γὰρ ἐπιζευχθῇ ἡ ΑΖ, ἡ ὑπὸ τῶν ΖΑΔ γωνία ἐλάσσων ἐστὶ τῆς ὑπὸ τῶν ΚΑΘ. τὸ Δ ἄρα τῷ Ζ τὸ αὐτὸ δόξει εἶναι. διὰ τὰ αὐτὰ δὴ καὶ τὸ Γ τῷ Ε δόξει τὸ αὐτὸ εἶναι· ὥστε καὶ ἡ ΓΔ τῇ ΕΖ ἀνεπαίσθητός ἐστιν. καὶ ὁ διορίζων ἄρα ἐν τῇ σελήνῃ τό τε σκιερὸν καὶ τὸ λαμπρὸν ἀνεπαίσθητός ἐστι τῷ μεγίστῳ.

And a magnitude seen from an angle of this size is imperceptible for our eye; and arc ΚΘ is equal to arc ΔΖ. Therefore, even less perceptible for our eye is arc ΔΖ. Since, if ΑΖ is joined, angle ΖΑΔ is smaller than angle ΚΑΘ [M] [Sch. 37 (18)]. Therefore, point Δ will seem be the same as point Ζ. For the same reason, also point Γ will seem the same as point Ε, so that also line ΓΔ is indistinguishable from line ΕΖ.[58] And, therefore, the [circle] delimiting the shaded and bright parts on the Moon is indistinguishable from the great [circle].

M. Pappus, in the same place.

---

58 As H:371 note 2 highlights, ἀνεπαίσθητός is used with dative, which is very unusual. Literally, it would mean *line ΓΔ is imperceptible from line ΕΖ*, but what it means is that *the difference between both lines is imperceptible* or that *line ΓΔ is indistinguishable for perception from line ΕΖ*, but in that case, it would seem more appropriate to use ἀδιάφορός instead of ἀνεπαίσθητός. Nevertheless, Pappus, in his *Commentary of the Almagest* (Rome 1930, p. 47, lines 19–20), uses ἀνεπαίσθητὰ in the same way.

ε' Ὅταν ἡ σελήνη διχότομος ἡμῖν φαίνηται, τότε ὁ μέγιστος κύκλος ὁ παρὰ τὸν διορίζοντα ἐν τῇ σελήνῃ τό τε σκιερὸν καὶ τὸ λαμπρὸν νεύει εἰς τὴν ἡμετέραν ὄψιν, τουτέστιν, ὁ παρὰ τὸν διορίζοντα μέγιστος κύκλος καὶ ἡ ἡμετέρα ὄψις ἐν ἑνί εἰσιν ἐπιπέδῳ.

* 3. positione.

** 4. huius.

Ἐπεὶ γὰρ διχοτόμου οὔσης τῆς σελήνης φαίνεται ὁ διορίζων τό τε λαμπρὸν καὶ τὸ σκιερὸν τῆς σελήνης κύκλος νεύων εἰς τὴν ἡμετέραν ὄψιν, καὶ αὐτῷ ἀδιάφορος ὁ παρὰ τὸν διορίζοντα μέγιστος κύκλος, ὅταν ἄρα ἡ σελήνη διχότομος ἡμῖν φαίνηται, τότε ὁ μέγιστος κύκλος ὁ παρὰ τὸν διορίζοντα νεύει εἰς τὴν ἡμετέραν ὄψιν.

5. When the Moon appears to us bisected, then, the great circle parallel to the one delimiting the shaded and bright parts on the Moon points towards our eye, that is, the great circle parallel to the delimiting one and our eye are in one plane.

Since, when the Moon is bisected, it is manifest that the circle delimiting the bright and shaded parts of the Moon[59] points towards our eye [*], and that the great circle, parallel to the delimiting one is indistinguishable from it [**], therefore, when the Moon appears to us bisected, the great circle parallel to the delimiting one points towards our eye.[60]

* From hypothesis 3.

** From [proposition] 4 of this [treatise].

---

59 This is the only time that the bright/shaded order is reversed. It is always τό τε σκιερὸν καὶ τὸ λαμπρὸν, but in this case the text says, τό τε λαμπρὸν καὶ τὸ σκιερὸν.
60 No:11 omits ὅταν ἄρα ἡ σελήνη διχότομος ἡμῖν φαίνηται.

ϛ'. Ἡ σελήνη κατώτερον φέρεται τοῦ ἡλίου,
καὶ διχότομος οὖσα ἔλασσον τεταρτημορίου
ἀπέχει ἀπὸ τοῦ ἡλίου.

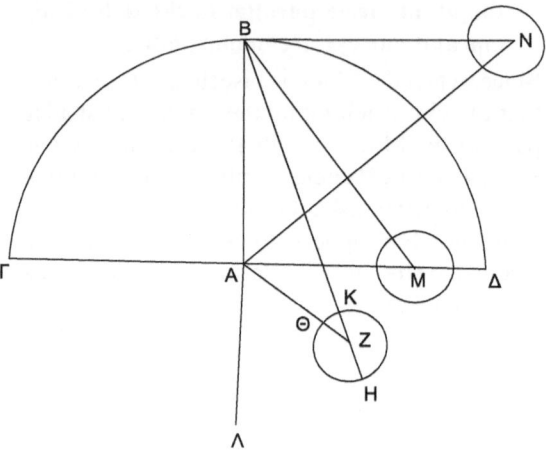

6. The Moon travels lower than the Sun, and, when it is bisected, it is distant[61] from the Sun by less than a quadrant.

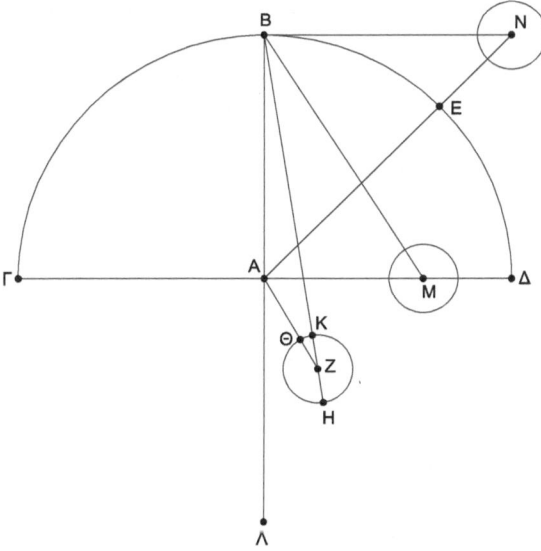

Figure 2.1.5 Diagram of proposition 6.

61 This refers to the angular distance, that is, to the elongation.

Ἔστω γὰρ ἡ ἡμετέρα ὄψις πρὸς τῷ Α, ἡλίου δὲ κέντρον τὸ Β, καὶ ἐπιζευχθεῖσα ἡ ΑΒ ἐκβεβλήσθω, καὶ ἐκβεβλήσθω διὰ τῆς ΑΒ καὶ τοῦ κέντρου τῆς σελήνης διχοτόμου οὔσης ἐπίπεδον· ποιήσει δὴ τομὴν ἐν τῇ σφαίρᾳ, καθ' ἧς φέρεται τὸ κέντρον τοῦ ἡλίου, κύκλον μέγιστον. ποιείτω οὖν τὸν ΓΒΔ κύκλον, καὶ ἀπὸ τοῦ Α τῇ ΑΒ πρὸς ὀρθὰς ἤχθω ἡ ΓΑΔ· τεταρτημορίου ἄρα ἐστὶν ἡ ΒΔ περιφέρεια. λέγω ὅτι ἡ σελήνη κατώτερον φέρεται τοῦ ἡλίου, καὶ διχότομος οὖσα ἔλασσον τεταρτημορίου ἀπέχει ἀπὸ τοῦ ἡλίου, τουτέστιν, ὅτι τὸ κέντρον ἐστὶν αὐτῆς μεταξὺ τῶν ΒΑ, ΑΔ εὐθειῶν καὶ τῆς ΔΕΒ περιφερείας.

Εἰ γὰρ μή, ἔστω τὸ κέντρον αὐτῆς τὸ Ζ μεταξὺ τῶν ΔΑ, ΑΛ εὐθειῶν, καὶ ἐπεζεύχθω ἡ ΒΖ.

Let our eye be at point A, and the center of the Sun be point B, and let AB be joined, and let it be produced,[62] and let a plane be produced through line AB and the center of the Moon when it is bisected. It will make a section on the sphere along which the center of the Sun travels[63] [that is] a great circle. So, let [the plane] make circle ΓBΔ, and let line ΓAΔ be drawn from point A and at right angles to line AB. Therefore, arc BΔ is one quadrant. I say that the Moon travels lower than the Sun,[64] and that when it is bisected, it is distant from the Sun by less than a quadrant, that is, that the center of it is between straight lines BA and AΔ and arc ΔEB.

If not, let the center of it, point Z, [be] between straight lines ΔA and AΛ, and let BZ be joined.

---

62 F:16 translates, *joints par une ligne AB, prolongée jusqu'en L* (joined by one line, AB, produced to L). G:140 in French and G:202 in Italian omits: ἐκβεβλήσθω.

63 M:48 translates, *este plano cortará a la esfera en un círculo máximo sobre el cual se mueve el centro del Sol* (this plane will cut the sphere in a great circle on which the center of the Sun travels). Therefore, according to her translation, it is the great circle and not the sphere that carries the center of the Sun. Accordingly, she holds that the expression καθ' ἧς φέρεται τὸ κέντρον τοῦ ἡλίου must be attributed to κύκλον μέγιστον and not to ἐν τῇ σφαίρᾳ. However, this is impossible since the relative particle ἧς is feminine, as σφαίρᾳ, but κύκλον is neutral. The meaning does not justify her translation either. The plane in which the Sun travels is the ecliptic, but Aristarchus knew that the Moon does not travel through the ecliptic. Consequently, the Sun, the Moon, and the Earth would not always be in the same plane, which is required by the argument. Massa translates the Latin of Commandino, which in this case is unambiguous because, in addition to the position of the clause, quam is feminine, as a sphere, while circulum is masculine: *faciet utiq; sectionem in sphaera, per quam fertur centrum solis circulum maximum* (C:13a).

64 H:373 translates, *I say that the moon moves (in an orbit) lower than (that of) the sun*, adding *orbit* into parentheses to clarify the translation; M:48 follows H, but she omits the parentheses.

A. Ex demonstratis in tertia
propositione huius.

B. Ex antecedente.

C. Ex tertia diffinitione
undecimi elementorum.

D. Essent enim trianguli ABD
tres anguli maiores duobus
rectis.

ἡ ΒΖ ἄρα ἄξων ἐστὶν τοῦ περιλαμβάνοντος
κώνου τόν τε ἥλιον καὶ τὴν σελήνην, καὶ γίνεται
ἡ ΒΖ ὀρθὴ πρὸς τὸν διορίζοντα ἐν τῇ σελήνῃ τό
τε σκιερὸν καὶ τὸ λαμπρὸν μέγιστον
κύκλον. Ἔστω οὖν ὁ μέγιστος κύκλος ἐν τῇ
σελήνῃ ὁ παρὰ τὸν διορίζοντα τό τε σκιερὸν καὶ
τὸ λαμπρὸν ὁ ΗΘΚ. καὶ ἐπεὶ διχοτόμου οὔσης
τῆς σελήνης ὁ μέγιστος κύκλος ὁ παρὰ τὸν
διορίζοντα ἐν τῇ σελήνῃ τό τε σκιερὸν καὶ τὸ
λαμπρὸν καὶ ἡ ἡμετέρα ὄψις ἐν ἑνί εἰσιν ἐπιπέδῳ,
ἐπεζεύχθω ἡ ΑΖ· ἡ ΑΖ ἄρα ἐν τῷ τοῦ ΚΗΘ
κύκλου ἐστὶν ἐπιπέδῳ. καὶ ἔστιν ἡ ΒΖ τῷ ΚΘΗ
κύκλῳ πρὸς ὀρθάς, ὥστε καὶ τῇ ΑΖ· ὀρθὴ ἄρα
ἐστὶν ἡ ὑπὸ ΒΖΑ γωνία. ἀλλὰ καὶ ἀμβλεῖα ἡ ὑπὸ
τῶν ΒΑΖ· ὅπερ ἀδύνατον. οὐκ ἄρα τὸ Ζ σημεῖον
ἐν τῷ ὑπὸ τὴν ΔΑΛ γωνίαν τόπῳ ἐστίν.[3]

---

3 This is the only time Aristarchus names an angle by the letters without preceding them by ὑπὸ τῶν.
Usually, he would write: οὐκ ἄρα τὸ Ζ σημεῖον ἐν τῷ ὑπὸ τὴν [ὑπὸ τῶν] ΔΑΛ γωνίᾳ τόπῳ ἐστίν. It
could be a transmission error, but ἐν τῷ ὑπὸ τὴν ὑπὸ τῶν ΔΑΛ γωνιαν would certainly look awkward,
so perhaps Aristarchus was taking a small liberty with the geometrical idiom for the sake of smooth
Greek.

Therefore, line BZ is the axis of the cone enveloping the Sun and the Moon, and line BZ is at right angles to the great circle[65] delimiting the shaded and bright parts on the Moon [A]. So, let the great circle on the Moon parallel[66] to the one delimiting the shaded and bright parts be HΘK. And since when the Moon is bisected, the great circle parallel to the one delimiting the shaded and bright parts on the Moon and our eye are in one plane [B], let AZ be joined. Therefore, line AZ is in the plane of circle KHΘ. And line BZ is at right angles to circle KΘH,[67] so that [it] also [is] with line AZ. Therefore, angle BZA[68] is right [C].[69] But also angle BAZ is obtuse, which is impossible. [D].[70] Therefore, point Z is not in the place [determined by] angle ΔAΛ.

A. From what has been demonstrated in the third proposition of this [treatise].

B. From above.

C. From the third definition of the eleventh [book] of the *Elements* [of Euclid].

D. For the three angles of triangle ABΔ would be greater than two right angles.

---

65 Strictly speaking, the circle delimiting the shaded and bright parts of the Moon is not a great circle, since the Moon is smaller than the Sun, although the difference between the two is so small that it is indistinguishable for our perception (proposition 4). There are, therefore, three possibilities: (1) We suppose that μέγιστον is spurious, and therefore we omit it, Ni:10 has chosen this option, as he had also omitted μέγιστον In the formulation of hypothesis 3 (see note 5). (2) Another possibility is to suppose that here Aristarchus refers to the maximum circle parallel to the one delimiting the shaded and bright parts of the Moon, and thus, "παρὰ τὸν" should be added between the τὸν and the διορίζοντα: ὀρθὴ πρὸς τὸν παρὰ τὸν διορίζοντα ἐν τῇ σελήνῃ τό τε σκιερὸν καὶ τὸ λαμπρὸν μέγιστον κύκλον. If this were the case, it could be a saut du même au même. This is the option followed by Nokk (No:12). (3) The third option is to assume that the text is correct and that Aristarchus allows himself the imprecision of calling great to the delimiting circle because the difference between them is imperceptible, as he proved in proposition 4. This option is explicitly followed by Heath (H:373, n.1) and implicitly by V:70, F:17, C:13b, and M:48. As usual, we prefer to keep the Greek as it is, but we do not find Heath's explanation satisfactory because, although they are indeed indistinguishable by perception, Aristarchus continues to treat them as distinct circles. The aim of showing that they are indistinguishable is not to treat them as if they were one and the same circle but to justify the geometric construction proposition 7 illustrates. Probably, by the omission of Aristarchus or a later copyist, παρὰ was lost, as Nokk suggests. However, we continue to believe that the μέγιστον in the enunciation of hypothesis 3 is original (see note 5), since in the formulation of the hypotheses can be allowed simplifications whose clarification would be very cumbersome at the beginning of the treatise. The same is true regarding hypothesis 5, where Aristarchus states that the width of the shadow is equal to two Moons when, in fact, it is equal to two diameters of the lunar circle of visibility, viewed from the center of the Earth, as becomes evident in proposition 13.

66 F:17 does not interpret παρὰ in a strict geometrical sense meaning *parallel to*, since he translates *auprès* (near to); V:70 translates it by *qui ad*. C:13b translates by *iuxta*. M:48 does not translate Commandinus's *iuxta*, but she translates by *paralelo* (parallel) the iuxta found in the following sentence. H:372 translates *parallel* in both cases. Since the meaning in this text, like typically in scientific texts, is clearly *parallel*, we prefer to translate it by *parallel* even if Aristarchus also uses another word (παράλληλος) for parallel.

67 No:12 *BHK* instead of *GHK*.

68 G:203, in Italian, says that angle BZA is obtuse. That is, he omits ὀρθὴ ἄρα ἐστὶν ἡ ὑπὸ BZA γωνία. ἀλλὰ καὶ and replaces BZA by BAZ. His French translation is correct.

69 F:17–18 omits ὀρθὴ ἄρα ἐστὶν ἡ ὑπὸ BZA γωνία. However, the sentence is not omitted in his Greek edition and Latin translation (FG:31–32).

70 V:70 translates *atqui etiam obtusus b sub baf*. The extra *b* does not make sense.

Λέγω ὅτι οὐδὲ ἐπὶ τῆς ΑΔ. εἰ γὰρ δυνατόν, ἔστω τὸ Μ, καὶ πάλιν ἐπεζεύχθω ἡ ΒΜ, καὶ ἔστω μέγιστος κύκλος ὁ παρὰ τὸν διορίζοντα, οὗ κέντρον τὸ Μ. κατὰ τὰ αὐτὰ δὴ δειχθήσεται ἡ ὑπὸ ΒΜΑ γωνία ὀρθὴ <πρὸς τὸν μέγιστον κύκλον>⁴· ἀλλὰ καὶ ἡ ὑπὸ τῶν ΒΑΜ· ὅπερ ἀδύνατον.

οὐκ ἄρα ἐπὶ τῆς ΑΔ τὸ κέντρον ἐστὶ τῆς σελήνης διχοτόμου οὔσης· μεταξὺ ἄρα τῶν ΑΒ, ΑΔ ἐστίν.

Λέγω δὴ ὅτι καὶ ἐντὸς τῆς ΒΔ περιφερείας. εἰ γὰρ δυνατόν, ἔστω ἐκτὸς κατὰ τὸ Ν, καὶ τὰ αὐτὰ κατεσκευάσθω. δειχθήσεται δὴ ἡ ὑπὸ τῶν ΒΝΑ γωνία ὀρθή· μείζων ἄρα ἐστὶν ἡ ΒΑ τῆς ΑΝ. ἴση δὲ ἡ ΒΑ τῇ ΑΕ· μείζων ἄρα ἐστὶν καὶ ἡ ΑΕ τῆς ΑΝ· ὅπερ ἀδύνατον. οὐκ ἄρα τὸ κέντρον τῆς σελήνης διχοτόμου οὔσης ἐκτὸς ἔσται τῆς ΒΕΔ περιφερείας.

ὁμοίως δειχθήσεται ὅτι οὐδὲ ἐπ' αὐτῆς τῆς ΒΕΔ περιφερείας· ἐντὸς ἄρα.

ἡ ἄρα σελήνη κατώτερον φέρεται τοῦ ἡλίου, καὶ διχότομος οὖσα ἔλασσον τεταρτημορίου ἀπέχει ἀπὸ τοῦ ἡλίου.

---

4 πρὸς τὸν μέγιστον κύκλον would seem to be an interpolation as suggested by H:374–375 in a note. No:12 does not translate it. F:18 translates *perpendiculaire* and includes πρὸς τὸν μέγιστον κύκλον in the Latin and Greek (FG:32–33). C:13b translates it as *angulus BMA rectus esse ad maximum circulum,* and M:49 translates accordingly: *el ángulo BMA es recto con respecto al círculo máximo.* G:142: *forme un angle droit avec le grand cercle.* G:203 incorporates it without arguing: *è retto rispetto al cerchi Massimo.* We have simply preferred to omit what might be an interpolation and not translate πρὸς τὸν μέγιστον κύκλον.

I say that it is also not on line AΔ. If possible, let point M [be the center of the Moon], and, again, let BM be joined, and let the great circle parallel to the delimiting one be [the circle] whose center is point M. In the same way, it will be proved that angle BMA is right. But also angle BAM [is right], which is impossible.

Therefore, the center of the Moon, when it is bisected, is not on line AΔ. Therefore, it is between lines AB and AΔ.[71]

I say that [it is] also inside arc BΔ.[72] If possible, let it be outside [arc BΔ], at point N[73] and let the same be constructed [as before]. It will be proved that angle BNA is right. Therefore, line BA is greater than line AN. But line BA is equal to line AE. Therefore, also line AE is greater than line AN, which is impossible. Therefore, the center of the Moon, when it is bisected, will not be outside arc BEΔ.

Similarly, it will be proved that [it is] also not on the same arc BEΔ. Therefore, [it is] inside.

Therefore, the Moon travels lower than the Sun, and when it is bisected, it is distant from the Sun by less than a quadrant.

71 M:49 *BD*; however, C:13b *AD*.

72 W:581 *BED*; it is correct, but it is not found in the manuscripts consulted by him (ms. A and W), although in the following references to the same arc, it is labeled *BEΔ*. C:13b, No:12, F:18, and FG:33 also have *BED*. FG:208 explains why he chooses *BED*. H:375, M:49, V:70, and Σ:81, 114 put *BD*.

73 M:49 translates: *pues si fuera posible que esté fuera, esté en N*. In this, she departs from the Greek text and from the edition of C:13b: *Nam si fieri potest, sit extra in puncto N*.

ζ'. Τὸ ἀπόστημα ὃ ἀπέχει ὁ ἥλιος ἀπὸ τῆς γῆς
τοῦ ἀποστήματος οὗ ἀπέχει ἡ σελήνη ἀπὸ τῆς
γῆς μεῖζον μέν ἐστιν ἢ ὀκτωκαιδεκαπλάσιον,
ἔλασσον δὲ ἢ εἰκοσαπλάσιον.

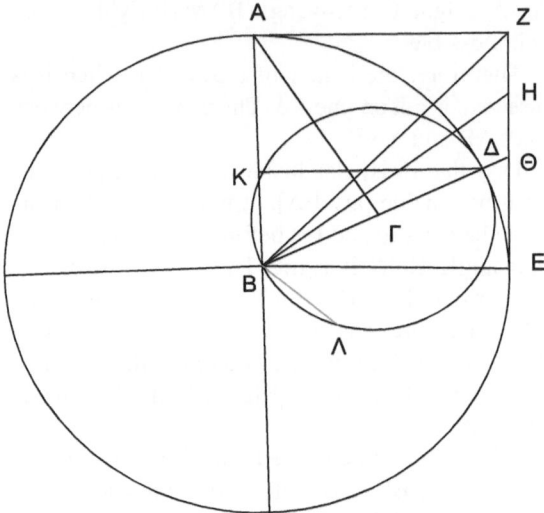

7. **The distance of the Sun from the Earth is greater than eighteen but smaller than twenty times the distance of the Moon from the Earth.**[74,75]

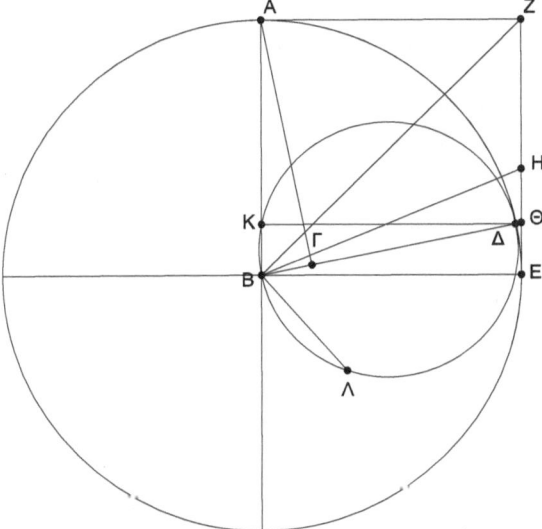

Figure 2.1.6 Diagram of proposition 7.

74 In the formulation of the proposition, V:70 uses *absentia* for *distance*, but then in the text, he uses *distantia*. However, we have not found a meaning for *absentia* consistent with distance. The participle of *absum* – which means "to be distant from" and that V uses to translate ἀπέχει – is *absens*, where the noun *absentia* comes from, so there is an etymological link between the two.
75 Line BΛ does not appear in the diagram of ms. A, the oldest, but it is mentioned in the text and appears in most manuscripts, so it is probably an omission in ms. A.

* 3. huius.

A. Hoc in figura ita esse ponatur, namque ob loci angustiam coacti sumus circumferentiam DE multo maiorem facere, quam sit trigesima pars circunferentiae EDA.

Ἔστω γὰρ ἡλίου μὲν κέντρον τὸ Α, γῆς δὲ τὸ Β, καὶ ἐπιζευχθεῖσα ἡ ΑΒ ἐκβεβλήσθω, σελήνης δὲ κέντρον διχοτόμου οὔσης τὸ Γ, καὶ ἐκβεβλήσθω διὰ τῆς ΑΒ καὶ τοῦ Γ ἐπίπεδον, καὶ ποιείτω τομὴν ἐν τῇ σφαίρᾳ, καθ᾽ ἧς φέρεται τὸ κέντρον τοῦ ἡλίου, μέγιστον κύκλον τὸν ΑΔΕ, καὶ ἐπεζεύχθωσαν αἱ ΑΓ, ΓΒ, καὶ ἐκβεβλήσθω ἡ ΒΓ ἐπὶ τὸ Δ. ἔσται δή, διὰ τὸ τὸ Γ σημεῖον κέντρον εἶναι τῆς σελήνης διχοτόμου οὔσης, ὀρθὴ ἡ ὑπὸ τῶν ΑΓΒ. ἤχθω δὴ ἀπὸ τοῦ Β τῇ ΒΑ πρὸς ὀρθὰς ἡ ΒΕ.

ἔσται δὴ ἡ ΕΔ περιφέρεια τῆς ΕΔΑ περιφερείας λ´· ὑπόκειται γάρ, ὅταν ἡ σελήνη διχότομος ἡμῖν φαίνηται, ἀπέχειν ἀπὸ τοῦ ἡλίου ἔλασσον τεταρτημορίου τῷ τοῦ τεταρτημορίου λ´· ὥστε καὶ ἡ ὑπὸ τῶν ΕΒΓ γωνία ὀρθῆς ἐστι λ´.

Let the center of the Sun [be] point A, that of the
Earth, point B, and let AB be joined, and let it be
produced,[76] and let the center of the Moon, when it
is bisected [be] point Γ, and let a plane be produced
through line AB and point Γ, and let it make a section on the sphere along which the center of the
Sun travels, [namely] the great circle AΔE,[77] and
let AΓ, ΓB be joined and let line BΓ be produced
until point Δ. Because point Γ is the center of the
Moon when it is bisected, angle AΓB will be right
[*]. Let line BE be drawn from point B at right
angles to line BA.[78]

Arc EΔ[79] will be the thirtieth [part] of arc EΔA[80]
[A]. Since it is hypothesized[81] that, when the Moon
appears to us bisected, it is distant from the Sun by
one quadrant minus one-thirtieth [part] of a quadrant,[82] hence also angle EBΓ is the thirtieth [part]
of a right angle.

* A. From [hypothesis] 3 of this
[treatise].

A. It is assumed to be so in the
diagram, for, because of the
narrow space, we are forced
to make arc ΔE much greater
than one-thirtieth of arc EΔA.

76 G:144, in French, and G:203, in Italian, omits ἐκβεβλήσθω. See note 62 on p. 93.
77 Again, M:49 attributes to the great circle and not to the sphere to carry the Sun. See note 63 on
p. 93.
78 V:70 adds *& producantur bc. in d* that was already included before where he translated καὶ
ἐκβεβλήσθω ἡ BΓ ἐπὶ τὸ Δ.
79 G:204, in Italian: AB; in French, it is correct.
80 G:144, in French: CDA; in Italian, it is correct.
81 M:50 omits ὑπόκειται.
82 G:204, in Italian, omits τῷ τοῦ τεταρτημορίου λ´; in French, it is correct.

B. Producatur etiam BD ad rectam lineam FE in H.

C. Illud nos hoc lemmate demonstrabimus. Sit triangulum orthogonium ABC rectum habens angulum ad C; et in recta linea AC sumatur quodvis punctum D, et BD iungatur. Dico rectam lineam AC ad rectam lineam CD maiorem proportionem habere, quam angulus ABC habeat ad DBC angulum.

Centro enim B et intervallo BD circuli circumferentia EDF describatur, et BC producatur ad F. Itaque quoniam triangulum quidem ABD maius est sectore EBD; triangulum vero DBC minus sectore DBF: habebit triangulum ABD ad triangulum DBC maiorem proportionem, quam sector EBD ad sectorem DBF. Ut autem triangulum ABD ad triangulum DBC, ita est recta linea AD ad ipsam DC {V. sexti.}: et ut sector ABD ad sectorem DBC, ita angulus ABD ad DBC angulum {Vlt. sexti.}. Ergo recta linea AD ad ipsam DC maiorem proportionem habet, quam angulus ABD ad angulum DBC: et componendo recta linea AC ad ipsam CD, maiorem habet proportionem quam angulus ABC ad DBC angulum.

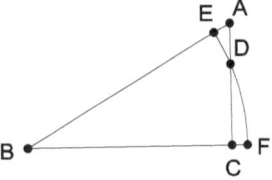

συμπεπληρώσθω δὴ τὸ ΑΕ παραλληλόγραμμον, καὶ ἐπεζεύχθω ἡ ΒΖ. ἔσται δὴ ἡ ὑπὸ τῶν ΖΒΕ γωνία ἡμίσεια ὀρθῆς. τετμήσθω ἡ ὑπὸ τῶν ΖΒΕ γωνία δίχα τῇ ΒΗ εὐθείᾳ· ἡ ἄρα ὑπὸ τῶν ΗΒΕ γωνία τέταρτον μέρος ἐστὶν ὀρθῆς. ἀλλὰ καὶ ἡ ὑπὸ τῶν ΔΒΕ γωνία λ΄ ἐστι μέρος ὀρθῆς· λόγος ἄρα τῆς ὑπὸ τῶν ΗΒΕ γωνίας πρὸς τὴν ὑπὸ τῶν ΔΒΕ γωνίαν <ἐστὶν> ὃν <ἔχει> τὰ ιε πρὸς τὰ δύο· οἵων γάρ ἐστιν ὀρθὴ γωνία ξ, τοιούτων ἐστὶν ἡ μὲν ὑπὸ τῶν ΗΒΕ ιε, ἡ δὲ ὑπὸ τῶν ΔΒΕ δύο. καὶ ἐπεὶ ἡ ΗΕ πρὸς τὴν ΕΘ μείζονα λόγον ἔχει ἤπερ ἡ ὑπὸ τῶν ΗΒΕ γωνία πρὸς τὴν ὑπὸ τῶν ΔΒΕ γωνίαν, ἡ ἄρα ΗΕ πρὸς τὴν ΕΘ μείζονα λόγον ἔχει ἤπερ τὰ ιε πρὸς τὰ β.

καὶ ἐπεὶ ἴση ἐστὶν ἡ ΒΕ τῇ ΕΖ, καὶ ἔστιν ὀρθὴ ἡ πρὸς τῷ Ε, τὸ ἄρα ἀπὸ τῆς ΖΒ τοῦ ἀπὸ ΒΕ διπλάσιόν ἐστι· ὡς δὲ τὸ ἀπὸ ΖΒ πρὸς τὸ ἀπὸ ΒΕ, οὕτως ἐστὶ τὸ ἀπὸ ΖΗ πρὸς τὸ ἀπὸ ΗΕ· τὸ ἄρα ἀπὸ ΖΗ τοῦ ἀπὸ ΗΕ διπλάσιόν ἐστι. τὰ δὲ μθ τῶν κε ἐλάσσονά ἐστιν ἢ διπλάσια, ὥστε τὸ ἀπὸ ΖΗ πρὸς τὸ ἀπὸ ΗΕ μείζονα λόγον ἔχει ἢ <ὃν τὰ> μθ πρὸς κε· καὶ ἡ ΖΗ ἄρα πρὸς τὴν ΗΕ μείζονα λόγον ἔχει ἢ <ὃν> τὰ ζ πρὸς τὰ ε· καὶ συνθέντι ἡ ΖΕ ἄρα πρὸς τὴν ΕΗ μείζονα λόγον ἔχει ἢ ὃν τὰ ιβ πρὸς τὰ ε, τουτέστιν, ἢ ὃν <τὰ> λς πρὸς τὰ ιε. ἐδείχθη δὲ καὶ ἡ ΗΕ πρὸς τὴν ΕΘ μείζονα λόγον ἔχουσα ἢ ὃν τὰ ιε πρὸς τὰ δύο· δι' ἴσου ἄρα ἡ ΖΕ πρὸς τὴν ΕΘ μείζονα λόγον ἔχει ἢ ὃν τὰ λς πρὸς τὰ δύο, τουτέστιν, ἢ ὃν τὰ ιη πρὸς α· ἡ ἄρα ΖΕ τῆς ΕΘ μείζων ἐστὶν ἢ ιη.

Let parallelogram AE be completed and let BZ be joined [B]. Angle ZBE will be half of a right angle. Let angle ZBE be cut into two parts by straight line BH. Therefore, angle HBE is one fourth part of a right angle. But also, angle ΔBE is 1/30[83] of a right angle. Therefore, the ratio that angle HBE has to angle ΔBE is the [ratio] that 15 has to 2; for of such parts as a right angle comprises 60, angle HBE comprises 15 and angle ΔBE two. And since line HE has to line EΘ a greater ratio than angle HBE [has] to angle ΔBE [C], therefore, line HE has to line EΘ a greater ratio than 15 [has] to 2.

And since line BE is equal to line EZ and the angle at point E is right, therefore, the square of line ZB is the double of the square of BE. And as the square of ZB is to the square of BE, so is the square of ZH to the square of HE [D] [Sch. 47 (30)]. Therefore, the square of ZH is the double of the square of HE. But 49 is smaller than the double of 25, so that the square of ZH has to the square of HE a ratio greater than the [ratio] that 49 has to 25.[84] And, therefore, line ZH has to line HE a ratio greater than the [ratio] that 7 has to 5. And therefore, by composition, line ZE has to line EH a ratio greater than the [ratio] that 12 has to 5, that is, than the [ratio] that 36 has to 15. But it has also been proved that line HE has to line EΘ a ratio greater than the [ratio] that 15 has to 2. Therefore, by equality of terms,[85] line ZE has to line EΘ a ratio greater than the [ratio] that 36 has to 2, that is, than the [ratio] that 18 has to 1. Therefore, line ZE is greater than 18 times line EΘ.

B. Let also BΔ be produced to straight line ZE in Θ.

C. We will demonstrate that in this lemma. Let it be triangle rectangle ABC with the right angle at C. And let add in straight line AC whatever point D, and let BD be joined. I say that straight line AC has to straight line CD a ratio greater than the one that angle ABC has to angle DBC.

For, with center at B and radius BD, let the arc of circle EDF be described, and let BC be produced to F. In the same manner, since triangle ABD is certainly greater than sector EBD, but triangle DBC is smaller than sector DBF, triangle ADB will have to triangle DBC a ratio greater than the one that sector EBD has to sector DBF. But as triangle ABD is to triangle DBC, so is straight line AD to DC itself {[*Elem.*] VI, 5.}: and as sector ABD is to sector DBC, so [is] angle ABD to angle DBC {[*Elem.*] VI, last [33].}.[86] Consequently, straight line AD has to DC itself a ratio greater than the one that angle ABD has to angle DBC. And, by composition, straight line AC has to CD itself a ratio greater than the one that angle ABC has to angle DBC.

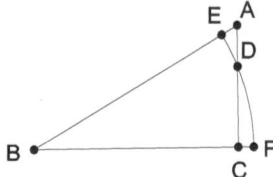

Figure 2.2.5 Diagram of Commandino's note C to proposition 7.

---

83 V:71 *pas*, presumably for *pars*.

84 V:71 *sed mh ipsorum Ke maior est q dupla* (but MH is greater than the double of the same KE), confusing the numbers μθ (49) and κε (25) with letters indicating lines. He also translates *maior* (greater) when it is minor (ἐλάσσονα).

85 G:146 omits δι' ἴσου. In his Italian translation, G:204 translates it by *egualmente*.

86 Here Commandino is applying the second part of the proposition, which links the angles and arcs to the sectors of the circles. It is an addition of Theon (cf. Heath 2002, 2: 274–275) that also appears in the translation of C of the *Elements* (p. 216), in which he includes *adhuc etiam et sectores* at the end.

D. Quoniam enim angulus DBE bifariam secatur recta linea BG, erit ex tertia sexti elementorum ut FB ad BE, ita FG ad GE: quare ex 22 eisudem, ut quadratum ex FB ad quadratum ex BE, ita quadratum ex FG ad quadratum ex GE.

E. Nam BH, quae maiori angulo, nempe recto subtenditur, maior est, quam ipsa BE.

F. Ducatur a puncto C, videlicet ab angulo recto trianguli ABC ad basim perpendicularis CM; fient triangula BCM ACM similia toti, et inter sese {8. sexti.}. Quare angulus BCM, hoc est angulus HBE est aequalis angulo BAC {29. primi.}, atque est ACB rectus aequalis recto BEH. Reliquus igitur ABC reliquo BHE est aequalis, et triangulum triangulo simile. Ergo ut BH ad HE, ita AB ad BC {4. sexti.}.

ἡ δὲ ZE ἴση ἐστὶν τῇ BE· καὶ ἡ BE ἄρα τῆς EΘ μείζων ἐστὶν ἢ ιη· πολλῷ ἄρα ἡ BΘ τῆς ΘE μείζων ἐστὶν ἢ ιη. ἀλλ᾽ ὡς ἡ BΘ πρὸς τὴν ΘE, οὕτως ἐστὶν ἡ AB πρὸς τὴν BΓ, διὰ τὴν ὁμοιότητα τῶν τριγώνων· καὶ ἡ AB ἄρα τῆς BΓ μείζων ἐστὶν ἢ ιη.

καὶ ἔστιν ἡ μὲν AB τὸ ἀπόστημα ὃ ἀπέχει ὁ ἥλιος ἀπὸ τῆς γῆς, ἡ δὲ ΓB τὸ ἀπόστημα ὃ ἀπέχει ἡ σελήνη ἀπὸ τῆς γῆς· τὸ ἄρα ἀπόστημα ὃ ἀπέχει ὁ ἥλιος ἀπὸ τῆς γῆς τοῦ ἀποστήματος, οὗ ἀπέχει ἡ σελήνη ἀπὸ τῆς γῆς, μεῖζόν ἐστιν ἢ ιη.

Λέγω δὴ ὅτι καὶ ἔλασσον ἢ κ.

ἤχθω γὰρ διὰ τοῦ Δ τῇ EB παράλληλος ἡ ΔK, καὶ περὶ τὸ ΔKB τρίγωνον κύκλος γεγράφθω ὁ ΔKB· ἔσται δὴ αὐτοῦ διάμετρος ἡ ΔB, διὰ τὸ ὀρθὴν εἶναι τὴν πρὸς τῷ K γωνίαν. καὶ ἐνηρμόσθω ἡ BΛ ἑξαγώνου.

And line ZE is equal to line BE.[87] Therefore, also line BE is greater than 18 times line EΘ.[88,89] Therefore, the more so is line BΘ greater than 18 times line ΘE[90] [E]. But, as line BΘ is to line ΘE, so is line AB to line BΓ, because of the similarity of the triangles [F]. Therefore, also line AB is greater than 18 times line BΓ.

And line AB is the distance of the Sun from the Earth, and line BΓ,[91] the distance of the Moon from the Earth. Therefore, the distance of the Sun from the Earth is greater than 18 times the distance of the Moon from the Earth.

I say that [this distance is] also smaller than 20.

Let line ΔK be drawn through[92] point Δ, parallel to line EB and let circle ΔKB be described around triangle ΔKB. Its diameter will be line ΔB, because the angle at point K is right [Sch. 50 (27)]. And let BΛ be fitted into a hexagon.

D. For, since angle ΔBE is cut in two parts by straight line BH, it will be – from the third [proposition] of the sixth [book of] the *Elements* [of Euclid] – [that] as ZB is to BE, so will ZH be to HE. Therefore – from [proposition] 22 of the same [book] – as the square of ZB is to the square of BE, so [is] the square of ZH to the square of HE.

E. For BΘ, which subtends a greater angle, that is, a right angle, is greater than BE itself.

F. Let ΓM be drawn from point Γ, that is, from the right angle of triangle ABΓ, perpendicular to the base. Triangles BΓM and AΓM will be similar to the whole and between each other {[*Elem.*] VI, 8.}. Therefore, angle BΓM, that is angle ΘBE, is equal to angle BAΓ {[*Elem.*] I, 29.}, and the right angle AΓB is equal to the right angle BEΘ. Accordingly, the remaining ABΓ is equal to the remaining BΘE, and one triangle is similar to the [other] triangle. Consequently, as BΘ is to ΘE, so [is] AB to BΓ {[*Elem.*] VI, 4.}.

---

87 G:146, in French, BF; but in Italian, G:204, BE.
88 V:71 *nam quadratum be. ergo ipso eh maior est q 18* (therefore, the square of BE is greater than 18 times the same EH).
89 G:146 BH, but in Italian, G:204, EH.
90 G:146 and G:204, DE.
91 G:148 and G:205, CB.
92 διὰ τοῦ Δ should be translated by *through*, but because the line begins at Δ, it would be better to translate it by *from*. See note 28.

G. Ex demonstratis a Ptolemeo in principio magnae constructionis.

H. Ex corollario quintae decimae quarti libri elementorum.

K. Ob triangulorum DBK ABC similitudinem. Rursus enim angulus MCB, hoc est BDK est aequalis angulo BAC, rectusque DKB recto ACB, et reliquus reliquo aequalis.

καὶ ἐπεὶ ἡ ὑπὸ τῶν ΔΒΕ γωνία λ΄ ἐστιν ὀρθῆς, καὶ ἡ ὑπὸ τῶν ΒΔΚ ἄρα λ΄ ἐστιν ὀρθῆς· ἡ ἄρα ΒΚ περιφέρεια ξ΄ ἐστιν τοῦ ὅλου κύκλου. ἔστιν δὲ καὶ ἡ ΒΛ ἕκτον μέρος τοῦ ὅλου κύκλου· ἡ ἄρα ΒΛ περιφέρεια τῆς ΒΚ περιφερείας ι ἐστίν. καὶ ἔχει ἡ ΒΛ περιφέρεια πρὸς τὴν ΒΚ περιφέρειαν μείζονα λόγον ἤπερ ἡ ΒΛ εὐθεῖα πρὸς τὴν ΒΚ εὐθεῖαν· ἡ ἄρα ΒΛ εὐθεῖα τῆς ΒΚ εὐθείας ἐλάσσων ἐστὶν ἢ ι. καὶ ἔστιν αὐτῆς διπλῆ ἡ ΒΔ· ἡ ἄρα ΒΔ τῆς ΒΚ ἐλάσσων ἐστὶν ἢ κ. ὡς δὲ ἡ ΒΔ πρὸς τὴν ΒΚ, ἡ ΑΒ πρὸς⁵ ΒΓ, ὥστε καὶ ἡ ΑΒ τῆς ΒΓ ἐλάσσων ἐστὶν ἢ κ. καὶ ἔστιν ἡ μὲν ΑΒ τὸ ἀπόστημα ὃ ἀπέχει ὁ ἥλιος ἀπὸ τῆς γῆς, ἡ δὲ ΒΓ τὸ ἀπόστημα ὃ ἀπέχει ἡ σελήνη ἀπὸ τῆς γῆς· τὸ ἄρα ἀπόστημα ὃ ἀπέχει ὁ ἥλιος ἀπὸ τῆς γῆς τοῦ ἀποστήματος, οὗ ἀπέχει ἡ σελήνη ἀπὸ τῆς γῆς, ἔλασσόν ἐστιν ἢ κ.

ἐδείχθη δὲ καὶ μεῖζον ἢ ιη.

---

5 H:380 inserts τὴν between πρὸς and ΒΓ. It is true that usually, the letters labeling lines are preceded by the article, but there are also many cases in which the article is omitted in the text. So either we assume that the omission could be original or we should modify every time it is omitted.

And since angle ΔBE is 1/30[93] [part] of a right angle, also angle BΔK is, therefore, 1/30 [part] of a right angle. Therefore, angle BK is 1/60 [part] of the whole circle [Sch. 52 (28)]. But also [arc] BΛ is one-sixth[94] part of the whole circle.[95] Therefore, arc BΛ is 10 times arc BK. And arc BΛ has to arc BK a greater ratio than straight line BΛ [has] to straight line BK [G] [Sch. 54 (29)]. Therefore, straight line BΛ is smaller than 10 times straight line BK. And line BΔ is the double of it[96] [H]. Therefore, line BΔ is smaller than 20 times line BK.[97] And, as line BΔ is to line BK, [so] is line AB to BΓ [K], so that also line AB is smaller than 20 times line BΓ. And line AB is the distance of the Sun from the Earth, and line BΓ is the distance of the Moon from the Earth. Therefore, the distance of the Sun from the Earth is smaller than 20 times the distance of the Moon from the Earth.

And it has also been proved that it is greater than 18 times. [Sch. 45 (23)]

G. As demonstrated by Ptolemy at the beginning of the *Great Construction* [*The Almagest*].[98]

H. From the corollary of the fifteenth [proposition] of the fourth book of the *Elements* [of Euclid].

K. Because of the similarity of triangles ΔBK and ABΓ. For, again, angle MΓB, that is BΔK, is equal to angle BAΓ, and the right angle ΔKB [is equal to] the right angle AΓB, and the remaining [angle] [is equal to] the remaining [angle].

---

93 V:71, *50*.
94 V:71 translates ἕκτον (sixth) by *quaelibet*. One of the meanings of ἕκτος is "qualities of a substance," and maybe that was the one he chose for his translation.
95 Although it is not made explicit that it is an arc and the feminine article only makes us presume a straight line, it is clear that it refers to an arc, since only an arc can be part of a circumference of a circle.
96 This refers to line BΛ.
97 G:205 omits ἡ ἄρα BΔ τῆς BK ἐλάσσων ἐστὶν ἢ κ. ὡς δὲ ἡ BΔ πρὸς τὴν BK, ἡ AB πρὸς BΓ and replaces AB by BD. The French translation is correct.
98 See note 54.

η′ Ὅταν ὁ ἥλιος ἐκλείπῃ ὅλος, τότε ὁ αὐτὸς κῶνος περιλαμβάνει τόν τε ἥλιον καὶ τὴν σελήνην, τὴν κορυφὴν ἔχων πρὸς τῇ ἡμετέρᾳ ὄψει.

Ἐπεὶ γάρ, ἐὰν ἐκλείπῃ ὁ ἥλιος, δι᾽ ἐπιπρόσθεσιν τῆς σελήνης ἐκλείπει, ἐμπίπτοι ἂν ὁ ἥλιος εἰς τὸν κῶνον τὸν περιλαμβάνοντα τὴν σελήνην τὴν κορυφὴν ἔχοντα πρὸς τῇ ἡμετέρᾳ ὄψει. ἐμπίπτων δὲ ἤτοι ἐναρμόσει εἰς αὐτόν, ἢ ὑπεραίροι, ἢ ἐλλείποι· εἰ μὲν οὖν ὑπεραίροι, οὐκ <ἂν> ἐκλείποι ὅλος, ἀλλὰ παραλλάττοι αὐτοῦ τὸ ὑπεραῖρον. εἰ δὲ ἐλλείποι, διαμένοι ἂν ἐκλελοιπὼς ἐν ὅσῳ διεξέρχεται τὸ ἐλλεῖπον. ὅλος δὲ ἐκλείπει καὶ οὐ διαμένει ἐκλελοιπώς· τοῦτο γὰρ ἐκ τῆς τηρήσεως φανερόν. ὥστε οὔτ᾽ ἂν ὑπεραίροι, οὔτε ἐλλείποι. ἐναρμόσει ἄρα εἰς τὸν κῶνον, καὶ περιληφθήσεται ὑπὸ τοῦ κώνου τοῦ περιλαμβάνοντος τὴν σελήνην τὴν κορυφὴν ἔχοντος πρὸς τῇ ἡμετέρᾳ ὄψει.

8. When the Sun is wholly eclipsed, one and the same cone envelops the Sun and the Moon, [the cone] having its vertex towards our eye.

Since, if the Sun is eclipsed, it is eclipsed by the interposition of the Moon, the Sun would fall within the cone enveloping the Moon and having its vertex towards our eye. If it falls within [the cone], either it fits, or it would exceed, or it would fall short of [the cone].[99] If it exceeds [the cone], it would not be wholly eclipsed, but what exceeds would pass beyond it[100,101] And, if it falls short of [the cone], it would remain eclipsed while it travels along what remains.[102] But it is wholly eclipsed and does not remain eclipsed; this is evident from observation.[103] So that it would not exceed [the cone], and it would not fall short of it. Therefore, it will fit the cone, and it will be enveloped by the cone enveloping the Moon and having its vertex towards our eye.

---

99 The Greek uses the future in the first case and the optative in the other two, pointing out that the first case is, in fact, the one that happens and thus anticipating the solution. In English, it is not possible to maintain that nuance without forcing the expression.

100 This refers to the cone.

101 Modern translations tend to render what the text seems to suggest rather than what it actually says, stating that what exceeds would be visible. For example, H:383 says, "If now it should overlap it, the sun would not be totally eclipsed, but the portion which overlaps would be unobstructed." In our translation, we have chosen to respect the sense of the Greek text and translate it in agreement with the Latin scholars: "but the surplus (or the surplus part) will protrude from itself" (V:71: "sed eminebit ipsius excedens"; C:18a: "Sed eminebit ipsius pars excedens"). FG:43 follows C.

102 It refers to the fact that it must travel the distance between the Sun and the limit of the cone.

103 F:23 omits γὰρ ἐκ τῆς τηρήσεως φανερόν, but it is in his Greek edition and Latin translation (FG:42–43).

θ' Ἡ τοῦ ἡλίου διάμετρος τῆς διαμέτρου τῆς σελήνης μείζων μέν ἐστιν ἢ ιη, ἐλάσσων δὲ ἢ κ.

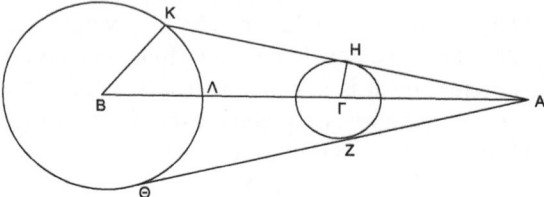

9. **The diameter of the Sun is greater than 18 times, but smaller than 20 times the diameter of the Moon.**

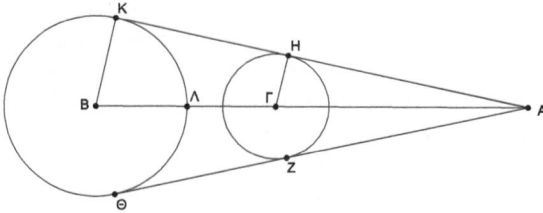

Figure 2.1.7 Diagram of proposition 9.

Ἔστω γὰρ ἡ μὲν ἡμετέρα ὄψις πρὸς τῷ Α, ἡλίου δὲ κέντρον τὸ Β, σελήνης δὲ κέντρον τὸ Γ, ὅταν ὁ περιλαμβάνων κῶνος τόν τε ἥλιον καὶ τὴν σελήνην τὴν κορυφὴν ἔχῃ πρὸς τῇ ἡμετέρᾳ ὄψει, τουτέστιν, ὅταν τὰ Α, Γ, Β σημεῖα ἐπ' εὐθείας ᾖ, καὶ ἐκβεβλήσθω διὰ τῆς ΑΓΒ ἐπίπεδον· ποιήσει δὴ τομάς, ἐν μὲν ταῖς σφαίραις μεγίστους κύκλους, ἐν δὲ τῷ κώνῳ εὐθείας. ποιείτω οὖν ἐν μὲν ταῖς σφαίραις μεγίστους κύκλους τοὺς ΖΗ, ΚΛΘ, ἐν δὲ τῷ κώνῳ εὐθείας τὰς ΑΖΘ, ΑΗΚ, καὶ ἐπεζεύχθωσαν αἱ ΓΗ, ΒΚ.

* 4. sexti

* 7. huius

ἔσται δή, ὡς ἡ ΒΑ πρὸς τὴν ΑΓ, ἡ ΒΚ⁶ πρὸς ΓΗ. ἡ δὲ ΒΑ τῆς ΑΓ⁷ ἐδείχθη μείζων μὲν ἢ ιη, ἐλάσσων δὲ ἢ κ. καὶ ἡ ΒΚ ἄρα τῆς ΓΗ μείζων μέν ἐστιν ἢ ιη, ἐλάσσων δὲ ἢ κ.

---

6 In mss. U: ΒΗ.
7 In mss. U: ΒΓ.

Let our eye be at point A, let the center of the Sun be point B, and let the center of the Moon be point Γ, when the cone enveloping the Sun and the Moon has its vertex towards our eye, that is, when points A, Γ, and B are on a straight line. And let a plane be produced through line AΓB. It will make sections on the spheres [that are] great circles, and on the cone, straight lines.[104] So let [the plane] make on the spheres great circles ZH and KΛΘ,[105] and on the cone, the straight lines AZΘ and AHK, and let ΓH and BK be joined.

As line BA is to line AΓ,[106] [so] will line BK be to ΓH [*]. And it has been proved that line BA is greater than 18 times, but smaller than 20 times line AΓ [**]. Therefore, also line BK is greater than 18 times, but smaller than 20 times line ΓH.

* [*Elem.*] VI, 4.

** From the seventh [proposition] of this [treatise].

---

104 V:72 translates: *ac in cono rectam lineam* (and in the cone, a straight line).
105 G:206, in Italian: KLM (and not KLH). The French is correct.
106 G:154, in French: CA; G:206, in Italian: CG.

ι' Ὁ ἥλιος πρὸς τὴν σελήνην μείζονα μὲν
λόγον ἔχει ἢ ὃν τὰ ͵εωλβ πρὸς α, ἐλάσσονα
δὲ ἢ ὃν τὰ ͵η πρὸς α.

A   B

10. The Sun has to the Moon a ratio greater than the [ratio] that 5832 has to 1, but smaller than the [ratio] that 8000 has to 1.

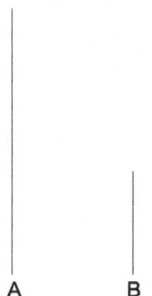

A   B

Figure 2.1.8 Diagram of proposition 10.

* 33. undecimi elmen. [1]8.[8]
       duo decimi.
** 11. quinti.

Ἔστω ἡ μὲν τοῦ ἡλίου διάμετρος ἡ Α, ἡ δὲ τῆς σελήνης ἡ Β. ἡ Α ἄρα πρὸς τὴν Β μείζονα λόγον ἔχει ἢ ὃν τὰ ιη πρὸς α, ἐλάσσονα δὲ ἢ ὃν τὰ κ πρὸς α. καὶ ἐπειδὴ ὁ ἀπὸ τῆς Α κύβος πρὸς τὸν ἀπὸ τῆς Β κύβον γ λόγον ἔχει ἤπερ ἡ Α πρὸς τὴν Β, ἔχει δὲ καὶ ἡ περὶ διάμετρον τὴν Α σφαῖρα πρὸς τὴν περὶ διάμετρον τὴν Β σφαῖραν γ λόγον ἤπερ ἡ Α πρὸς τὴν Β, ἔστιν ἄρα, ὡς ἡ περὶ διάμετρον τὴν Α σφαῖρα πρὸς τὴν περὶ διάμετρον τὴν Β σφαῖραν, οὕτως ὁ ἀπὸ τῆς Α κύβος πρὸς τὸν ἀπὸ τῆς Β κύβον. ὁ δὲ ἀπὸ τῆς Α κύβος πρὸς τὸν ἀπὸ τῆς Β κύβον μείζονα λόγον ἔχει ἢ ὃν τὰ ͵εωλβ πρὸς α, ἐλάσσονα δὲ ἢ ὃν τὰ ͵η πρὸς α, ἐπειδὴ ἡ Α πρὸς τὴν Β μείζονα λόγον ἔχει ἢ ὃν τὰ ιη πρὸς α, ἐλάσσονα δὲ ἢ ὃν τὰ κ πρὸς ἕν· ὥστε ὁ ἥλιος πρὸς τὴν σελήνην μείζονα λόγον ἔχει ἤπερ τὰ ͵εωλβ πρὸς α, ἐλάσσονα δὲ ἢ ὃν τὰ ͵η πρὸς α.

---

8   In C:18b, 28 instead of 18, but it is clear that he wanted to refer to proposition 18.

Let the diameter of the Sun be line A, and that of the Moon, line B. Therefore, line A has to line B a ratio greater than the [ratio] that 18 has to 1, but smaller than the [ratio] that 20[107] has to 1.[108] And because the cube of line A has to the cube of line B the triplicate ratio[109] that line A [has] to line B [*], and also the sphere [constructed] around diameter A has to the sphere [constructed] around diameter B the triplicate ratio[110] that line A [has] to line B, therefore, as the sphere [constructed] around diameter A is to the sphere [constructed] around diameter B, so is the cube of line A to the cube of line B [**]. But the cube of line A has to the cube of line B a ratio greater than the [ratio] that 5832 has to 1, but smaller than the [ratio] that 8000 has to 1, since line A has to line B a ratio greater than the [ratio] that 18 has to 1,[111] but smaller than the [ratio] that 20 has to one. So that the Sun has to the Moon a greater ratio than 5832 [has] to 1, but smaller than the [ratio] that 8000 has to 1. [Sch. 59; Sch. 60 (33)]

* [*Elem.*] XI, 33 and XII, 18.

** [*Elem.*] V, 11.

---

107 V:72: *8*.

108 V:72 translates α (1) by *a*.

109 V:72 translates γ (3) by *c*. *Triplicate ratio* must be understood as *the ratio multiplied three times by itself*, that is, cubed. See the first note of FG:216, where he explains that.

110 V:72 translates γ (3) by *c*.

111 V:72 omits ἐλάσσονα δὲ ἢ ὃν τὰ ͵η πρὸς α, ἐπειδὴ ἡ Α πρὸς τὴν Β μείζονα λόγον ἔχει ἢ ὃν τὰ ιη πρὸς α. It could be a case of *saut du même au même*.

ια' Ἡ τῆς σελήνης διάμετρος τοῦ ἀποστήματος,
οὗ ἀπέχει τὸ κέντρον τῆς σελήνης ἀπὸ τῆς
ἡμετέρας ὄψεως, ἐλάσσων μέν ἐστιν ἢ δύο
με΄, μείζων δὲ ἢ λ΄.

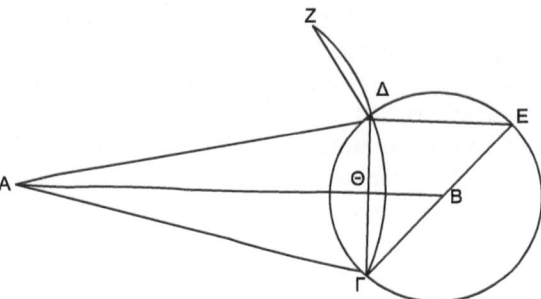

11. **The diameter of the Moon is smaller than 2/45, but greater than 1/30 [parts] the distance of the center of the Moon from our eye.**

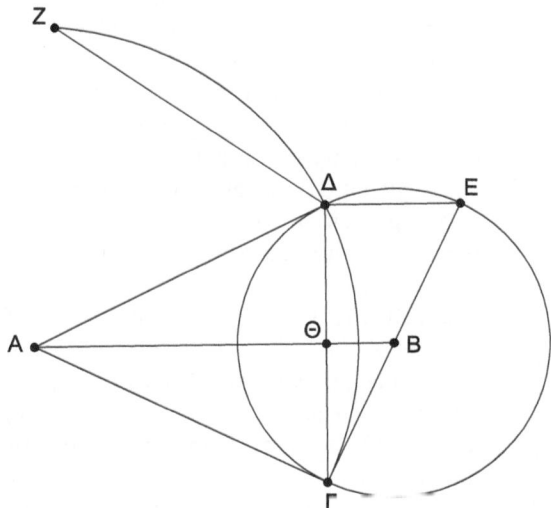

Figure 2.1.9 Diagram of proposition 11.

Ἔστω γὰρ ἡ μὲν ἡμετέρα ὄψις πρὸς τῷ Α, σελήνης δὲ κέντρον τὸ Β, ὅταν ὁ περιλαμβάνων κῶνος τόν τε ἥλιον καὶ τὴν σελήνην τὴν κορυφὴν ἔχῃ πρὸς τῇ ἡμετέρᾳ ὄψει.

λέγω ὅτι γίγνεται τὰ διὰ τῆς προτάσεως.

Ἐπεζεύχθω γὰρ ἡ ΑΒ, καὶ ἐκβεβλήσθω τὸ[9] διὰ τῆς ΑΒ ἐπίπεδον· ποιήσει δὴ τομ{ἢν}<ὰς>[10] ἐν μὲν τῇ σφαίρᾳ κύκλον, ἐν δὲ τῷ κώνῳ εὐθείας.

ποιείτω οὖν ἐν μὲν τῇ σφαίρᾳ κύκλον τὸν ΓΕΔ, ἐν δὲ τῷ κώνῳ εὐθείας τὰς ΑΔ, ΑΓ, καὶ ἐπεζεύχθω ἡ ΒΓ, καὶ διήχθω ἐπὶ τὸ Ε.

φανερὸν δὴ ἐκ τοῦ προδεδειγμένου ὅτι ἡ ὑπὸ τῶν ΒΑΓ γωνία ἡμισείας ὀρθῆς ἐστι με΄· καὶ κατὰ τὰ αὐτὰ ἡ ΒΓ τῆς ΓΑ ἐλάσσων ἐστὶν ἢ με΄. πολλῷ ἄρα ἡ ΒΓ τῆς ΒΑ ἐλάσσων ἐστὶν ἢ με΄ μέρος. καὶ ἔστι τῆς ΒΓ διπλῆ ἡ ΓΕ· ἡ ΓΕ ἄρα τῆς ΑΒ ἐλάσσων ἐστὶν ἢ δύο με΄. καὶ ἔστιν ἡ μὲν ΓΕ ἡ τῆς σελήνης διάμετρος, ἡ δὲ ΒΑ τὸ ἀπόστημα ὃ ἀπέχει τὸ κέντρον τῆς σελήνης ἀπὸ τῆς ἡμετέρας ὄψεως· ἡ ἄρα διάμετρος τῆς σελήνης τοῦ ἀποστήματος, οὗ ἀπέχει τὸ κέντρον τῆς σελήνης ἀπὸ τῆς ἡμετέρας ὄψεως, ἐλάσσων ἐστὶν ἢ δύο με΄.

A. Demonstratum est hoc in quarta huius.

B. Est enim BA maior, quam AC, cum maiori angulo subtendatur.

---

9 This is the only time the plane is specified by the article in this formula: ἐκβεβλήσθω τὸ διὰ τῆς ΑΒ ἐπίπεδον·. Even if the plane is specified by the article, there are infinite planes that go through line AB.

10 The manuscripts and editions have τομὴν. We justify the use of the plural from the use that Aristarchus makes in the treatise. He always uses the plural when the cuts are made in more than one sphere (ποιήσει δὴ τομὰς ἐν ταῖς σφαίραις μεγίστους κύκλους) or in more than one sphere and in the cone (δὴ τομάς, ἐν μὲν ταῖς σφαίραις μεγίστους κύκλους, ἐν δὲ τῷ κώνῳ εὐθε). Instead, he uses the singular when the cut is made in a single sphere and not in the cone (ποιήσει δὴ τομὴν ἐν τῇ σφαίρᾳ). Here the plane makes two cuts, one on a sphere and one on the cone. Aristarchus himself uses the plural when referring to the same case at the beginning of proposition 12 (ποιήσει δὴ τομὰς ἐν μὲν τῇ σφαίρᾳ κύκλον, ἐν δὲ τῷ κώνῳ εὐθείας). All translators, following the manuscripts or their editions, translate in the singular.

Let our eye be at point A, let the center of the Sun be point B when the cone enveloping the Sun and the Moon has its vertex towards our eye.

I say that the things [specified] in the enunciation happen.

Let AB be joined, and let the plane through line AB be produced. It will make sections: on the sphere, a circle, and on the cone, straight lines.

So, let it make, on the sphere circle ΓΕΔ, and on the cone, the straight lines AΔ and AΓ, and let BΓ be joined, and let [line BΓ] be produced to point E.

It is evident from what has been proved that angle BAΓ is 1/45 [part] of half a right angle [A], and, by the same [reasoning], that line BΓ is smaller than 1/45[112] [part] of line ΓA. Therefore, the more so is line BΓ smaller than 1/45 part of line BA [B]. And line ΓE is double line BΓ. Therefore, line ΓE is smaller than two 45th [part] of line AB.[113] And line ΓE is the diameter of the Moon, and line BA is the distance of the center of the Moon from our eye.[114] Therefore, the diameter of the Moon is smaller than 2/45 [parts] of the distance of the center of the Moon from our eye.

A. This has been demonstrated in the fourth [proposition] of this [treatise].

B. For BA is greater than AΓ, because it subtends a greater angle.

---

112 V:72, *48*.

113 V:72 omits ΓE.

114 G:207, in Italian, ἡ ἄρα διάμετρος τῆς σελήνης τοῦ ἀποστήματος, οὗ ἀπέχει τὸ κέντρον τῆς σελήνης ἀπὸ τῆς ἡμετέρας ὄψεως, ἐλάσσων ἐστὶν ἢ δύο με´ is omitted. The French is correct.

C. Ex 8. sexti elementorum.
Quoniam enim ab angulo
recto ACB perpendicularis
ducta est CH, fiunt triangula
ACH HCB similia toti, et
inter sese. quare angulus
BCH, videlicet ECD est
aequalis angulo BAC.

D. Hoc demonstratum est in
quarta huius.

E. Angulus enim rectus
consistit in quarta
parte circumferentiae
totius circuli, hoc est
in gradibus nonaginta,
cuius circumferentiae
pars quadragesima quinta
sunt duo gradus, videlicet
centesima, et octogesima
pars totius circuli.

F. Nam circumferentia
DF, quae maior est
circumferentia CD ad
ipsam CD circumferentiam
maiorem proportionem
habet, quam recta DF
ad rectam CD, quod
demonstravit Ptolemaeus
in principio magnae
constructionis. Quare
convertendo ex 26
quinti circumferentia
CD ad circumferentiam
DF minorem habet
proportionem, quam recta
linea CD ad DF rectam.

Λέγω δὴ ὅτι καὶ μείζων ἐστὶν ἡ ΓΕ τῆς ΒΑ ἢ λ΄ μέρος.

ἐπεζεύχθω γὰρ ἡ ΔΕ καὶ ἡ ΔΓ, καὶ κέντρῳ μὲν τῷ Α, διαστήματι δὲ τῷ ΑΓ, κύκλος γεγράφθω ὁ ΓΔΖ, καὶ ἐνηρμόσθω εἰς τὸν ΓΔΖ κύκλον τῇ ΑΓ ἴση ἡ ΔΖ.

καὶ ἐπεὶ ὀρθὴ ἡ ὑπὸ τῶν ΕΔΓ ὀρθῇ τῇ ὑπὸ τῶν ΒΓΑ ἐστὶν ἴση, ἀλλὰ καὶ ἡ ὑπὸ τῶν ΒΑΓ τῇ ὑπὸ τῶν ΘΓΒ ἐστὶν ἴση, λοιπὴ ἄρα ἡ ὑπὸ τῶν ΔΕΓ λοιπῇ τῇ ὑπὸ τῶν ΘΒΓ ἐστὶν ἴση· ἰσογώνιον ἄρα ἐστὶν τὸ ΓΔΕ τρίγωνον τῷ ΑΒΓ τριγώνῳ. ἔστιν ἄρα, ὡς ἡ ΒΑ πρὸς ΑΓ, οὕτως ἡ ΕΓ πρὸς ΓΔ· καὶ ἐναλλάξ, ὡς ἡ ΑΒ πρὸς ΓΕ, οὕτως ἡ ΑΓ πρὸς ΓΔ, τουτέστιν, ἡ ΔΖ πρὸς ΓΔ. ἀλλ᾽ ἐπεὶ πάλιν ἡ ὑπὸ τῶν ΔΑΓ γωνία με΄ μέρος ἐστὶν ὀρθῆς, ἡ ΓΔ ἄρα περιφέρεια ρπ΄ μέρος ἐστὶ τοῦ κύκλου· ἡ δὲ ΔΖ περιφέρεια ἕκτον μέρος ἐστὶν τοῦ ὅλου κύκλου· ὥστε ἡ ΓΔ περιφέρεια τῆς ΔΖ περιφερείας λ΄ μέρος ἐστίν. καὶ ἔχει ἡ ΓΔ περιφέρεια, ἐλάσσων οὖσα τῆς ΔΖ περιφερείας, πρὸς αὐτὴν τὴν ΔΖ περιφέρειαν ἐλάσσονα λόγον ἤπερ ἡ ΓΔ εὐθεῖα πρὸς τὴν ΖΔ εὐθεῖαν· ἡ ἄρα ΓΔ εὐθεῖα τῆς ΔΖ μείζων ἐστὶν ἢ λ΄.

I say that also line ΓΕ is greater than 1/30 part of line BA.

Let ΔΕ and ΔΓ be joined and, with center at point A and radius AΓ let circle ΓΔΖ be described, and let line ΔΖ, equal to line AΓ, be fitted into circle ΓΔΖ.[115]

And, since right angle ΕΔΓ is equal to right angle ΒΓΑ and, also, angle ΒΑΓ is equal to angle ΘΓΒ [C],[116] therefore, the remaining angle ΔΕΓ is equal to the remaining angle ΘΒΓ.[117] Therefore, triangle ΓΔΕ is equiangular to triangle ΑΒΓ. Therefore, as line BA is to line AΓ, so is line ΕΓ to ΓΔ and, by alternation, as line AB is to ΓΕ, so is line AΓ to ΓΔ, that is, line ΔΖ to ΓΔ. But, since, again, angle ΔΑΓ[118] is 1/45 part of a right angle [D], therefore, arc ΓΔ is 1/180 part of the circle [E]. And arc ΔΖ is one sixth of the whole circle,[119] so that arc ΓΔ is 1/30 part of arc ΔΖ. And arc ΓΔ, which is smaller than arc ΔΖ, has to the same arc ΔΖ a smaller ratio than straight line ΓΔ [has] to straight line ΖΔ [F].[120] Therefore, straight line ΓΔ is greater than 1/30 [part] of line ΔΖ.[121]

C. From [proposition] 8 of the sixth [book] of the *Elements* [of Euclid]. For, since from right angle AΓB, line ΓΘ has been drawn perpendicular [to AB], triangles AΓΘ and ΘΓB become similar to the whole and to each other. Therefore, angle ΒΓΘ, that is ΕΓΔ, is equal to angle ΒΑΓ.

D. This has been demonstrated in the fourth [proposition] of this [treatise].

E. For a right angle consists of the fourth part of the circumference of a whole circle, that is, in ninety degrees. From that circumference, the forty-fifth part is two degrees, that is, 1/180 part of the whole circle.

F. For arc ΔΖ, which is greater than arc ΓΔ, has to arc ΓΔ itself a ratio greater than the [ratio] that straight line ΛΖ has to straight line ΓΔ, what Ptolemy demonstrated in the beginning of the *Great Construction* [*Almagest*].[122] Therefore, by inversion – from [proposition] 26 of the fifth [book of the *Elements* of Euclid][123] – arc ΓΔ has to arc ΔΖ a ratio smaller than the [ratio] that straight line ΓΔ has to straight line ΔΖ.

115 G:160, in French, and 107, in Italian, omits καὶ ἐνηρμόσθω εἰς τὸν ΓΔΖ κύκλον τῇ ΑΓ ἴση ἡ ΔΖ.

116 V:72 replaces ΘΓΒ with DCE. As seen in the figure, the two letterings refer to the same angle, but the manuscripts have ΘΓΒ. C:20a, W:585, No:18, and M:57 (ECD) adopt the change. W:585 has ΕΓΔ. H:386–388 and F:26 translate it correctly. However, it is curious that Aristarchus introduces point Θ, only to indicate angles that could be better indicated without using it and thus using only the points of the triangles that he intends to show are equal. So ΘΓΒ would be better expressed as ΔΓΕ or, in any case, as ΔΓΒ, and ΘΒΓ as ΑΒΓ.

117 No:18 translates ΘΒΓ by ABC (it should be HCB).

118 V:72, *dkc* (instead of *dac*) and omits ὀρθῆς.

119 V:73 omits ἡ δὲ ΔΖ περιφέρεια ἔκτον μέρος ἐστὶν τοῦ ὅλου κύκλου.

120 G:160, in French, omits ἡ ἄρα ΓΔ εὐθεῖα τῆς ΔΖ μείζων ἐστὶν ἢ λʹ. ἴση δὲ ἡ ΖΔ τῇ ΑΓ. In Italian, it is not omitted (G:208).

121 G:160: CA.

122 See note 56.

123 See note 50.

G. Superius namque demonstratum est, ut AB ad CE, ita esse AC ad CD. Quare convertendo ut CE ad AB, ita DC ad CA. Quod cum DC maior sit, quam trigesima pars ipsius CA, et CE ipsius AB, quam trigesima pars maior erit.

ἴση δὲ ἡ ΖΔ τῇ ΑΓ· ἡ ἄρα ΔΓ τῆς ΓΑ μείζων ἐστὶν ἢ λ΄, ὥστε καὶ ἡ ΓΕ τῆς ΒΑ μείζων ἐστὶν ἢ λ΄. ἐδείχθη δὲ καὶ ἐλάσσων οὖσα ἢ δύο με΄.

And line ΖΔ is equal to line ΑΓ. Therefore, line ΔΓ is greater than 1/30 [part] of line ΓΑ,[124] so that also line ΓΕ is greater than 1/30 [part] of line ΒΑ [G].[125] But it has also been proved that it is smaller than two 45th [parts].[126]

G. Since above it has been demonstrated that, as AB is to ΓE, so is ΑΓ to ΓΔ. Therefore, by inversion, as ΓΕ is to AB, so [is] ΔΓ to ΓΑ. Because, since ΔΓ is greater than one-thirtieth part of ΓΑ itself, also ΓΕ will be greater than one-thirtieth part of AB itself.

---

124 No:19 omits ἴση δὲ ἡ ΖΔ τῇ ΑΓ· ἡ ἄρα ΔΓ τῆς ΓΑ μείζων ἐστὶν ἢ λ΄.
125 V:73 omits λ΄.
126 C:20b adds *ipsius BA. Quod ostendendum proponebatur* (in the same way as BA. What we proposed to prove). M:58 translates "quintas partes (del mismo BA)" (fifth parts (of the same BA)), but while "línea" (line) in Spanish is feminine, "del mismo" is masculine. So it should say, "(de la misma)."

ιβ' Ἡ διάμετρος τοῦ διορίζοντος ἐν τῇ σελήνῃ
τό τε σκιερὸν καὶ τὸ λαμπρὸν τῆς
διαμέτρου τῆς σελήνης ἐλάσσων μέν ἐστι,
μείζονα δὲ λόγον ἔχει πρὸς αὐτὴν ἢ ὂν τὰ
πθ πρὸς ϙ.

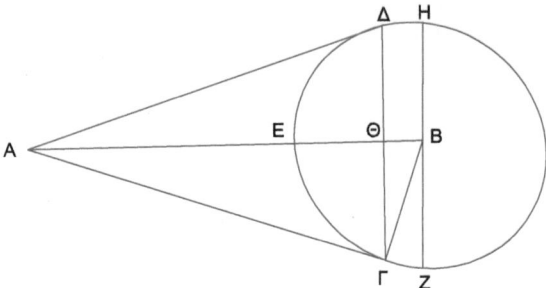

12. **The diameter of the [circle] delimiting the shaded and bright parts on the Moon is smaller than the diameter of the Moon, but it has to [that diameter] a ratio greater than the [ratio] that 89[127]has to 90.**

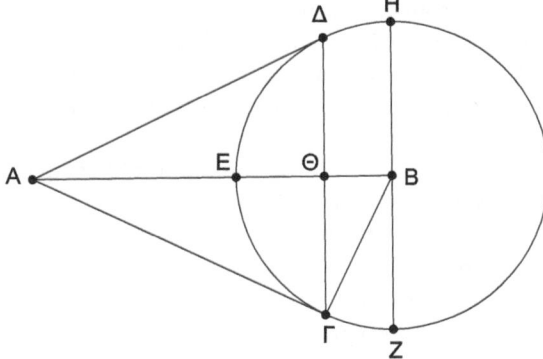

Figure 2.1.10 Diagram of proposition 12.

Ἔστω γὰρ ἡ μὲν ἡμετέρα ὄψις πρὸς τῷ Α,
σελήνης δὲ κέντρον τὸ Β, ὅταν ὁ περιλαμβάνων
κῶνος τόν τε ἥλιον καὶ τὴν σελήνην τὴν κορυφὴν
ἔχῃ πρὸς τῇ ἡμετέρᾳ ὄψει, καὶ ἐπεζεύχθω ἡ ΑΒ,
καὶ ἐκβεβλήσθω διὰ τῆς ΑΒ ἐπίπεδον· ποιήσει
δὴ τομὰς ἐν μὲν τῇ σφαίρᾳ κύκλον, ἐν δὲ τῷ
κώνῳ εὐθείας. ποιείτω <ἐν μὲν τῇ σφαίρᾳ
κύκλον τὸν ΔΕΓ, ἐν δὲ τῷ κώνῳ εὐθείας>[11] τὰς
ΑΔ, ΑΓ, ΓΔ. ἡ ΓΔ ἄρα διάμετρός ἐστι τοῦ κύκλου
τοῦ διορίζοντος ἐν τῇ σελήνῃ τὸ σκιερὸν καὶ τὸ
λαμπρόν.

λέγω δὴ ὅτι ἡ ΓΔ τῆς διαμέτρου τῆς σελήνης
ἐλάσσων μέν ἐστι, μείζονα δὲ λόγον ἔχει <πρὸς
αὐτὴν> ἢ ὃν τὰ πθ πρὸς ϙ. Ὅτι μὲν οὖν ἡ ΓΔ
ἐλάσσων ἐστὶ τῆς διαμέτρου τῆς σελήνης,
φανερόν. λέγω δὴ ὅτι καὶ μείζονα λόγον ἔχει
<πρὸς αὐτὴν> ἢ ὃν τὰ πθ πρὸς ϙ.

---

11 The text in angle brackets was added by W:586, probably inspired by the translation of C:21b–22a, who adds *faciat in sphaera circulum DEC, & in cono rectas lineas AD AC CD*. However, in ms. U – presumably used by C (cf. Noack 1992: 61) – it does not appear. V:73 does not add it. FG:54 adds it and, in a note on page 219, clarifies that he agrees with W in the addition. In W there is no τῇ before σφαίρᾳ, but H:388 and others add it.

Let our eye be at point A, let point B be the center of the Moon when the cone enveloping the Sun and the Moon has its vertex towards our eye, and let AB be joined and let a plane be produced through line AB. It will make sections: on the sphere, a circle, and on the cone, straight lines. Let it make on the sphere, circle ΔΕΓ, and on the cone, the straight lines ΑΔ, ΑΓ, and ΓΔ. Therefore, line ΓΔ is a diameter of the circle delimiting[128] the shaded and bright parts on the Moon.

I say that line ΓΔ is smaller than the diameter of the Moon, but it has to [that diameter] a ratio greater than the [ratio] that 89 has to 90.[129] Now, it is evident that line ΓΔ is smaller than the diameter of the Moon. I say that it also has to [that diameter] a ratio greater than the [ratio] that 89 has to 90.[130]

---

128 V:73, Cd igitur diameter est circuli dispescens in luna opacum & luminosum.

129 No:19–20 omits Ὅτι μὲν οὖν ἡ ΓΔ ἐλάσσων ἐστὶ τῆς διαμέτρου τῆς σελήνης, φανερόν. λέγω δὴ ὅτι καὶ μείζονα λόγον ἔχει <πρὸς αὐτὴν> ἢ ὃν τὰ πθ πρὸς ρ.

130 V:73, *60*.

A. Anguli enim eundem habe[n]t proportionem quam circunferentiae, in quibus insistunt, ex ultima sexti elementorum.

B. Ex 15 quinti elementorum.

C. Ex demonstratis a Ptolemaeo. Nam circumferentia GEF ad circumferentiam DEC maiorem habet proportionem, quam GF recta ad rectam DC. Ergo convertendo circumferentia DEC ad circumferentiam GEF minorem proportionem habet, quam recta DC ad rectam GF, ideoque recta DC ad rectam GF maiorem proportionem habebit, quam circumferentia DEC ad GEF circumferentiam.

Ἤχθω γὰρ διὰ τοῦ Β τῇ ΓΔ παράλληλος ἡ ΖΗ, καὶ ἐπεζεύχθω ἡ ΒΓ. ἔσται δὴ πάλιν κατὰ τὰ αὐτὰ ἡ ὑπὸ τῶν ΔΑΓ γωνία ὀρθῆς μεʹ μέρος, ἡ δὲ ὑπὸ τῶν ΒΑΓ ὀρθῆς ϙʹ μέρος. καὶ ἔστιν ἡ ὑπὸ τῶν ΒΑΓ γωνία ἴση τῇ ὑπὸ τῶν ΓΒΖ· καὶ ἡ ὑπὸ τῶν ΓΒΖ ἄρα γωνία ὀρθῆς ἐστιν ϙʹ, τουτέστιν, τῆς ὑπὸ τῶν ΖΒΕ γωνίας ϙʹ, ὥστε καὶ ἡ ΓΖ περιφέρεια τῆς ΖΓΕ περιφερείας ἐστὶν ϙʹ. ἡ ΓΕ ἄρα περιφέρεια πρὸς τὴν ΕΓΖ περιφέρειαν λόγον ἔχει ὃν τὰ πθ πρὸς ϙ. καὶ ἔστι τῆς ΓΕ β ἡ ΔΕΓ, τῆς δὲ ΕΓΖ β ἡ ΗΕΖ· ἡ ἄρα ΔΕΓ περιφέρεια πρὸς τὴν ΗΕΖ περιφέρειαν λόγον ἔχει ὃν τὰ πθ πρὸς ϙ. καὶ ἔχει ἡ ΔΓ εὐθεῖα πρὸς <τὴν> ΗΖ εὐθεῖαν μείζονα λόγον ἤπερ ἡ ΔΕΓ περιφέρεια πρὸς τὴν ΗΕΖ περιφέρειαν· καὶ ἡ ΔΓ ἄρα εὐθεῖα πρὸς τὴν ΗΖ εὐθεῖαν μείζονα λόγον ἔχει ἢ ὃν τὰ πθ πρὸς ϙ.

Let line ZH be drawn parallel to line ΓΔ through point B and let BΓ be joined.[131] Again, by the same [reasoning], angle ΔΑΓ will be 1/45 part of a right angle, and angle ΒΑΓ will be 1/90 part of a right angle. And angle ΒΑΓ is equal to angle ΓΒΖ. And therefore, angle ΓΒΖ is 1/90 [part] of a right angle,[132] that is, 1/90 [part] of angle ΖΒΕ, so that also arc ΓΖ is 1/90 [part] of arc ΖΓΕ [Α].[133,134] Therefore, arc ΓΕ has to arc ΕΓΖ the ratio that 89 has to 90. And [arc] ΔΕΓ is the double of [arc] ΓΕ [Β][135] and [arc] ΗΕΖ, the double of [arc] ΕΓΖ.[136,137] Therefore, arc ΔΕΓ has to arc ΗΕΖ[138] the ratio that 89 has to 90 [C]. And straight line ΔΓ has to straight line ΗΖ a greater ratio than arc ΔΕΓ [has] to arc ΗΕΖ [Sch. 77 (47)]. And, therefore, straight line ΔΓ has to straight line ΗΖ a ratio greater than the [ratio] that 89 has to 90.

A. For the angles has the same ratio as the arcs in which they stand – from the last [33] [proposition] of the sixth [book] of the *Elements* [of Euclid].

B. From [proposition] 15 of the fifth [book] of the *Elements* [of Euclid].

C. From what has been demonstrated by Ptolemy. For arc ΗΕΖ has to arc ΔΕΓ a ratio greater than the [ratio] that straight line ΗΖ has to straight line ΔΓ. Consequently, by inversion, arc ΔΕΓ has to arc ΗΕΖ a ratio smaller than the [ratio] that straight line ΔΓ has to straight line ΗΖ. And, for that reason, straight line ΔΓ will have to straight line ΗΖ a ratio greater than the [ratio] that arc ΔΕΓ has to arc ΗΕΖ.

---

131 G:208, in Italian, ABC; the French is correct.
132 M:59 omits καὶ ἡ ὑπὸ τῶν ΓΒΖ ἄρα γωνία ὀρθῆς ἐστιν ϙ´.
133 G:164, in French, omits ΖΓΕ. The Italian is correct (G:209).
134 M:59 omits ΓΖ.
135 V:73 translates β (2) by *b*.
136 V:73 translates τῆς δὲ ΕΓΖ β ἡ ΗΕΖ by *ipisus decfb ipse gef*. Ms. L, which follows V, says τῆς δ ΕΓΖ β ἡ ΗΕΖ, so he translates the δ by another d, and again, the β by b and not by 2. And so he obtains *decfb*. The elision of the ὲ in δὲ could have provoked the first mistake. However, it must be considered that the δ is separated from ΕΓΖ and there is not a line above it.
137 Even if περιφέρεια is omitted and the feminine article usually stands for a *line*, by context and by the diagram, it is evident that they refer to arcs.
138 G:209, in Italian, CEF (it should be GEF); the French is correct.

ιγ′ Ἡ ὑποτείνουσα εὐθεῖα ὑπὸ τὴν
ἀπολαμβανομένην ἐν τῷ σκιάσματι τῆς
γῆς περιφέρειαν τοῦ κύκλου, καθ' οὗ
φέρεται τὰ ἄκρα τῆς διαμέτρου τοῦ
διορίζοντος ἐν τῇ σελήνῃ τό τε σκιερὸν
καὶ τὸ λαμπρόν, τῆς μὲν διαμέτρου τῆς
σελήνης ἐλάσσων ἐστὶν ἢ διπλῆ, μείζονα
δὲ λόγον ἔχει πρὸς αὐτὴν ἢ ὃν τὰ πη πρὸς
με, τῆς δὲ τοῦ ἡλίου διαμέτρου ἐλάσσων
μέν ἐστιν ἢ ἔνατον μέρος, μείζονα δὲ λόγον
ἔχει πρὸς αὐτὴν ἢ ὃν κβ πρὸς σκε, πρὸς δὲ
τὴν ἀπὸ τοῦ κέντρου τοῦ ἡλίου ἡγμένην
πρὸς ὀρθὰς τῷ ἄξονι, συμβάλλουσαν δὲ
ταῖς τοῦ κώνου πλευραῖς, μείζονα λόγον
ἔχει ἢ ὃν τὰ ϡοθ πρὸς Μ̅ρκε.

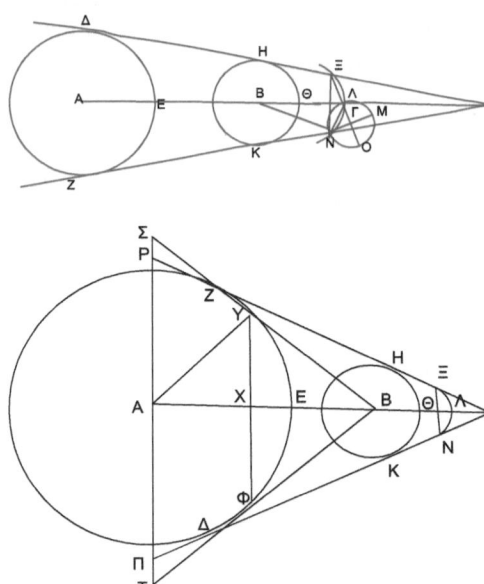

13. **The straight line subtending the arc cut off in the shadow of the Earth from the circle on which travel the extremities of the diameter of the [circle] delimiting the shaded and bright parts on the Moon, in the first place, is smaller than the double of the diameter of the Moon, but it has to [that diameter] a ratio greater than the [ratio] that 88 has to 45; in the second place, it is smaller than one ninth part of the diameter of the Sun, but it has to [that diameter] a ratio greater than the [ratio] that 22 has to 225;[139] and in the third place, it has to the line drawn from the center of the Sun that is at right angles to the axis and joins both sides of the cone, a ratio greater than the [ratio] that 979[140] has to 10125.[141]**

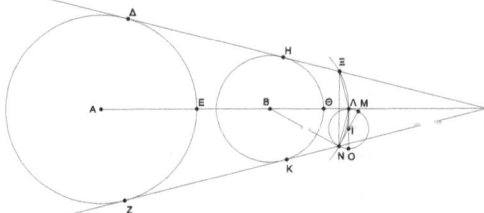

Figure 2.1.11 First diagram of proposition 13.

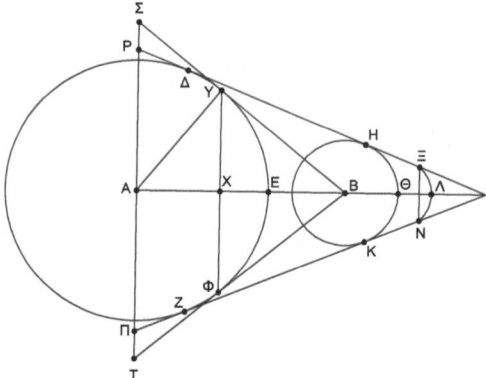

Figure 2.1.12 Second diagram of proposition 13.

139 V:73: *225.*
140 V:73: *379.*
141 V:73: *41125.*

Ἔστω γὰρ ἡλίου μὲν κέντρον πρὸς τῷ Α, γῆς δὲ κέντρον τὸ Β, σελήνης δὲ τὸ Γ, τελείας οὔσης τῆς ἐκλείψεως καὶ πρώτως ὅλης ἐμπεπτωκυίας εἰς τὸ τῆς γῆς σκίασμα, καὶ ἐκβεβλήσθω διὰ τῶν Α, Β, Γ ἐπίπεδον· ποιήσει δὴ τομὰς ἐν μὲν ταῖς σφαίραις κύκλους, ἐν δὲ τῷ κώνῳ εὐθείας τῷ περιλαμβάνοντι τόν τε ἥλιον καὶ τὴν γῆν. ποιείτω ἐν μὲν ταῖς σφαίραις μεγίστους κύκλους τοὺς ΔΕΖ, ΗΘΚ, ΛΜΝ, ἐν δὲ τῷ σκιάσματι τῆς γῆς κύκλον, καθ᾿ οὗ φέρεται τὰ ἄκρα τῆς διαμέτρου τοῦ διορίζοντος ἐν τῇ σελήνῃ τό τε σκιερὸν καὶ τὸ λαμπρόν, τὸν ΞΛΝ, ἐν δὲ τῷ κώνῳ εὐθείας τὰς ΔΗΞ, ΖΚΝ· ἄξων δὲ ἔστω ὁ ΑΒΛ. φανερὸν δὴ ὅτι ὁ ΑΒΛ ἄξων ἐφάπτεται τοῦ ΛΜΝ κύκλου, διὰ τὸ τὸ σκίασμα τῆς γῆς σεληνῶν εἶναι δύο, καὶ δίχα διαιρεῖσθαι τὴν ΝΛΞ περιφέρειαν ὑπὸ τοῦ ΑΒΛ ἄξονος, καὶ ἔτι τὴν σελήνην πρώτως ἐμπεπτωκέναι εἰς τὸ τῆς γῆς σκίασμα. ἐπεζεύχθωσαν δὴ αἱ ΞΝ, ΝΛ, ΒΝ, ΛΞ.

ἡ ΛΝ ἄρα ἐστὶν ἡ διάμετρος τοῦ διορίζοντος ἐν τῇ σελήνῃ τό τε σκιερὸν καὶ τὸ λαμπρόν, καὶ ἡ ΒΝ ἐφάπτεται τοῦ ΛΝΟΜ κύκλου, διὰ τὸ εἶναι τὸ Β πρὸς τῇ ἡμετέρᾳ ὄψει, καὶ τὴν ΛΝ διάμετρον τοῦ διορίζοντος ἐν τῇ σελήνῃ τό τε σκιερὸν καὶ τὸ λαμπρόν.

Let the center of the Sun be at point A, let the center of the Earth [be] point B, let that of the Moon [be] point Γ, when the eclipse[142] is total and for the first time the entire [Moon] has fallen upon the shadow of the Earth, and let a plane be produced through points A, B and Γ. It will make sections on the spheres [that are] circles, and on the cone enveloping the Sun and the Earth,[143] [that are] straight lines. Let [the plane] make on the spheres great circles ΔEZ, HΘK, and ΛMN; and on the shadow of the Earth, the circle ΞΛN on which travel the extremities of the diameter of the [circle] delimiting the shaded and bright parts on the Moon; and straight lines ΔHΞ and ZKN on the cone. Let there be axis ABΛ. It is evident that axis ABΛ is tangent to circle ΛMN,[144] because the shadow of the Earth is two moons and arc NΛΞ[145] is divided into two by axis ABΛ[146] and, besides, the Moon has fallen for the first time upon the shadow of the Earth[147] [Sch. 81 (51)]. Let ΞN, NΛ, BN, and ΛΞ be joined.

Therefore, line ΛN is the diameter [of the circle] delimiting the shaded and bright parts on the Moon, and line BN is tangent to circle ΛNOM, because point B [points] towards our eye, and line ΛN is a diameter of the [circle] delimiting the shaded and bright parts on the Moon. [Referring to the diagram: Sch. 83]

---

142 V:73 adds, after eclipse, *quem defectum dicimus* (which we call eclipse). It seems that he wants to clarify the meaning of the word *eclipsi* by introducing a synonym (even if *eclipsis* is feminine and does not agree with the masculine *quem*).

143 G:166, in French, *lune*; and G:209, in Italian, *luna*.

144 V:74 omits Λ.

145 G:166, in French, and G:209, in Italian: ONL.

146 V:74 omits Λ.

147 G:166 in French omits πρώτως; in Italian, it is correct (G:210).

A. Sunt enim trianguli LXN duo latera XL LN reliquo XN m[a]iora, ex 20 primi elementorum.

B. Namque LO cum sit lunae diameter, maior est, quam LN diameter circuli, qui in luna opacum, et splendidum determinat.

C. Ex 18 tertii elementorum, quod recta linea BL circulum LMN contingat.

D. Ex 28 primi elmentorum, est n. BL et ad XN perpendicularis, cum ipsam bifariam secet. {3. tertii.}

E. Quoniam n. LO XN parallelae sunt, erit angulus LNX aequalis angulo CLN {29. primi.}. Sed angulus LXN est aequalis angulo LNX, et angulus CNL ipsi CLN, quod XL LN aequales sint, itemque aequales LC CN {5. primi.}. Ergo et reliquus angulus XLN est aequalis reliquo LCN, et triangulum triangulo simile.

F. Habet enim NL ad lunae diametrum LO maiorem proportionem, quam 89 ad 90, quod in antecedente demonstratum est.

G. Est enim 7921 numerus quadratus, qui sit ex 89 et 2025 quadratus, qui ex 45.

καὶ ἐπεὶ αἱ ΞΛ, ΛΝ ἴσαι εἰσίν, διπλασίονες ἄρα εἰσὶ τῆς ΛΝ, ὥστε ἡ ΞΝ τῆς ΛΝ ἐλάσσων ἐστὶν ἢ διπλῆ. ἐπεζεύχθωσαν δὴ αἱ ΛΓ, ΓΝ, καὶ διήχθω ἡ ΛΓ ἐπὶ τὸ Ο· πολλῷ ἄρα ἡ ΞΝ τῆς ΛΟ ἐλάσσων ἐστὶν ἢ β. καὶ ἐπεὶ ἡ ΓΛ κάθετός ἐστιν ἐπὶ τὴν ΒΛ, παράλληλος ἄρα ἐστὶ τῇ ΞΝ· ἴση ἄρα ἐστὶν ἡ ὑπὸ τῶν ΛΞΝ τῇ ὑπὸ τῶν ΓΛΝ γωνίᾳ. καὶ ἔστιν ἴση μὲν ἡ ΝΛ τῇ ΛΞ, ἡ δὲ ΛΓ τῇ ΓΝ· ὅμοιον ἄρα ἐστὶν τὸ ΞΝΛ τρίγωνον τῷ ΛΝΓ τριγώνῳ· ἔστιν ἄρα, ὡς ἡ ΞΝ πρὸς τὴν ΝΛ, οὕτως ἡ ΝΛ πρὸς τὴν ΛΓ. ἀλλ' ἡ ΝΛ πρὸς τὴν ΛΓ μείζονα λόγον ἔχει ἢ ὃν τὰ πθ πρὸς τὰ με, τουτέστι, τὸ ἀπὸ ΝΛ πρὸς τὸ ἀπὸ ΛΓ μείζονα λόγον ἔχει ἤπερ τὰ ͵ζϡκα πρὸς τὰ ͵βκε· καὶ τὸ ἀπὸ ΞΝ ἄρα πρὸς τὸ ἀπὸ ΝΛ μείζονα λόγον ἔχει ἤπερ τὰ ͵ζϡκα πρὸς τὰ ͵βκε, καὶ ἡ ΞΝ πρὸς τὴν ΛΟ μείζονα λόγον ἔχει ἤπερ τὰ ͵ζϡκα πρὸς ͵δν. ἔχει δὲ καὶ τὰ ͵ζϡκα πρὸς ͵δν μείζονα λόγον ἤπερ τὰ πη πρὸς με· ἡ ΝΞ ἄρα πρὸς ΛΟ μείζονα λόγον ἔχει ἢ ὃν τὰ πη πρὸς τὰ με.

And, since lines ΞΛ and ΛΝ are equal, therefore, [both added together] are the double of line ΛΝ, so that line ΞΝ is smaller than the double of line ΛΝ [A]. Let ΛΓ and ΓΝ be joined,[148] and produce line ΛΓ to point O. Therefore, the more so is line ΞΝ smaller than 2 [times] line ΛΟ [B].[149] And, since line ΓΛ is perpendicular to line ΒΛ [C], therefore, it is parallel to line ΞΝ [D]. Therefore, angle ΛΞΝ is equal to angle ΓΛΝ [E]. And line ΝΛ[150] is equal to line ΛΞ, and line ΛΓ to line ΓΝ. Therefore, triangle ΞΝΛ is similar to triangle ΛΝΓ. Therefore, as line ΞΝ is to line ΝΛ, so is line ΝΛ to line ΛΓ. But line ΝΛ has to line ΛΓ a ratio greater than the [ratio] that 89 has to 45 [F], that is, the square of ΝΛ has to the square of ΛΓ[151] a greater ratio than 7921[152] [has] to 2025 [G]. Therefore, the square of ΞΝ has to the square of ΝΛ a greater ratio than 7921[153] [has] to 2025; and line ΞΝ has to line ΛΟ a greater ratio than 7921[154] [has] to 4050 [H]. And also 7921[155] has to 4050 a greater ratio than 88 [has] to 45 [K]. Therefore, line ΝΞ has to ΛΟ a ratio greater than the [ratio] that 88 has to 45 [L].

A. For sides ΞΛ and ΛΝ of triangle ΛΞΝ are greater than the remining [side] ΞΝ, from [proposition] 20 of the first [book of] the *Elements* [of Euclid].

B. For ΛΟ, by being the diameter of the Moon, is greater than ΛΝ, the diameter of the circle delimiting the shaded and the bright on the Moon.

C. From [proposition] 18 of the third [book of] the *Elements* [of Euclid, it follows] that straight line ΒΛ be tangent to circle ΛΜΝ.

D. From [proposition] 28 of the first [book of] the *Elements* [of Euclid], because ΒΛ is also perpendicular to ΞΝ, for it cuts [that line] itself into two parts {[*Elem.*] III, 3}.

E. Since, because ΛΟ and ΞΝ are parallel, angle ΛΝΞ will be equal to angle ΓΛΝ {[*Elem.*] I, 29.}. But angle ΛΞΝ is equal to angle ΛΝΞ, and angle ΓΝΛ to ΓΛΝ itself, because ΞΛ and ΛΝ are equal, and likewise, ΛΓ and ΓΝ [are] equal {[*Elem.*] I, 5.}. Consequently, also the remaining angle ΞΛΝ is equal to the remaining ΛΓΝ, and one triangle is similar to the [other] triangle.

F. For ΝΛ has to the diameter of the Moon ΛΟ a ratio greater than the [ratio] that 89 has to 90, as it has been demonstrated earlier.

G. For the number 7921 is the square of 89, and 2025 is the square of 45.

---

148 V:74 translates ΛΓ and ΓΝ by *lc* and *on* instead of by *lc* and *cn*.

149 V:74 translates β by *b* instead of 2 and translates ΛΟ by *no*. Thus, he translates: πολλῷ ἄρα ἡ ΞΝ τῆς ΛΟ ἐλάσσων ἐστὶν ἢ β by *multo igitur xn non minor est quam b* (Therefore, XN is not too much smaller than B).

150 G:168, in French, LN; in Italian, it is correct (G:210).

151 F:30 translates ΝΛ (NL) by *NX* and (LC) by *LN*.

152 V:74: *7321*.

153 V:74: *7321*.

154 V:74: *7321*.

155 V:74: *7321*.

H. Nam cum XN ad
NL maiorem habeat
proportionem, quam 89
ad 45, hoc est quam 7921
ad 4005; et NL ad LO
maiorem, quam 89 ad 90,
hoc est quam 4005 ad
4050; habebit ex aequali
XN ad LO multo maiorem
proportionem, quam 7291
ad 4050, ex iis quae nos
demonstravimus ad 13
quinti elementorum.

K. Est enim 88 ad 45, ut 7921
ad 4050 $^{45}/_{88}$. Sed 7921
ad 4050 maiorem habet
proportionem, quam ad
4050 $^{45}/_{88}$. Ergo 7921 ad
4050 maiorem proportionem
habebit, quam 89 ad 45.

L. Immo vero longe maiorem
ex ante dictis.

H. For, because ΞN has to NΛ a
ratio greater than the [ratio]
that 89 has to 45, that is, to the
[ratio] that 7921 has to 4005;
and NΛ [has] to ΛO [a ratio]
greater than the [ratio] that 89
has to 90, that is, the [ratio] that
4005 has to 4050; by equality of
terms, the more so ΞN will have
to ΛO a greater ratio than the
[ratio] that 7291 has to 4050,
from what we demonstrated in
[the commentary to proposition]
13 of the fifth [book of] the
Elements [of Euclid].[156]

K. For 88 is to 45 as 7921 to 4050
$^{45}/_{88}$. But 7921 has to 4050 a
ratio greater than to 4050 $^{45}/_{88}$.
Consequently, 7921 will have
to 4050 a ratio greater than the
[ratio] that 89 has to 45.

L. But rather much greater, from
what it has been said before.

---

156 Commandino (1572: 64b). In the same way, it will be shown that if the first has to the second the
same ratio as the third to the fourth but the third has to the fourth a smaller ratio than the fifth to
second, also the first has to the second a smaller ratio than the fifth to the sixth. That if the first
has to the second a greater ratio than the third to the fourth but the third has to the fourth a greater
ratio than the fifth to the sixth, also the first will have to the second a greater ratio than the fifth to
the sixth (*Eodem modo demonstrabitur si prima ad secundam eandem habeat proportionem, quam
tertia ad quartam, tertia autem ad quartam minorem proportionem habeat, quam quinta ad secundam: et prima ad secundam minorem proportionem habere quam quintam ad sextam. Quod si
prima ad secundam maiorem habeat proportionem quam tertia ad quartam, tertia autem ad
quartam maiorem habeat, quam quinta ad sextam: et prima ad secundam maiorem proportionem
habebit, quam quinta ad sextam*).

M. Ita ut secet rectam lineam NKF in puncto P, et rectam lineam XGD in R.

N. Ex 9. huius; solis enim diameter maior est, quam duodevigintupla diametri lunae.

O. Ex nona huius. Nam cum solis diameter minor sit, quam vigintupla diametri lunae, habebit diameter lunae ad solis diametrum maiorem proportionem, quam 1 ad 20, hoc est 45 ad 900, ex 15 quinti.

P. Immo vero longe maiorem.

Q. Secent autem rectam lineam PAR in punctis S T

R. Illud nos hoc lemmate demonstrabimus.
Sit noster visus ad A, solis centrum B, lunae vero centrum C, quando conus solem et lunam comprehendens ad visum nostrum verticem habeat. Erunt ACB puncta in eadem recta linea. Ducatur per ACB planum, quod faciat sectiones, in sphaeris quidem circulos maximos DEF, GHK, in cono autem rectas lineas DGA FKA: iunganturque BD, CG et a punctis D G ad BA ducantur ad rectos angulos DLF GMK: et DB GC ad puncta N O producantur. Dico ut KG ad GO, ita esse FD ad DN.

ἡ ἄρα ὑποτείνουσα ὑπὸ τὴν ἀπολαμβανομένην ἐν τῷ σκιάσματι τῆς γῆς περιφέρειαν τοῦ κύκλου, καθ᾽ οὗ φέρεται τὰ ἄκρα τῆς διαμέτρου τοῦ διορίζοντος ἐν τῇ σελήνῃ τό τε σκιερὸν καὶ τὸ λαμπρόν, τῆς διαμέτρου τῆς σελήνης ἐλάσσων μέν ἐστιν ἢ β, μείζονα δὲ λόγον ἔχει <πρὸς αὐτὴν> ἢ ὃν τὰ πη πρὸς με.

Τῶν αὐτῶν ὑποκειμένων, ἤχθω ἀπὸ τοῦ Α τῇ ΑΒ πρὸς ὀρθὰς ἡ ΠΑΡ· λέγω ὅτι ἡ ΞΝ τῆς διαμέτρου τοῦ ἡλίου ἐλάσσων μέν ἐστιν ἢ θ´ μέρος, μείζονα δὲ λόγον ἔχει πρὸς αὐτὴν ἢ ὃν τὰ κβ πρὸς τὰ σκε, πρὸς δὲ τὴν ΠΡ μείζονα λόγον ἔχει ἢ ὃν τὰ ϡοθ πρὸς Μ̅ρκε.

ἐπεὶ γὰρ ἐδείχθη ἡ ΞΝ τῆς διαμέτρου τῆς σελήνης ἐλάσσων οὖσα ἢ β, ἡ δὲ διάμετρος τῆς σελήνης τῆς διαμέτρου τοῦ ἡλίου ἐλάσσων ἐστὶν ἢ ιη´ μέρος, ἡ ἄρα ΞΝ τῆς διαμέτρου τοῦ ἡλίου ἐλάσσων ἐστὶν ἢ θ´ μέρος.

πάλιν ἐπεὶ ἡ ΞΝ πρὸς τὴν διάμετρον τῆς σελήνης μείζονα λόγον ἔχει ἢ ὃν τὰ πη πρὸς τὰ με, ἡ δὲ διάμετρος τῆς σελήνης πρὸς τὴν τοῦ ἡλίου διάμετρον μείζονα λόγον ἔχει ἢ ὃν τὰ με πρὸς ϡ· ἐπεὶ γὰρ ἡ τῆς σελήνης διάμετρος πρὸς τὴν τοῦ ἡλίου μείζονα λόγον ἔχει ἢ ὃν α πρὸς κ, καὶ πάντα τεσσαρακοντάκις καὶ πεντάκις· ἕξει ἄρα ἡ ΞΝ πρὸς τὴν διάμετρον τοῦ ἡλίου μείζονα λόγον ἢ ὃν τὰ πη πρὸς τὰ ϡ, τουτέστιν, ἢ ὃν τὰ κβ πρὸς τὰ σκε.

Therefore, the line subtending the arc cut off in the shadow of the Earth, from the circle[157] on which travel the extremities of the diameter of the [circle] delimiting the shaded and bright parts on the Moon is smaller than the double of the diameter of the Moon,[158] but it has to [that diameter] a ratio greater than the [ratio] that 88 has to 45.

Hypothesizing the same [as before], let line ΠAP be drawn from point A at right angles to line AB [M]. I say that line ΞN is smaller than 1/9 part of the diameter of the Sun, but it has to [that diameter] a ratio greater than the [ratio] that 22 has to 225, and it has to line ΠP a ratio greater than the [ratio] that 979[159] has to 10125.[160]

Since it has been proved that line ΞN is smaller than the double[161] of the diameter of the Moon and that the diameter of the Moon is smaller than 1/18 part of the diameter of the Sun [N], therefore, line ΞN is smaller than 1/9 part of the diameter of the Sun.

Again, since line ΞN has to the diameter of the Moon a ratio greater than the [ratio] that 88 has to 45 and the diameter of the Moon has to the diameter of the Sun a ratio greater than the [ratio] that 45 has to 900[162] [O] – since the diameter of the Moon has to that of the Sun a ratio greater than the [ratio] that 1 has to 20, and all [multiplied] times forty five [gives 45 to 900][163] – therefore, line ΞN will have to the diameter of the Sun a ratio greater than the [ratio] that 88 has to 900 [P],[164] that is, the [ratio] that 22 has to 225.

M. So that it would cut straight line NKZ in point Π and straight line ΞHΔ in point P.

N. From [proposition] 9 of this [treatise]; for the diameter of the Sun is greater than eighteenth times the diameter of the Moon.

O. From the ninth [proposition] of this [treatise]. For, because the diameter of the Sun is smaller than twenty times the diameter of the Moon, the diameter of the Moon will have to the diameter of the Sun a ratio greater than the [ratio] that 1 has to 20, that is, to the [ratio] that 45 has to 900, from [proposition] 15 of the fifth [book of the *Elements* of Euclid].

P. But rather much greater.

Q. But they would cut straight line ΠAP in points Σ and T.

R. We will demonstrate that in this lemma. Let our sight be at A, and let the center of the Sun be B, and [let] C be the center of the Moon, when the cone enveloping the Sun and the Moon has its vertex towards our sight. Points A, C, and B will be on the same straight line. Let it be drawn through ACB a plane making sections, on the spheres, indeed, the great circles DEF and GHK, but on the cone, the straight lines DGA and FKA. And let BD and CG be joined and, from points D and G, at right angles with respect to BA, let DLF and GMK be drawn. And let DB and GC be produced to points N and O. I say that, as KG is to GO, so is FD to DN.

157 V:74 translates ἡ ἄρα ὑποτείνουσα ὑπὸ τὴν ἀπολαμβανομένην ἐν τῷ σκιάσματι τῆς γῆς περιφέρειαν τοῦ κύκλου by *ergo extensa sub compraehendentibus in opaco terrae ambitum circuli*. It seems senseless using the dative plural of *compraehendentibus*, translating ἀπολαμβανομένην.

158 V:74 translates β (double) by *b*.

159 V:74: *379*.

160 V:74: *mo.125* taking only the three last letters, of M̅ρκε (10125), that is, ρκε as the number and the two first M̅ as the name of a line (*mo*), confusing, besides, the α by *o*.

161 V:74 translates β (double) by *b*.

162 V:74: *3000*.

163 No:22 adds *als 45 zu 900* (as 45 to 900). F:32 adds *on trouvera les deux nombres que je viens de donner* (we will find the two numbers I just gave).

164 V:74: *300*.

Quoniam enim recta linea AGD circulos DEF GHK contingit: et a centris B C ad contactus ducuntur BD, CG, erunt anguli ADB AGC recti {18. tertii.}. Quare trianguli ABD angulus ADB est aequalis angulo AGC trianguli ACG: atque est angulus DAB utrique communis. Reliquus igitur DBA est aequalis reliquo GCA. Rursus trianguli BDL angulus DLB rectus est aequalis recto GMC, et angulus DBL aequalis ipsi GCM. Ergo et reliquus reliquo aequalis, et triangulum triangulo simile {4. sexti.}. Ut igitur MG ad LD, ita GC ad DB: permutandoque ut MG ad GC, ita LD ad DB {15. quinti.}. Et eorum dupla, ut KG ad GO, ita FD ad DN. Est autem GK diameter circuli, qui in luna opacum et splendidum determinat, et GO lunae diameter. Ergo ut diameter circuli in luna opacum et splendidum determinantis ad diametrum lunae, ita FD ad DN, hoc est ad solis diametrum.

ἤχθωσαν δὴ ἀπὸ τοῦ Β τοῦ ΔΕ κύκλου ἐφαπτόμεναι αἱ ΒΥΣ, ΒΦΤ, καὶ ἐπεζεύχθω ἡ ΥΦ καὶ ἡ ΥΑ.

ἔσται δή, ὡς ἡ διάμετρος τοῦ διορίζοντος ἐν τῇ σελήνῃ τό τε σκιερὸν καὶ τὸ λαμπρὸν πρὸς τὴν διάμετρον τῆς σελήνης, οὕτως ἡ ΥΦ πρὸς τὴν διάμετρον τοῦ ἡλίου, διὰ τὸ τὸν αὐτὸν κῶνον περιλαμβάνειν τόν τε ἥλιον καὶ τὴν σελήνην τὴν κορυφὴν ἔχοντα πρὸς τῇ ἡμετέρᾳ ὄψει.

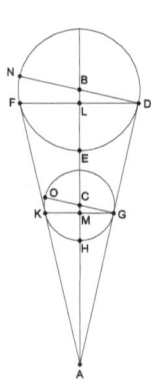

From point B and tangent to circle ΔE,[165] let lines BYΣ and BΦT[166] be drawn [Q], and let YΦ and YA be joined.[167]

As the diameter of the [circle] delimiting the shaded and bright part on the Moon will be to the diameter of the Moon, so will line YΦ be to the diameter of the Sun, because one and the same cone having its vertex towards our eye, envelops the Sun and the Moon [R] [Sch. 99 (66)].

For, since straight line AGB is tangent to circles DEF and GHK, and, from the centers B and C to the points of contact are drawn BG an CG, angles ADB and AGC will be right {[*Elem.*] III, 18.}. Therefore, angle ADB of triangle ABD is equal to angle AGC of triangle ACG. And angle DAB is common to both. Accordingly, the remaining [angle] DBA is equal to the remaining [angle] GCA. Again, right angle DLB of triangle BDL is equal to right angle GMC, and angle DBL is equal to angle GCM itself. Consequently, also the remaining [angle] is equal to the [other] remaining [angle] and one triangle is similar to the [other] triangle {[*Elem.*] VI, 4.}. Accordingly, as MG is to LD, so [is] GC to DB and, by permutation, as MG is to GC, so [is] LD to DB {[*Elem.*] V, 15.}. And [the same happens] to the doubles of them, as KG is to GO, so [is] FD to DN. But GK is the diameter of the circle delimiting the shaded and the bright on the Moon, and GO is the diameter of the Moon. Consequently, as the diameter of the circle delimiting the shaded and the bright on the Moon is to the diameter of the Moon, so [is] FD to DN, that is, to the diameter of the Sun.

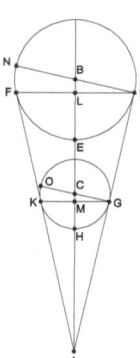

Figure 2.2.6 Diagram of Commandino's note R to proposition 13.

---

165 F:32: *DEF*, the same in his Greek edition (ΔEZ) and Latin translation (FG:66–67); FG:224 justifies his choice.

166 No:22 translates BΦT (BVT) by *SVT*. V:74 omits BΦT, but he introduces *contingentes* (in plural).

167 G:172, in French, and G:211, in Italian: VA (it should be UA).

S. Ut enim YV ad solis
    diametrum, ita QY ad YA, cum
    sint earum dimidiae, ex 15
    quinti.

T. Quoniam enim parallelae
    sunt SA YQ, erit angulus
    YAS aequalis angulo AYQ:
    atque, est angulus AYS rectus
    aequalis recto AQY {29.
    primi.}. Ergo et reliquus
    reliquo aequalis, et triangulum
    simile triangulo {4. sexti.}. Ut
    igitur QY ad YA, ita est YA ad
    AS.

V. Ex 8 quinti. Est enim AR
    minor, quam AS. Quare et
    dupla ipsius YA ad duplam
    ipsius AR, hoc est solis
    diameter ad PR maiorem
    habebit proportionem, quam
    89 ad 90.

ἡ δὲ διάμετρος τοῦ διορίζοντος ἐν τῇ σελήνῃ τό τε
σκιερὸν καὶ τὸ λαμπρὸν πρὸς τὴν διάμετρον τῆς
σελήνης μείζονα λόγον ἔχει ἢ ὃν τὰ πθ πρὸς τὰ ϙ·
καὶ ἡ ΥΦ ἄρα πρὸς τὴν τοῦ ἡλίου διάμετρον
μείζονα λόγον ἔχει ἢ ὃν τὰ πθ πρὸς ϙ· καὶ ἡ ΧΥ ἄρα
πρὸς τὴν ΥΑ μείζονα λόγον ἔχει ἢ ὃν τὰ πθ πρὸς ϙ.
ὡς δὲ ἡ ΧΥ πρὸς τὴν ΥΑ, οὕτως ἡ ΥΑ πρὸς τὴν ΑΣ,
διὰ τὸ παραλλήλους εἶναι τὰς ΣΑ, ΥΧ· καὶ ἡ ΥΑ
ἄρα πρὸς τὴν ΑΣ μείζονα λόγον ἔχει ἢ ὃν τὰ πθ
πρὸς τὰ ϙ· πολλῷ ἄρα ἡ ΥΑ πρὸς τὴν ΑΡ μείζονα
λόγον ἔχει ἢ ὃν τὰ πθ πρὸς τὰ ϙ.

And the diameter of the [circle] delimiting the shaded and bright parts on the Moon has to the diameter of the Moon a ratio greater than the [ratio] that 89 has to 90. And therefore, line YΦ has to the diameter of the Sun a ratio greater than the [ratio] that 89 has to 90. And therefore, line XY[168] has to line YA a ratio greater than the [ratio] that 89 has to 90 [S].[169] And, as line XY is to line YA, so is line YA to line AΣ, because lines ΣA and YX are parallel [T]. And therefore, line YA has to line AΣ a ratio greater than the [ratio] that 89 has to 90. Therefore, the more so has line YA to line AP a greater ratio than the [ratio] that 89 has to 90 [V].

S. For, as YΦ is to the diameter of the Sun, so [is] XY to YA, because they are the halves of them,[170] from [proposition] 15 of the fifth [book of the *Elements* of Euclid].

T. For, since ΣA and YX are parallel, angle YAΣ will be equal to angle AYX. And right angle AYΣ is equal to right angle AXY {[*Elem.*] I, 29.}. Consequently, also the remaining [angle] is equal to the remaining [angle] and one triangle is similar to the [other] triangle {[*Elem.*] VI, 4.}. Accordingly, as XY is to YA, so is YA to AΣ.

V. From [proposition] 8 of the fifth [book of the *Elements* of Euclid]. For AP is smaller than AΣ. Therefore, also the double of AY itself, will have to the double of AP itself, that is, the diameter of the Sun to ΠP, a ratio greater than the [ratio] than 89 has to 90.

168 G:174 and G:211: WV (it should be WU).
169 V:74 omits καὶ ἡ XY ἄρα πρὸς τὴν YA μείζονα λόγον ἔχει ἢ ὃν τὰ πθ πρὸς ϙ.
170 That is, because XY and YA are the halves of YΦ and the diameter of the Sun.

X. Quoniam enim XN ad diametrum solis maiorem habet proportionem quam 22 ad 225. Et diameter solis ad PR maiorem habet, quam 89 ad 90, fiat ut 225 ad 22, ita 89 ad alium. Erit 8 $^{158}/_{225}$. Cum igitur XN ad diametrum solis maiorem habeat proportionem, quam 22 ad 225, hoc est quam 8 $^{158}/_{225}$ ad 89; et solis diameter ad PR habeat maiorem, quam 89 ad 90; habebit ex aequali XN ad PR multo maiorem proportionem, quam 8 $^{158}/_{225}$ ad 90. Sed 8 $^{158}/_{225}$ hoc est 1958/225 ad 90 est ut 1958 ad 20250. Quod ita manifestum erit. Dispositis enim numeris in hunc modum, et decussatim multiplicatis

	20250
1958	90
225	1

videlicet 225 in 90. Fient 20250 et 1 in 1958, fient 1958. Habebit 1958/225 ad 90 eandem proportionem, quam 1958 ad 20250. Quod nos demonstravimus in commentariis in tertiam propositionem. libri Archimedis de circuli dimensione, propositione septima. Quare XN ad PR multo maiorem proportionem habebit, quam numerus productus ex 22 et 89, hoc est 1958 ad eum qui producitur ex 90 et 225, videlicet ad 20250.

καὶ τὰ β· ἡ ἄρα διάμετρος τοῦ ἡλίου πρὸς τὴν ΠΡ μείζονα λόγον ἔχει ἢ ὃν τὰ πθ πρὸς τὰ ϙ. ἐδείχθη δὲ καὶ ἡ ΞΝ πρὸς τὴν διάμετρον τοῦ ἡλίου μείζονα λόγον ἔχουσα ἢ ὃν τὰ κβ πρὸς τὰ σκε. δι' ἴσου πολλῷ ἄρα ἡ ΞΝ πρὸς τὴν ΠΡ μείζονα λόγον ἔχει ἢ <ὃν> ὁ συνηγμένος ἔκ τε τῶν κβ καὶ πθ πρὸς τὸν ἐκ τῶν ϙ καὶ σκε, τουτέστιν, τὰ ͵αϡνη πρὸς τὰ Μ̅β̅σν· καὶ τὰ ἡμίση, τουτέστιν, τὰ ϡοθ πρὸς τὰ Μ̅α̅ρκε.

And [the same holds for] the double [of them]. Therefore, the diameter of the Sun has to line ΠΡ[171] a ratio greater than the [ratio] that 89 has to 90. And it has also been proved that line ΞΝ has to the diameter of the Sun a ratio greater than the [ratio] that 22 has to 225. Therefore, by equality of terms, the more so has line ΞΝ to line ΠΡ a greater ratio than the [ratio] that the product of 22 and 89 has to the product of 90 and 225[172] [X], that is, than the [ratio] that 1958 has to 20250;[173] and to the halves [of these numbers], that is, to the [ratio] that 979[174] has to 10125.[175]

X. For, since ΞΝ has to the diameter of the Sun a ratio greater than the [ratio] that 22 has to 225. And the diameter of the Sun has to ΠΡ a ratio greater than the [ratio] that 89 has to 90, let it be that as 225 is to 22, so [is] 89 to another [number]. [This number] will be $8\,^{158}/_{225}$. Accordingly, since ΞΝ has to the diameter of the Sun a ratio greater than the [ratio] that 22 has to 225, that is, the [ratio] that $8\,^{158}/_{225}$ [has] to 89, and the diameter of the Sun has to ΠΡ a [ratio] greater than the [ratio] that 89 has to 90, by equality of terms, the more so ΞΝ will have to ΠΡ a greater ratio than the [ratio] that $8\,^{158}/_{225}$ has to 90. But $8\,^{158}/_{225}$, that is, $^{1958}/_{225}$ is to 90 as 1958 to 20250. What will be manifest so, for, if the numbers are arranged in this way and multiplied crosswise:

		20250
	1958	90
	225	1

that is, 225 by 90. It will result 20250; and 1 by 1958, it will result 1958. $^{1958}/_{225}$ will have to 90 the same ratio that 1958 has to 20250. What we demonstrated in the commentaries to the seventh proposition of the third proposition of the book of Archimedes, *On the dimension of the Circle*,[176] therefore the more so ΞΝ will have to ΠΡ a greater ratio than the [ratio] that the number [resulting from the] product of 22 and 89, that is 1958 has to what is the product of 90 and 225, that is, 20250.

171 V:74 omits πρὸς τὴν ΠΡ.
172 G:211, in Italian: 22 × 225; in French, it is correct. See Berggren and Sidoli (2007: 246) for a discussion of the term συνηγμένος that is not the typical technical term for the product: ὁ περιεχόμενος ἀριθμὸς (that Aristarachus uses in proposition 15).
173 V:75 translates αϧνη πρὸς τὰ Μσν (1958 to 20250) by *knfyg*.
174 V:75: *479*.
175 V:75 translates Μσν (10125) by *41 r y go*.
176 Commandino (1558: 9) (of the commentaries starting with a new numeration): *Numeris datarun proportionum decussatim multiplicatis, quam proportionum habuerit productum ex antecedente priori, & consequente posterioris, ad productum ex consequente prioris, & antecedente posterioris, eandem habere invenietur quantitas a ad quantitatem b.*

$$\frac{8}{2}\;\frac{9}{3}\;\frac{\overline{}}{8} $$
$$\frac{2}{3}\;\frac{3}{4}=\frac{8}{9}$$

ιδ' Ἡ ἀπὸ τοῦ κέντρου τῆς γῆς ἐπὶ τὸ κέντρον
τῆς σελήνης ἐπιζευγνυμένη εὐθεῖα πρὸς τὴν
εὐθεῖαν, ἣν ἀπολαμβάνει ἀπὸ τοῦ ἄξονος
πρὸς τῷ κέντρῳ τῆς σελήνης ἡ ὑπὸ τὴν ἐν
τῷ σκιάσματι τῆς γῆς ὑποτείνουσα εὐθεῖα,
μείζονα λόγον ἔχει ἢ ὃν τὰ χοε πρὸς α.

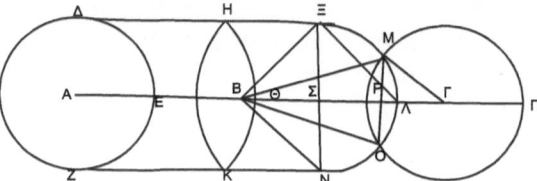

14. **The straight line [BΓ] joined [from] the center of the Earth [B] to the center of the
Moon [Γ] has to the straight line [ΣΓ] cut off
from the axis [AΠ] towards the center of the
Moon [Γ] by the straight line [ΞN] subtending [the arc ΞMΛON cut off] in the shadow
of the Earth a ratio greater than the [ratio]
that 675 has to 1.[177]**

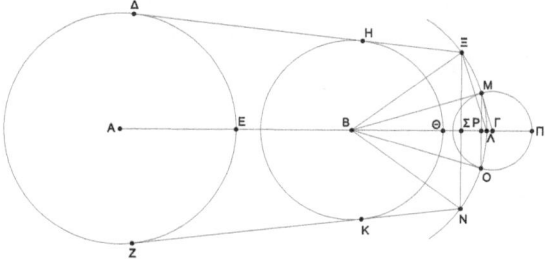

Figure 2.1.13 Diagram of proposition 14

177 We added the letters referring to the diagram of the proposition in order to make easier the meaning
of the paragraph.

A. Ex 28 tertii elementorum.
   In aequalis enim circulis
   aequales rectae lineae aequales
   circumferentias auferunt.

B. Quoniam enim circumferentia
   XML est aequalis
   circumferentiae MLO, dempta
   circumferentia ML utrique
   communi, erit reliqua XM
   reliquae LO aequalis.

Ἔστω τὸ αὐτὸ σχῆμα τῷ πρότερον, καὶ ἡ σελήνη οὕτως ἔστω ὥστε τὸ κέντρον αὐτῆς εἶναι ἐπὶ τοῦ ἄξονος τοῦ κώνου τοῦ περιλαμβάνοντος τόν τε ἥλιον καὶ τὴν γῆν, καὶ ἔστω τὸ Γ, μέγιστος δὲ τῶν ἐν τῇ σφαίρᾳ κύκλος ὁ ΠΟΜ ἐν τῷ αὐτῷ ἐπιπέδῳ ὢν αὐτοῖς, καὶ ἐπεζεύχθω ἡ ΜΟ· ἡ ΜΟ ἄρα διάμετρός ἐστι τοῦ διορίζοντος ἐν τῇ σελήνῃ τό τε σκιερὸν καὶ τὸ λαμπρόν. ἐπεζεύχθωσαν δὴ αἱ ΜΒ, ΒΟ, ΛΞ, ΞΒ, ΜΓ· ἐφάπτονται ἄρα τοῦ ΜΟΠ κύκλου αἱ ΜΒ, ΒΟ, διὰ τὸ τὴν ΟΜ διάμετρον εἶναι τοῦ διορίζοντος ἐν τῇ σελήνῃ τὸ σκιερὸν καὶ τὸ λαμπρόν.

καὶ ἐπεὶ ἴση ἐστὶν ἡ ΞΛ τῇ ΜΟ· ἑκατέρα γὰρ αὐτῶν διάμετρός ἐστι τοῦ διορίζοντος ἐν τῇ σελήνῃ τό τε σκιερὸν καὶ τὸ λαμπρόν· ἴση ἄρα καὶ ἡ ΞΜΛ περιφέρεια τῇ ΜΛΟ περιφερείᾳ, καὶ ἡ ΞΜ ἄρα ἴση ἐστὶν τῇ ΛΟ.

Let there be the same diagram as before,[178] and let the Moon be so that its center is on the axis of the cone enveloping the Sun and the Earth;[179] and let [its center] be point Γ; and of the great circles on the sphere,[180] let ΠOM be [the one] in the same plane as them;[181] and let MO be joined. Therefore, line MO is a diameter of the [circle] delimiting the shaded and bright parts on the Moon. Let MB, BO, ΛΞ, ΞB, and MΓ be joined.[182] Therefore, lines MB and BO[183] are tangent to circle[184] MOΠ,[185] because line OM is a diameter of the [circle] delimiting the shaded and bright parts on the Moon.

And, since line ΞΛ is equal to line MO[186] – for each of them is a diameter [of the circle] delimiting the shaded and bright parts on the Moon – therefore, also arc ΞMΛ is equal to arc MΛO [A], and, therefore, [arc] ΞM is equal to [arc] ΛO [B].

A. From [proposition] 28 of the third [book of] the *Elements* [of Euclid]. For, in equal circles, equal straight lines divide equal arcs.

B. For, since arc ΞMΛ is equal to arc MΛO, once removed arc MΛ, common to both, the remaining [arc] ΞM will be equal to the remaining [arc] ΛO.

---

178 On the differences and similarities of both diagrams and the fact that, nevertheless, Aristarchus asks that the same diagram be drawn, see Netz (2007: 40–43) and Carman (2018b).

179 FG:71 translates *in axe coni comprehendentis solem et lunam* (in the axis enveloping the Sun and the Moon), omitting the reference to point C and putting *Lunam* (Moon) instead of *Terram* (Earth).

180 This refers to the sphere of the Moon, that is, to the Moon itself, and not to the sphere in which the Moon travels.

181 This refers to all the cuts of the plane of the previous diagram. V:75 translates *eodem plano ex istis cum ipsis*, and *ex istis* seems to be unnecessary.

182 V:75 omits αἱ MB, BO, ΛΞ, ΞB y MΓ.

183 V:75 omits αἱ MB, BO.

184 G:178, in French: *aux cercles* (to the circles); the Italian (G:212) is correct.

185 G:178, in French, and G:212, in Italian: MBQ (it should be MPQ).

186 V:75 omits: τῇ MO.

C. A centro enim B ad circumferentiam ducuntur.

D. Ex 3 tertii elementorum, nam recta linea BM ex centro ducta circumfrentiam XML, et ob id rectam lineam XL bifariam secat.

E. Ex 18 tertii. Ducta est enim recta linea ex centro C ad punctum, in quo BM circulum POM contingit.

F. Ex 28 primi elementorum

G. Namque angulus LXS aequalis est angulo CMR, et angulus LSX rectus aequalis recto CRM {29. primi.}. Ergo et reliquus reliquo aequalis, et triangulum triangulo simile.

H. Ex 15 quinti elmentorum.

K. Ex demonstratis in antecedente.

L. Est enim RS minor, q[u]<n>am SL.

M. Ex 8 quinti elmentorum.

ἀλλ' ἡ ΛΟ τῇ ΛΜ ἴση ἐστίν· καὶ ἡ ΞΜ ἄρα ἴση ἐστὶν τῇ ΛΜ. ἔστι δὲ καὶ ἡ ΞΒ ἴση τῇ ΒΛ, διὰ τὸ τὸ Β σημεῖον κέντρον εἶναι τῆς γῆς, καὶ <τὴν γῆν> σημείου καὶ κέντρου λόγον ἔχειν πρὸς τὴν τῆς σελήνης σφαῖραν, καὶ τὸν ΜΟΠ κύκλον ἐν τῷ αὐτῷ ἐπιπέδῳ εἶναι· ἡ ἄρα ΒΜ κάθετός ἐστιν ἐπὶ τὴν ΞΛ. ἔστιν δὲ καὶ ἡ ΓΜ κάθετος ἐπὶ τὴν ΒΜ· παράλληλος ἄρα ἐστὶν ἡ ΓΜ τῇ ΞΛ. ἔστι δὲ καὶ ἡ ΣΞ τῇ ΜΡ παράλληλος· ὅμοιον ἄρα ἐστὶ τὸ ΛΞΣ τρίγωνον τῷ ΜΡΓ τριγώνῳ· ἔστιν ἄρα, ὡς ἡ ΣΞ πρὸς τὴν ΜΡ, οὕτως ἡ ΣΛ πρὸς τὴν ΡΓ. ἀλλ' ἡ ΣΞ τῆς ΜΡ ἐστὶν ἐλάσσων ἢ β, ἐπεὶ καὶ ἡ ΞΝ τῆς ΜΟ ἐλάσσων ἐστὶν ἢ β· καὶ ἡ ΣΛ ἄρα τῆς ΓΡ ἐλάσσων ἐστὶν ἢ β· ὥστε ἡ ΣΡ τῆς ΡΓ πολλῷ ἐλάσσων ἐστὶν ἢ β. ἡ ΣΓ ἄρα τῆς ΓΡ ἐλάσσων ἐστὶν ἢ τριπλασίων· ἡ ΓΡ ἄρα πρὸς τὴν ΓΣ μείζονα λόγον ἔχει ἢ ὃν α πρὸς γ. καὶ ἐπεί ἐστιν, ὡς ἡ ΒΓ πρὸς ΓΜ, οὕτως ἡ ΓΜ πρὸς τὴν ΓΡ, ἡ δὲ ΒΓ πρὸς τὴν ΓΜ μείζονα λόγον ἔχει ἢ ὃν με πρὸς α, καὶ ἡ ΓΜ ἄρα πρὸς ΓΡ μείζονα λόγον ἔχει ἢ ὃν με πρὸς α. ἔχει δὲ καὶ ἡ ΓΡ πρὸς τὴν ΓΣ μείζονα λόγον ἢ ὃν α πρὸς γ· δι' ἴσου ἄρα ἡ ΓΜ πρὸς τὴν ΓΣ μείζονα λόγον ἔχει ἢ ὃν με πρὸς γ, τουτέστιν, <ἢ> ὃν ιε πρὸς α. ἐδείχθη δὲ καὶ ἡ ΒΓ πρὸς τὴν ΓΜ μείζονα λόγον ἔχουσα ἢ ὃν με πρὸς α· δι' ἴσου ἄρα ἡ ΒΓ πρὸς τὴν ΓΣ μείζονα λόγον ἔχει ἢ ὃν τὰ χοε πρὸς α.

But [arc] ΛΟ is equal to [arc] ΛΜ.[187] And, there-
fore, [arc] ΞΜ is equal to [arc] ΛΜ. And also line
ΞΒ is equal to line ΒΛ [C], because point B is the
center of the Earth, and the Earth has a ratio of a
point and center with respect to the sphere of the
Moon,[188,189] and circle ΜΟΠ is on the same plane.
Therefore, line ΒΜ is perpendicular to line ΞΛ
[D]. And also line ΓΜ is perpendicular to line ΒΜ
[E]. Therefore, line ΓΜ is parallel to line ΞΛ [F].
And also line ΣΞ is parallel to line ΜΡ.[190] There-
fore, triangle ΛΞΣ is similar to triangle ΜΡΓ [G].
Therefore, as line ΣΞ is to line ΜΡ, so is line ΣΛ to
line ΡΓ. But line ΣΞ is smaller than 2 times line ΜΡ
[H], since also line ΞΝ is smaller than 2 times line
ΜΟ [K]. And, therefore, also line ΣΛ is smaller[191]
than 2 times line ΓΡ.[192] So that the more so is line
ΣΡ[193] smaller than 2 times line ΡΓ [L]. Therefore,
line ΣΓ is smaller than three times line ΓΡ. Therefore, line ΓΡ[194] has to line ΓΣ a ratio greater than
the [ratio] that 1 has to 3 [M]. And, since, as line
ΒΓ is to line ΓΜ, so is line ΓΜ to line ΓΡ [N], and
line ΒΓ has to line ΓΜ a ratio greater than the
[ratio] that 45 ha to 1 [O]; therefore, also line ΓΜ
has to ΓΡ a ratio greater than the [ratio] that 45 has
to 1.[195] And also, line ΓΡ has to line ΓΣ a ratio
greater than the [ratio] that 1 has to 3. Therefore,
by equality of terms, line ΓΜ has to line ΓΣ a ratio
greater than the [ratio] that 45 has to 3, that is, the
[ratio] that 15 has to 1. And it has also been proved
that line ΒΓ[196] has to line ΓΜ a ratio greater than
the [ratio] that 45 has to 1. Therefore, by equality
of terms, line ΒΓ has to line ΓΣ a ratio greater than
the [ratio] that 675 has to 1 [P].

C. For they are drawn from center
B to the arc.

D. From [proposition] 3 of the
third [book of] the *Elements* [of
Euclid] for straight line BM,
drawn from the center, bisects
arc ΞΜΛ and, because of that,
[it also bisects] straight line ΞΛ.

E. From [proposition] 18 of the
third [book of the *Elements* of
Euclid]. For a straight line has
been drawn from the center Γ
to the point in which BM is
tangent to circle ΠΟΜ.

F. From [proposition] 28 of the
first [book of] the *Elements* [of
Euclid].

G. For angle ΛΞΣ is equal to
angle ΓΜΡ, and angle ΛΣΞ
is equal to right angle ΓΡΜ
{[*Elem.*] I, 29.}. Consequently,
also the remaining [angle] is
equal to the remaining [angle],
and one triangle is similar to
the [other] triangle.

H. From [proposition] 15 of the
fifth [book of] the *Elements* [of
Euclid].

K. From what has been
demonstrated above.

L. For ΡΣ is smaller than ΣΛ.

M. From [proposition] 8 of the
fifth [book of] the *Elements* [of
Euclid].

---

187 V:75 translates ΛΜ (LM) by *LB*.
188 G:178 *qu'on peut se représenter comme un point au centre de la sphère de la Lune* (which can be
represented by a point in the center of the sphere of the Moon). The translation in Italian is similar
(G:212).
189 This refers to the sphere in which the Moon travels, that is, to the orbit of the Moon.
190 V:75 omits τῇ ΜΡ.
191 V:75: *maior* (greater).
192 V:75 omits τῆς ΓΡ.
193 C:30b: R. In W:590 and M:65, it is corrected.
194 V:75 translates ΓΣ (*cs*) by *co*.
195 C:31a omits καὶ ἡ ΓΜ ἄρα πρὸς ΓΡ μείζονα λόγον ἔχει ἢ ὃν με πρὸς α. W:590 corrects it (cfr. note m).
196 V:10 omits Γ.

N. Ex 4 sexti nam triangula BMC,
   MCR simila sunt ex 8 eiusdem
   quod ab angulo recto trianguli
   BMC ad basim perpendicularis
   ducta est MR.

O. Ex undecima huius.

P. Si enim fiat, ut 1 ad 45, ita 15
   ad alium, erit ad 675. Itaque
   quoniam BC ad CM maiorem
   proportionem habet, quam 45
   ad 1, hoc est, quam 675 ad 15;
   et MC ad CS maiorem, quam
   15 ad 1, habebit ex aequali BC
   ad CS, maiorem proportionem,
   quam 675 ad 1.

N. From [proposition] 4 of the sixth [book of the *Elements* of Euclid], for triangles BMΓ and MΓP are similar from [proposition] 8 of the same [book], because MP has been drawn from the right angle of triangle BMΓ, perpendicular to the base.

O. From [proposition] 11 of this [treatise].

P. For, let it be that as 1 is to 45, so [is] 15 to another [number, then that number] will be 675. In the same manner, since BC has to CM a ratio greater than the [ratio] that 45 has to 1, that is [the ratio] that 675 has to 15; and MC has to CS [a ratio] greater than [the ratio] that 15 has to 1, by equality of terms, BC will have to CS a ratio greater than the [ratio] that 675 has to 1.

ιε' Ἡ τοῦ ἡλίου διάμετρος πρὸς τὴν τῆς γῆς
διάμετρον μείζονα λόγον ἔχει ἢ ὃν τὰ ιθ
πρὸς γ, ἐλάσσονα δὲ ἢ ὃν τὰ μγ πρὸς τὰ ς.

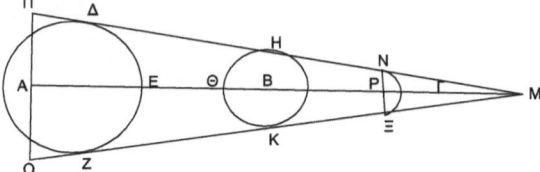

15. **The diameter of the Sun has to the diameter of the Earth[197]a ratio greater than the
[ratio] than 19 has to 3, but smaller than
the [ratio] that 43 has to 6.**

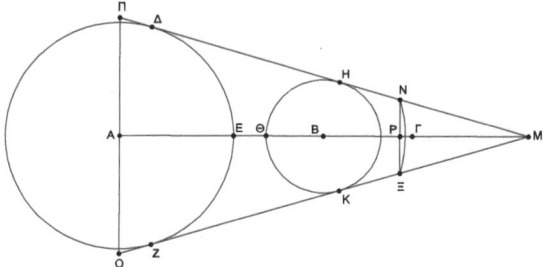

Figure 2.1.14 Diagram of proposition 15.

---

197 G:213, in Italian: *luna*; the French is correct.

Ἔστω γὰρ ἡλίου μὲν κέντρον τὸ Α, γῆς δὲ κέντρον
τὸ Β, σελήνης δὲ κέντρον τὸ Γ, τελείας οὔσης τῆς
ἐκλείψεως, τουτέστιν, ἵνα τὰ Α, Β, Γ ἐπ᾽ εὐθείας ᾖ,
καὶ ἐκβεβλήσθω διὰ τοῦ ἄξονος ἐπίπεδον, καὶ
ποιείτω τομὰς ἐν μὲν τῷ ἡλίῳ τὸν ΔΕΖ κύκλον, ἐν
δὲ τῇ γῇ τὸν ΗΘΚ, ἐν δὲ τῷ σκιάσματι τὴν ΝΞ
περιφέρειαν, ἐν δὲ τῷ κώνῳ εὐθείας τὰς ΔΜ, ΖΜ,
καὶ ἐπεζεύχθω ἡ ΝΞ, καὶ ἀπὸ τοῦ Α τῇ ΑΜ πρὸς
ὀρθὰς ἤχθω ἡ ΟΑΠ.

Let the center of the Sun be point A; the center of
the Earth, point B, and the center of the Moon, point
Γ, when it is totally eclipsed, that is, so that points
A, B, and Γ[198] are on a straight line, and let a plane
be produced through the axis; and let [the plane]
make sections,[199] [that are] circle ΔEZ[200] on the Sun,
and circle HΘK on the Earth, arc NΞ on the shadow,
and straight lines ΔM[201] and ZM[202] on the cone, and
let NΞ be joined and from point A and at right
angles to line AM let line OAΠ be drawn.

198 M:67 translates *para asegurar así* (so as to secure that), following Heath's version. Indeed, C:32b
    translates *hoc est ita ut puncta ABC in eadem recta linea constituantur* and H:403 *so as to secure
    that A, B, C may be in a straight line*. This passage is ambiguous and could mean that, when the
    eclipse is total, points A, B y Γ are on the same straight line, but this is not correct. The three points
    are on the same straight line only at the middle of the total eclipse. So this passage asks to place
    the center of the Moon at point Γ, which is already aligned with A and B, so that the center of the
    Moon be aligned with the centers of the Sun and the Earth.
199 M:67 translates "y córtese" (and let it cut), making ποιείτω have an undefined subject when the
    subject is clearly the plane. The Latin translation of C:32b attributes the action to the plane in an
    even clearer way than the Greek, because it is in a subordinate clause: "planum, quod faciat sec-
    tiones . . ."
200 V:11 omits E.
201 G:213, in Italian: OM (instead of DM); the French is correct.
202 V:11 translates ΔM, ZM by dgn.fkx.

A. Ex 12 huius. Ex quo sequitur
ex 8 quinti NX ad diametrum
solis minorem habere
proportion[e]m quam 1 ad
9. Quare convertendo ex 26
quinti diameter solis ad NX
maiorem habet proportionem,
quam 9 ad 1. Et OP quae
maior est, quam solis diameter,
ad NX multo maiorem
proportionem habet, quam
9 ad 1 {8. quinti.}. Sed ut
AO ad RN, hoc est ut earum
duplae OP ad NX, ita erit
AM ad MR ob similitudinem
triangulorum AMO RMN {15.
quinti.}. Ergo et AM ad MR
multo maiorem proportionem
habebit, quam 9 ad 1.

B. Ex 30 quinti.

καὶ ἐπεὶ ἡ ΝΞ τῆς διαμέτρου τοῦ ἡλίου ἐλάσσων
ἐστὶν ἢ θ΄ μέρος, ἡ ΟΠ ἄρα πρὸς τὴν ΝΞ πολλῷ
μείζονα λόγον ἔχει ἢ ὃν τὰ θ πρὸς α· καὶ ἡ ΑΜ ἄρα
πρὸς τὴν ΜΡ μείζονα λόγον ἔχει ἢ ὃν τὰ θ πρὸς α.
καὶ ἀναστρέψαντι ἡ ΜΑ πρὸς ΑΡ ἐλάσσονα λόγον
ἔχει ἢ ὃν τὰ θ πρὸς η. πάλιν ἐπεὶ ἡ ΑΒ τῆς ΒΓ
μείζων ἐστὶν ἢ ιη, πολλῷ ἄρα τῆς ΒΡ μείζων ἐστὶν
ἢ ιη· ἡ ΑΒ ἄρα πρὸς τὴν ΒΡ μείζονα λόγον ἔχει ἢ
ὃν τὰ ιη πρὸς α.

And since line NΞ is smaller than 1/9 part of the diameter of the Sun [A],[203] therefore, the more so has line OΠ to line NΞ a greater ratio than the [ratio] that 9 has to 1. And, therefore, line AM has to line MP a ratio greater than the [ratio] that 9 has to 1.[204] And, by conversion, line MA has to AP a ratio smaller than the [ratio] that 9 has to 8 [B]. Again, since line AB[205] is greater than 18 times line BΓ [C], therefore, the more so is it greater than 18 times line BP [D]. Therefore, line AB has to line BP a ratio greater than the [ratio] that 18 has to 1.

A. From [proposition] 12 of this [treatise]. Therefore, from [proposition] 8 of the fifth [book of the *Elements* of Euclid], it follows that NΞ has to the diameter of the Sun a ratio smaller than the [ratio] that 1 has to 9. Therefore, by inversion – from the [proposition] 26 of the fifth [book of the *Elements* of Euclid][206]– the diameter of the Sun has to NΞ a ratio greater than the [ratio] that 9 has to 1. And the more so[207] OΠ, which is greater than the diameter of the Sun, has to NΞ a greater ratio than the [ratio] that 9 has to 1 {[*Elem.*] V, 8.}. But, as AO is to PN, that is, as the double of them, OΠ to NΞ, so will AM be to MP, because of the similarity of triangles AMO and PMN {[*Elem.*] V, 15.}. Consequently, also the more so AM will have to MP a greater ratio than the [ratio] that 9 has to 1.

B. From [proposition] 30 of the fifth [book of the *Elements* of Euclid].[208]

203 W:591 adds ἡ ἄρα διάμετρος τοῦ ἡλίου μείζονα λόγον ἔχει πρὸς τὴν ΝΞ ἢ ὃν τὰ θ πρὸς α· καὶ (therefore, the diameter of the Sun has to line NΞ a ratio greater than the [ratio] that 9 has to 1, and . . .), adding in a note that the text is not in the manuscript or in Commandino's edition: *suppleo (duobus in locis) uncis inclusa, quae exciderant, tum in nostro codice Graeco, tum in codice Commandini, eiusque versione*. This text does not appear in any of the other editions or in any of the consulted manuscripts. This is the first of the two *loci*; the second corresponds to the next note. FG:228 justifies why he does not introduce the modification of W.

204 W:591 adds here the second of the mentioned texts (duobus in locis): ἀλλ᾽ ὡς ἡ ΟΠ πρὸς τὴν ΝΞ, τουτέστιν, ὡς ἡ ΑΟ πρὸς τὴν ΡΝ, οὕτως ἡ ΑΜ πρὸς τὴν ΡΜ, δι᾽ ὁμοιότητα τριγώνων, καὶ ἡ ΑΜ ἄρα πρὸς τὴν ΜΡ μείζονα λόγον ἔχει ἢ ὃν τὰ θ πρὸς α (but as line OΠ is to line NΞ, that is, like line AO is to line PN, so is line AM to line PM, because of the similarity of the triangles, and therefore, line AM has to line MP a ratio greater than the [ratio] that 9 has to 1). In mss. A, B, and F, is found only the last part of this sentence (from the καὶ), that we, as well as H and all the modern editions, have included. Instead, mss. K, P, U, Y, W, and 1 do not include the text. V:76 does not translate the sentence added by W. M:67 translates the missing sentence of C without mentioning its source. The text in FG:77 and No:25 is equal to that of H. Ni:18 asserts in note (d) that he will follow No and F, omitting the text introduced by W. Accordingly, his edition is similar to that of H. FG:228–229 justifies his decision not to include the modification of W. We want to highlight the acuteness of W, who not only realized that a text was missing but also knew how to restore it, at least in the part preserved in the manuscripts in a style very close to that of Aristarchus.

205 V:11: *abc*.

206 See note 50.

207 We translate *multo maiorem proportionem* by *the more so . . .* , following the way we translate πολλῷ μείζων in Aristarchus's text.

208 Here again, as in note 50, C refers to propositions that are towards the end of Book V, which appear in Commandino's translation, but not in the others. In this case, proposition 30 states: "If the first and the second have to the second a greater ratio than the [ratio] that the third and the fourth have

C. Ex 7 huius.

D. Est enim BR minor, quam BC.

  * 26. quinti

** 28. quinti

E. Quoniam enim MA ad AR minorem propo[r]tionem habet quam 9 ad 8, hoc est, quam eorum undevigintupla, videlicet 171 ad 152: habet autem RA ad AB proportionem minorem, quam 19 ad 18. Fiat ut 19 ad 18, ita 152 ad alium; erit ad 144. Cum igitur MA ad AR minorem habeat proportionem, quam 171 ad 152; habeatque RA ad AB proportionem minorem, quam 152 ad 144: ex aequali MA ad AB minorem proportionem habebit quam 171 ad 144; hoc est quam 19 ad 16.

*** 15. Quinti

**** 30. quinti.

ἀνάπαλιν ἄρα ἡ ΒΡ πρὸς τὴν ΒΑ ἐλάσσονα λόγον ἔχει ἢ ὃν α πρὸς ιη. καὶ συνθέντι ἡ ΡΑ ἄρα πρὸς τὴν ΑΒ ἐλάσσονα λόγον ἔχει ἢ ὃν τὰ ιθ πρὸς τὰ ιη. ἐδείχθη δὲ καὶ ἡ ΜΑ πρὸς τὴν ΑΡ ἐλάσσονα λόγον ἔχουσα ἢ ὃν τὰ θ πρὸς τὰ η· ἕξει ἄρα δι' ἴσου ἡ ΜΑ πρὸς τὴν ΑΒ ἐλάσσονα λόγον ἢ ὃν τὰ ροα πρὸς ρμδ.[12] καὶ ὃν τὰ ιθ πρὸς ις, τὰ γὰρ μέρη τοῖς ὡσαύτως πολλαπλασίοις τὸν αὐτὸν ἔχει λόγον· ἀναστρέψαντι ἄρα ἡ ΑΜ πρὸς ΒΜ μείζονα λόγον ἔχει ἢ ὃν τὰ ιθ πρὸς τὰ γ. ὡς δὲ ἡ ΑΜ πρὸς ΜΒ, οὕτως ἡ διάμετρος τοῦ ΔΕΖ κύκλου πρὸς τὴν διάμετρον τοῦ ΗΘΚ κύκλου· ἡ ἄρα τοῦ ἡλίου διάμετρος πρὸς τὴν τῆς γῆς διάμετρον μείζονα λόγον ἔχει ἢ ὃν τὰ ιθ πρὸς γ.

---

12 We have changed the punctuation so that τὰ γὰρ . . . keeps unequivocally linked to the sentences starting with καί.

Therefore, by inversion,[209] line BP has to line BA[210] a ratio smaller than the [ratio] that 1 has to 18. [*] And, therefore, by composition, line PA has to line AB a ratio[211] smaller than the [ratio] that 19 has to 18. [**] And it has also been proved that line MA has to line AP a ratio smaller than the [ratio] that 9 has to 8. Therefore, by equality of terms, line MA will have to line AB a ratio smaller than the [ratio] that 171 has to 144 [E], and than the [ratio] that 19 has to 16, since the parts have the same ratio as their multiples [***]. Therefore, by conversion, line AM has to BM a ratio greater than the [ratio] that 19 has to 3 [****]. And, as line AM is to MB, so is the diameter of circle ΔEZ[212] to the diameter of circle $H\Theta K$[213] [F]. Therefore, the diameter of the Sun has to the diameter of the Earth a ratio greater than the [ratio] that 19[214] has to 3.

C. From [proposition] 7 of this [treatise].

D. For BP is smaller than BΓ.

* [*Elem.*] V, 26[215]
** [*Elem.*] V, 28[216]

E. For, since MA has to AP a ratio smaller than the [ratio] that 9 has to 8, that is, than the [ratio] that 19 times those numbers have, that is 171 to 152: but, PA has to AB a ratio smaller than the [ratio] that 19 has to 18. Let it be that as 19 is to 18, so [is] 152 to another [number]. [This number] will be 144. Accordingly, since MA has to AP a ratio smaller than the [ratio] that 171 has to 152, and PA has to AB a ratio smaller than the [ratio] that 152 has to 144: by equality of terms, MA will have to AB a ratio smaller than the [ratio] that 171 has to 144, that is, the [ratio] that 19 has to 16.

*** [*Elem.*] V, 15.
**** [*Elem.*] V, 30.[217]

to the fourth, by conversion of the ratio, the first and the second will have to first a ratio smaller than the [ratio] that the third and the fourth have to the third" (*Si prima & secunda ad secundam maiorem proportionem habeat, quam tertia & quarta ad quarta, per conversionem rationis prima & secunda ad primam minorem habebit proportionem, quam tertia & quarta ad tertiam*).

209 F always translates ἀνάπαλιν and ἀναστρέψαντι by *convertendo*. C translates ἀνάπαλιν by *convertendo*, and ἀναστρέψαντι by *per conversionem rationis*. H translates ἀνάπαλιν by *inversely*, and ἀναστρέψαντι by *convertendo*. M:67 affirms in note 75 that here she decided to follow H and not C and translate "al invertir" (*inversely*). We translate ἀνάπαλιν by *by inversion* and ἀναστρέψαντι by *by conversion*. See Appendix I.

210 G:184, in French: AB; the Italian is correct (G:213).

211 C.32b omits λόγον.

212 V:76 omits ΔEZ.

213 G:214, in Italian: GHL (instead of GHK); the French is correct.

214 V:76: *9*.

215 See note 50.

216 Here, like in notes 50 and 208, C refers to a proposition that appears in his translation but not in the other editions. Proposition XXVIII asserts that if the first has to the second a ratio greater than the third to the fourth, also, by composition, the first and the second will have to the second a ratio greater than the third and the fourth to the fourth (*si prima ad secundam maiorem proportionem habeat, quam tertia ad quartam, etiam componendo prima et secunda ad secundam maiorem proportionem habebit, quam tertia et quarta ad quartam*).

217 See note 208.

F. Iungantur AD BG. Erit
    trianguli MDA angulus
    ADM rect[us] aequalis recto
    BGM trianguli MGB. Sed
    angulus DMA est communis
    utrique. Ergo et reliquus
    reliquo aequalis, et triangulum
    triangulo simile: ut igitur
    AM ad MB, ita AD ad BG,
    et ita earum duplae, videlicet
    diameter circuli DEF ad circuli
    GHK diametrum [4. sexti; 15.
    quinti].

G. Ex 13 huius.

*   30. quinti.

H. Ex 7 huius.

K. Nam cum AB ad BC minorem
    habeat proportionem, quam
    20 ad 1, hoc est quam 13500
    ad 675, et CB ad BR habeat
    minorem proportionem quam
    675 ad 674; habebit ex a[e]
    quali AB ad BR minorem
    proportionem, quam 13500
    ad 674, hoc est, quam eorum
    dimidia 6750 ad 337.

Λέγω δὴ ὅτι ἐλάσσονα λόγον ἔχει <πρὸς αὐτὴν> ἢ
ὃν τὰ μγ πρὸς ϛ.

  ἐπεὶ γὰρ ἡ ΒΓ πρὸς τὴν ΓΡ μείζονα λόγον ἔχει ἢ
ὃν τὰ χοε πρὸς α, ἀναστρέψαντι ἄρα ἡ ΓΒ πρὸς τὴν
ΒΡ ἐλάσσονα λόγον ἔχει ἢ ὃν τὰ χοε πρὸς τὰ χοδ.
ἔχει δὲ καὶ ἡ ΑΒ πρὸς τὴν ΒΓ ἐλάσσονα λόγον ἢ
ὃν τὰ κ πρὸς α· ἕξει ἄρα δι' ἴσου ἡ ΑΒ πρὸς τὴν ΒΡ
ἐλάσσονα λόγον ἢ ὃν τὰ $\overset{α}{Μ}$γφ πρὸς τὰ χοδ,
τουτέστιν, ἢ ὃν τὰ ͵ϛψν πρὸς τὰ τλζ·

I say that [the diameter of the Sun] has with [the diameter of the Earth] a ratio smaller[218] than the [ratio] that 43 has to 6.

Since line ΒΓ has to line ΓΡ a ratio greater than the [ratio] that 675 has to 1 [G], therefore, by conversion, line ΓΒ has to line ΒΡ a ratio smaller than the [ratio] that 675 has to 674 [*]. And also line ΑΒ has[219] to line ΒΓ a ratio smaller than the [ratio] that 20 has to 1 [H]. Therefore, by equality of terms, line ΑΒ will have to line ΒΡ a ratio smaller than the [ratio] that 13500[220] has to 674, that is, than the [ratio] that 6750 has to 337 [K].

F. Let ΑΔ and ΒΗ be joined. Right angle ΑΔΜ of triangle ΜΔΑ will be equal to right angle ΒΗΜ of triangle ΜΗΒ. But angle ΔΜΑ is common to both. Consequently, also the remaining [angle of one triangle] is equal to the remaining [angle of the other triangle], and one triangle is similar to [the other] triangle. Accordingly, as ΑΜ is to ΜΒ, so [is] ΑΔ to ΒΗ, and their doubles, that is, the diameter of circle ΔΕΖ to the diameter of circle ΗΘΚ {[*Elem.*] VI, 4 and V, 15.}.

G. From [proposition] 13 of this [treatise].

* [*Elem.*] V, 30.[221]

H. From [proposition] 7 of this [treatise].

K. For, because ΑΒ has to ΒΓ a ratio smaller than the [ratio] that 20 has to 1, that is, than the [ratio] that 13500 has to 675, and ΓΒ has to ΒΡ a ratio smaller than the [ratio] that 675 has to 674, by equality of terms, ΑΒ will have to ΒΡ a ratio smaller than the [ratio] that 13500 has to 674, that is, than the [ratio] of their halves, 6750 to 337.

218 V:11: *maiorem* (greater).
219 V:11 duplicates ἔχει: *habet autem et ab ad bc minorem habet rationem quam quae 20 ad 1.*
220 V:11 translates τὰ Μ̅γφ (13500) by *ma. 3500.*
221 See note 208.

* 26. quinti.
** 28. quinti.

L. Ex 12 huius.

* 26. quinti.

M. Sunt enim triangula AMO RMN inter se simila, ut superius dictum est.

N. Quoniam eni[m] MA ad AR maiorem habet proportionem, quam 10125 ad 9146, et RA ad AB habet maiorem, quam 7087 ad 6750, fiat ut 9146 ad 10125, ita 7087 ad alium. Erit ad 7845 $^{5500}/_{9146}$; si enim multiplicemus 10125 per 7087, et quod producitur, videlicet 71755875 dividamus per 9146, exibunt 7845$^{5500}/_{9146}$. Itaque cum MA ad AR, maiorem habeat proportionem, quam 10125 ad 9146, hoc est quam 7845$^{5500}/_{9146}$ ad 7085, et RA ad AB habeat maiorem, quam 7087 ad 6750: habebit ex aequali MA ad AB maiorem proportionem, quam 7845 $^{5500}/_{9146}$ ad 6750. Sed 7845 $^{5500}/_{9146}$ hoc est $^{71755875}/_{9146}$; ad 6750 est ut 71755875 ad 61735500, quod quidem numeris decussatim multiplicatis perspicuum erit, ex iis

	61735500
71755875	6750
9149	1

ἀνάπαλιν ἄρα καὶ συνθέντι ἡ ΡΑ πρὸς τὴν ΑΒ μείζονα λόγον ἔχει ἢ ὃν τὰ ͵ζπζ πρὸς ͵ϛψν.

καὶ ἐπεὶ ἡ ΝΞ πρὸς τὴν ΟΠ μείζονα λόγον ἔχει ἢ <ὃν τὰ> ͵ϡοθ πρὸς Μ̅ρκε, ἀνάπαλιν ἄρα ἡ ΟΠ πρὸς ΝΞ ἐλάσσονα λόγον ἔχει ἢ <ὃν τὰ> Μ̅ρκε πρὸς ͵ϡοθ· ὡς δὲ ἡ ΟΠ πρὸς ΝΞ, οὕτως ἡ ΑΜ πρὸς ΜΡ· καὶ ἡ ΑΜ ἄρα πρὸς ΜΡ ἐλάσσονα λόγον ἔχει ἢ <ὃν τὰ> Μ̅ρκε πρὸς ͵ϡοθ· ἀναστρέψαντι ἡ ΜΑ ἄρα πρὸς τὴν ΑΡ μείζονα λόγον ἔχει ἢ ὃν τὰ Μ̅ρκε πρὸς τὰ ͵θρμϛ. ἔχει δὲ καὶ ἡ ΡΑ πρὸς ΑΒ μείζονα λόγον ἢ ὃν τὰ ͵ζπζ πρὸς τὰ ͵ϛψν·

δι' ἴσου ἄρα ἕξει ἡ ΜΑ πρὸς τὴν ΑΒ μείζονα λόγον ἢ ὃν ὁ περιεχόμενος ἀριθμὸς ὑπὸ τῶν Μ̅ρκε καὶ τῶν ͵ζπζ πρὸς τὸν περιεχόμενον ἀριθμὸν ὑπό τε τῶν ͵θρμϛ καὶ τῶν ͵ϛψν, τουτέστιν, ὁ Μ̅ ͵ζροε ͵εωοε πρὸς Μ̅ ͵ϛρόγ ͵εφ.

Therefore, by inversion [*], and by composition [**], line PA has to line AB a ratio greater than the [ratio] that 7087 has to 6750.

And since line NΞ has to line OΠ a ratio greater than the [ratio] that 979[222] has to 10125 [L],[223] therefore, by inversion [*], line OΠ has to NΞ a ratio smaller than the [ratio] that 10125[224] has to 979.[225] And, as line OΠ is to NΞ, so is line AM to MP [M]. And, therefore, line AM has to MP a ratio smaller than the [ratio] that 10125 has to 979.[226] Therefore, by conversion, line MA has to line AP a ratio greater than the [ratio] that 10125 has to 9146. And also[227] line PA has to AB a ratio greater than the [ratio] that 7087 has to 6750.

Therefore, by equality of terms, line MA will have to line AB a ratio greater than the [ratio] that the number product[228] of 10125[229] and 7087 has to the number product of 9146 and 6750, that is, 71755875 to 61735500 [N].[230]

* [*Elem.*] V, 26[231]
** [*Elem.*] V, 28[232]
L. From [proposition] 12 of this [treatise].
* [*Elem.*] V, 26[233]
M. For triangles AMO and PMN are similar between each other, as it has previously been said.
N. For, since MA has to AP a ratio greater than the [ratio] that 10125 has to 9146, and PA has to AB a [ratio] greater than the [ratio] that 7087 has to 6750, and let it be as 9146 is to 10125, so [is] 7087 to another [number]. [This number] will be 7845 $^{5500}/_{9146}$; for if we multiply 10125 by 7087 and we divide what is produced, that is 71755875, by 9146, it results 7845 $^{5500}/_{9146}$. In the same manner, since MA has to AP a ratio greater than the [ratio] that 10125 has to 9146, that is, than the [ratio] that 7845 $^{5500}/_{9146}$ has to 7085, and PA has to AB a ratio greater than the [ratio] that 7087 has to 6750, by equality of terms, MA will have to AB a ratio greater than the [ratio] that 7845 $^{5500}/_{9146}$ has to 6750. But 7845 $^{5500}/_{9146}$, that is, $^{71755875}/_{9146}$ to 6750 is as 71755875 to 61735500, what certainly will be clear if the numbers are multiplied crosswise:

	61735500
71755875	6750
9149	1

222 V:11 translates ϡοθ (979) by 579.
223 V:11 omits *rationem* and translates M̅ρκε (10125) by 10.000.
224 V:11 translates M̅ρκε by 10000.yrke.
225 V:11 translates ϡοθ (979) by 379.
226 V:11 omits ὡς δὲ ἡ ΟΠ πρὸς ΝΞ, οὕτως ἡ ΑΜ πρὸς ΜΡ· καὶ ἡ ΑΜ ἄρα πρὸς ΜΡ ἐλάσσονα λόγον ἔχει ἢ <ὃν τὰ> M̅ρκε πρὸς ϡοθ.
227 M:69 translates *Sin embargo, se demostró que . . .* (However, it has been demonstrated that . . .).
228 M:69 translates *el número que representa el producto*, following again H:407: *the number representing the product*.
229 V:11 translates M̅ρκε (10125) by 10425.
230 V:11 translates ὁ M̅͵εωοε πρὸς M̅͵εφ (71.755.875 to 61.735.500) by 15875 and 5500, and the second time that they appear, the first number again by 15875 and the second by 15500.
231 See note 50.
232 See note 216.
233 See note 50.

quae nos demonstravimus
in commentariis in tertiam
propositionem libri
Archimedis de circuli
dimensione, propositione
septima, ut proxime diximus.
Ergo MA ad AB maiorem
habet proportionem, quam
numerus productus ex 10125
et 7087 ad eum, qui ex 9146 et
6750 producitur.

O. Si enim fiat ut 43 ad 37, ita
71755875 ad alium. Erit ad
61743427 qui maior est, quam
61735500. Ergo 71755875
ad 6173500 maiorem habebit
proportionem, quam ad
61743427, hoc est quam 43 ad
37 {8. quinti.}.

ἔχει δὲ καὶ ὁ Μ ͵ζϱοε ͞εωοε πρὸς Μ ͵ϛϱογ ͞εφ μείζονα λόγον ἢ
ὃν τὰ μγ πρὸς λζ· καὶ ἡ ΜΑ ἄρα πρὸς τὴν ΑΒ
μείζονα λόγον ἔχει ἢ ὃν μγ πρὸς λζ· ἀναστρέψαντι
ἄρα ἡ ΑΜ πρὸς τὴν ΜΒ ἐλάσσονα λόγον ἔχει ἢ ὃν
τὰ μγ πρὸς ϛ. ὡς δὲ ἡ ΑΜ πρὸς τὴν ΒΜ, οὕτως
ἐστὶν ἡ διάμετρος τοῦ ἡλίου πρὸς τὴν διάμετρον
τῆς γῆς· ἡ ἄρα διάμετρος τοῦ ἡλίου πρὸς τὴν
διάμετρον τῆς γῆς ἐλάσσονα λόγον ἔχει ἢ ὃν μγ
πρὸς ϛ.

ἐδείχθη δὲ καὶ μείζονα λόγον <πρὸς αὐτὴν>[13]
ἔχουσα ἢ ὃν τὰ ιθ πρὸς τὰ γ.

---

13 πρὸς αὐτὴν has been added by us.

But also 71755875 has to 61735500 a ratio greater than the [ratio] that 43 has to 37 [O]. And, therefore, line MA has to line AB a ratio greater than the [ratio] that 43 has to 37.[234,235] Therefore, by conversion, line AM has to line MB a ratio smaller than the [ratio] that 43 has to 6. And, as line AM is to line BM so is the diameter of the Sun to the diameter of the Earth. Therefore, the diameter of the Sun has to the diameter of the Earth a ratio smaller than the [ratio] that 43 has to 6.

And it has been proved that it also has [to the diameter of the Earth] a ratio greater than the [ratio] that 19 has to 3.

from what we have demonstrated in the comments to the seventh proposition of the third proposition of the book of Archimedes, *On the dimension of the circle*,[236] as we recently said. Consequently, MA has to AB a ratio greater than the [ratio] that the number product of 10125 and 7087 has to the [number] that is the product of 9146 and 6750.

O. For, let it be that as 43 is to 37, so [is] 71755875 to another [number]. [This number] will be 61743427, which is greater than 61735500. Consequently, 71755875 will have to 6173500 a ratio greater than to 61743427, that is, than the [ratio] that 43 has to 37 {[*Elem.*] V, 8.}.

---

234 V:11 omits καὶ ἡ ΜΑ ἄρα πρὸς τὴν ΑΒ μείζονα λόγον ἔχει ἤ ὄν μγ πρὸς λζ.
235 G:188 and G:214 omit καὶ ἡ ΜΑ ἄρα πρὸς τὴν ΑΒ μείζονα λόγον ἔχει ἤ ὄν μγ πρὸς λζ
236 See note 176.

ις' Ὁ ἥλιος πρὸς τὴν γῆν μείζονα λόγον ἔχει ἢ

ὃν ͵ϛωνθ πρὸς κζ, ἐλάσσονα δὲ ἢ ὃν Ṁθφζ

πρὸς σις.

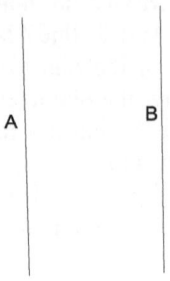

16. The Sun has to the Earth a ratio greater than the [ratio] that 6859 has to 27, but smaller than the [ratio] that 79507[237] has to 216.

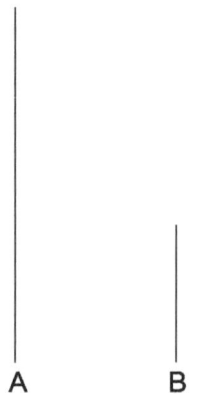

A   B

Figure 2.1.15 Diagram of proposition 16.

---

237 V:11 translates M͵θφζ (79507) by 19507.

* In decima enim propositione
huius demonstratum est ut
cubus qui sit ex diametro
solis ad cubum qui ex
diametro lunae, ita esse
sphaeram solis ad lunae
sphaeram, quod similiter in
terra demonstrabitur.

Ἔστω γὰρ ἡλίου μὲν διάμετρος ἡ Α, γῆς δὲ ἡ Β.
ἀποδείκνυται δὲ ὅτι ἐστίν, ὡς ἡ τοῦ ἡλίου σφαῖρα
πρὸς τὴν τῆς γῆς σφαῖραν, οὕτως ὁ ἀπὸ τῆς
διαμέτρου τοῦ ἡλίου κύβος πρὸς τὸν ἀπὸ τῆς
διαμέτρου τῆς γῆς κύβον, ὥσπερ καὶ ἐπὶ τῆς
σελήνης·[14] ὥστε ἐπεί ἐστιν, ὡς ὁ ἀπὸ τῆς Α κύβος
πρὸς τὸν ἀπὸ τῆς Β κύβον, οὕτως ὁ ἥλιος πρὸς τὴν
γῆν, ὁ δὲ ἀπὸ τῆς Α κύβος πρὸς τὸν ἀπὸ τῆς Β
<κύβον> μείζονα λόγον ἔχει ἢ ὃν τὰ ͵ϛωνθ πρὸς
κζ, ἐλάσσονα δὲ ἢ ὃν Μ̅ ͵θφϛ πρὸς σιϛ· καὶ γὰρ ἡ Α
πρὸς τὴν Β μείζονα λόγον ἔχει ἢ ὃν ιθ πρὸς γ,
ἐλάσσονα δὲ ἢ ὃν μγ πρὸς ϛ· ὥστε ὁ ἥλιος πρὸς
τὴν γῆν μείζονα λόγον ἔχει ἢ ὃν ͵ϛωνθ πρὸς κζ,
ἐλάσσονα δὲ ἢ ὃν Μ̅ ͵θφϛ πρὸς σιϛ.

14 We changed the comma by high dot.

Let line A be a diameter of the Sun, and line B, a diameter of the Earth. It is demonstrated that, as the sphere of the Sun is to the sphere of the Earth, so is the cube[238] of the diameter of the Sun to the cube of the diameter of the Earth, as also in the case of the Moon [*]. Therefore, as the cube of A is to the cube of B, so is the Sun to the Earth. But, the cube of A has to the cube of B a ratio greater than the [ratio] that 6859 has to 27,[239] but smaller than the [ratio] that 79507[240] has to 216 (since indeed, line A has to line B a ratio greater than the [ratio] that 19 has to 3, but smaller than the [ratio] that 43 has to 6). Hence the Sun has to the Earth a ratio greater than the [ratio] that 6859 has to 27, but smaller than the [ratio] that 79507[241,242] has to 216.

* For in the tenth proposition of this [treatise] it has been demonstrated that as the cube of the diameter of the Sun is to that of the diameter of the Moon, so is the sphere of the Sun to the sphere of the Moon, as it will be similarly demonstrated for the Earth.

238 V:77 translates *circulus* (circle).
239 V:77 introduces one extra &: *habet rationem quam quae 6859 & ad 27.*
240 V:77 translates Μ̅,θϙζ (79507) by 19507.
241 V:77 translates Μ̅,θϙζ (79507) by 19507.
242 G:190, in French, and G:215, in Italian: 79057.

ιζ' Ἡ διάμετρος τῆς γῆς πρὸς τὴν διάμετρον
τῆς σελήνης ἐν μείζονι μὲν λόγῳ ἐστὶν ἢ ὃν
<ἔχει> ρη πρὸς μγ, ἐν ἐλάσσονι δὲ ἢ ὃν ξ
πρὸς ιθ.

A   B   Γ

17. **The diameter of the Earth is in a ratio to the diameter of the Moon greater than the [ratio] that 108 has to 43, but smaller than the [ratio] that 60 has to 19.**[243]

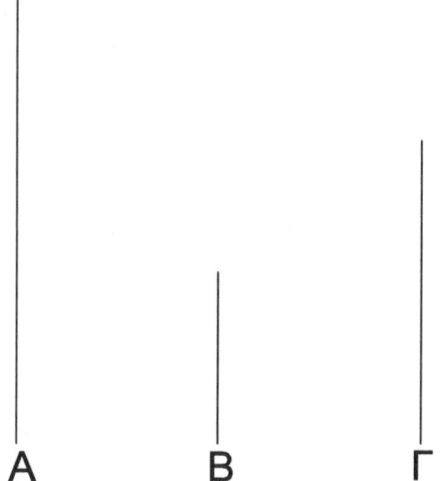

Figure 2.1.16 Diagram of proposition 17.

243 This is the first time that Aristarchus uses ἐν μείζονι μὲν λόγῳ ἐστὶν instead of μείζονα λόγον ἔχει (used 51 times until here).

A. Ex 14 huius.

B. Ex 9 huius.

C. Quoniam enim C ad
A maiorem habet
proportionem quam 6 a[d]
43; et A ad B maiorem quam
18 ad 1, fiat ut 18 ad 1,
ita 43 ad alium. Erit 2 $^7/_{18}$.
Cum igitur C ad A maiorem
proportionem habeat, quam
6 ad 43, et A ad B maiorem,
quam 43 ad 2 $^7/_{18}$, habebit
ex aequali A ad B maiorem
proportionem quam 6 ad 2
$^7/_{18}$, hoc est quam 108 ad 43,
quod numeris decussatim
multiplicatis manifeste
constat, ex iis, quae superius
dicta sunt:

$$\frac{108}{\quad}$$
$$\frac{6 \quad 43}{1 \quad 18}$$

D. Ex 14 huius.

E. Ex 9 huius.

F. Fiat ut 20 ad 1, ita 19 ad
alium. Erit ad $^{19}/_{20}$. Quare
cum C ad A mi[n]orem
proportionem habeat, quam
3 ad 19, et A ad B minorem,
quam 19 ad $^{19}/_{20}$, ex aequali
C ad B minorem habebit
proportionem, quam 3 ad
$^{19}/_{20}$, hoc est, quam 60 ad 19.

$$\frac{60}{\quad}$$
$$\frac{3 \quad 19}{1 \quad 20}$$

Ἔστω γὰρ ἡλίου μὲν διάμετρος ἡ Α, σελήνης δὲ
ἡ Β, γῆς δὲ ἡ Γ.
καὶ ἐπεὶ ἡ Α πρὸς τὴν Γ ἐλάσσονα λόγον ἔχει ἢ
ὃν τὰ μγ πρὸς ϛ, ἀνάπαλιν ἄρα ἡ Γ πρὸς τὴν Α
μείζονα λόγον ἔχει ἢ ὃν ϛ πρὸς μγ. ἔχει δὲ καὶ ἡ
Α πρὸς τὴν Β μείζονα λόγον ἢ ὃν τὰ ιη πρὸς α·
δι' ἴσου ἄρα ἡ Γ πρὸς τὴν Β μείζονα λόγον ἔχει
ἢ ὃν τὰ ρη πρὸς τὰ μγ. πάλιν ἐπεὶ ἡ Α πρὸς τὴν Γ
μείζονα λόγον ἔχει ἢ ὃν τὰ ιθ πρὸς τὰ γ, ἀνάπαλιν
ἄρα ἡ Γ πρὸς τὴν Α ἐλάσσονα λόγον ἔχει ἢ ὃν τὰ
γ πρὸς τὰ ιθ. ἔχει δὲ ἡ Α πρὸς τὴν Β ἐλάσσονα
λόγον ἢ ὃν τὰ κ πρὸς α· δι' ἴσου ἄρα ἡ Γ πρὸς τὴν
Β ἐλάσσονα λόγον ἔχει ἢ ὃν ξ πρὸς ιθ.

Let line A be a diameter of the Sun; line B, a diameter of the Moon, and line Γ, a diameter of the Earth.

And since line A has to line Γ a ratio smaller than the [ratio] that 43 has to 6 [A], therefore, by inversion, line Γ has to line A a ratio greater than the [ratio] that 6 has to 43. And also line A has to line B a ratio greater than the [ratio] that 18 has to 1 [B].[244] Therefore, by equality of terms, line Γ[245] has to line B a ratio greater than the [ratio] that 108 has to 43 [C]. Again, since line A has to line Γ a ratio greater than the [ratio] that 19 has to 3 [D], therefore, by inversion, line Γ has to line A a ratio smaller than the [ratio] that 3 has to 19.[246] And line A has to line B a ratio smaller than the [ratio] that 20 has to 1 [E]. Therefore, by equality of terms, line Γ[247] has to line B a ratio smaller than the [ratio] that 60 has to 19 [F].

A. From [proposition] 14 of this [treatise].

B. From [proposition] 9 of this [treatise].

C. For, since Γ has to A a ratio greater than the [ratio] that 6 has to 43; and A has to B a [ratio] greater than the [ratio] that 18 has to 1, let it be that as 18 is to 1, so [is] 43 to another [number]. [This number] will be $2\,7/_{18}$. Accordingly, since Γ has to A a ratio greater than the [ratio] that 6 has to 43, and A has to B a [ratio] greater than the [ratio] that 43 has to $2\,7/_{18}$, by equality of terms, A will have to B a ratio greater than the [ratio] that 6 has to $2\,7/_{18}$, that is, than the ratio that 108 has to 43, what is manifest if the numbers are multiplied crosswise, from what it has been explained above:

$$\frac{108}{6 \quad 43}$$
$$1 \quad 18$$

D. From [proposition] 14 of this [treatise].

E. From [proposition] 9 of this [treatise].

F. Let it be that as 20 is to 1, so [is] 19 to another [number]. [This number] will be $19/_{20}$. Therefore, since Γ has to A a ratio smaller than the [ratio] that 3 has to 19 and A has to B a [ratio] smaller than the [ratio] that 19 has to $19/_{20}$, by equality of terms, Γ will have to B a ratio smaller than the [ratio] that 3 has to $19/_{20}$, that is, that the [ratio] that 60 has to 19.

$$\frac{60}{3 \quad 19}$$
$$1 \quad 20$$

244　V:77 translates A (1) by *a*.

245　V:77 omits ἡ Γ.

246　V:77 translates ιθ (19) by 9.

247　V:12 translates ἡ Γ πρὸς τὴν B by *est ad b*, translating ἡ Γ by *est*.

ιη' Ἡ γῆ πρὸς τὴν σελήνην ἐν μείζονι μὲν
λόγῳ ἐστὶν ἢ ὃν <ἔχει> Μ̅θψιβ πρὸς
Μ̅‚θφζ, ἐν ἐλάσσονι δὲ ἢ ὃν Μ̅ς πρὸς
‚ςωνθ.

A   B

**18. The Earth is in a ratio to the Moon greater
than the [ratio] that 1259712 has to 79507,
but smaller than the [ratio] that 216000 has
to 6859.**[248]

A   B

Figure 2.1.17 Diagram of proposition 18

248 V:12 translates M̅,θψιβ (1,259,712) by 19712, M̅,θφζ (79,507) by 19760, and M̅,ϛ (216,000) by
16100.

Ἔστω γὰρ γῆς μὲν διάμετρος ἡ Α, σελήνης δὲ ἡ Β·

ἡ Α ἄρα πρὸς τὴν Β μείζονα λόγον ἔχει ἢ ὂν τὰ ρη πρὸς τὰ μγ, ἐλάσσονα δὲ ἢ ὂν τὰ ξ πρὸς ιθ· καὶ ὁ ἀπὸ τῆς Α ἄρα κύβος πρὸς τὸν ἀπὸ τῆς Β κύβον μείζονα λόγον ἔχει ἢ ὂν $\overset{\rho\kappa\varepsilon}{\text{Μ}}$‚θψιβ πρὸς $\overset{\zeta}{\text{Μ}}$‚θφζ, ἐλάσσονα δὲ ἢ ὂν $\overset{\kappa\alpha}{\text{Μ}}$‚ς πρὸς ‚ςωνθ. ὡς δὲ ὁ ἀπὸ τῆς Α κύβος πρὸς τὸν ἀπὸ τῆς Β κύβον, οὕτως ἐστὶν ἡ γῆ πρὸς τὴν σελήνην· ἡ γῆ ἄρα πρὸς τὴν σελήνην μείζονα μὲν λόγον ἔχει ἢ ὂν $\overset{\rho\kappa\varepsilon}{\text{Μ}}$‚θψιβ πρὸς $\overset{\zeta}{\text{Μ}}$‚θφζ, ἐλάσσονα δὲ ἢ ὂν $\overset{\kappa\alpha}{\text{Μ}}$‚ς πρὸς ‚ςωνθ.

Let line A be a diameter of the Earth, and line B, a diameter of the Moon.

Therefore, line A has to line B a ratio greater than the [ratio] that 108 has to 43, but smaller than the [ratio] that 60[249] has to 19. And, therefore, the cube of line A has to the cube of line B a ratio greater than the [ratio] that 1259712 has to 79507, but smaller than the [ratio] that 216000[250] has to[251] 6859. And, as the cube of line A is to the cube of line B, so is the Earth to the Moon. Therefore, the Earth has to the Moon a ratio greater than the [ratio] that 1259712 has to 79507,[252] but smaller than the [ratio] that 216000 has to 6859.[253]

---

249 V:12 translates ιθ (60) by 30.

250 V:77 translates $\overset{\rho\kappa\epsilon}{\text{M}}$,θψιβ (1,259,712) by 19712, $\overset{\varsigma}{\text{M}}$θφζ (79,507) by 19507, and $\overset{\kappa\alpha}{\text{M}}$,ς (216,000) by 16,000.

251 No:28 omits πρὸς.

252 C:38b translates $\overset{}{\text{M}}$,θφζ (79,507) by 795071.

253 V:77 omits ἡ γῆ ἄρα πρὸς τὴν σελήνην and translates $\overset{\rho\kappa\epsilon}{\text{M}}$,θψιβ (1,259,712) by 19712, $\overset{\varsigma}{\text{M}}$,θφζ (79,507) by 19507, and $\overset{\kappa\alpha}{\text{M}}$,ς (216,000) by 6000.

3 Analysis of *On the Sizes and Distances of the Sun and Moon*

In this chapter, we present in detail the calculations made by Aristarchus in *On Sizes*. First, we briefly explain why reading the treatise can be difficult for a modern reader. Second, we describe its general structure. Third, we analyze each one of the six starting hypotheses. Finally, we go into the analysis of the calculations of each proposition.

Comprehension difficulties for a modern reader

Several reasons undoubtedly make reading the treatise difficult. As we already discussed (see p. 39), the book uses a scientific vocabulary that tends to be standardized. Nevertheless, there is nothing close to the current mathematical formalism in it. If we wrote it as

$$HE/E\Theta > HBE/\Delta BE \rightarrow HE/E\Theta > 15/2$$

this would be understandable at first sight. However, it is much more difficult to understand this same material when expressed in prose:

> And since line HE has to line EΘ a greater ratio than angle HBE has to angle ΔBE, therefore, line HE has to line EΘ a greater ratio than 15 has to 2.

Nonetheless, we did not translate Aristarchus's expressions into current mathematical formalisms because we decided to follow as faithfully as possible the style of the original text.

Second, trigonometry was not available for Aristarchus. It only emerges at least a few centuries later (van Brummelen 2009). Consequently, Aristarchus devotes whole pages – sometimes even entire propositions – full of labyrinthine steps going from sides to angles and back only to calculate something that just one trigonometric function could solve.

A consequence of the aforesaid is the use of inequalities, which imply a third difficulty. Usually, Aristarchus cannot establish the proportion between sides and angles. Still, he knows that, under defined configurations, the ratio between angles is greater or smaller than the ratio between sides. Consequently, the lack of

DOI: 10.4324/9781003184553-3

trigonometry gives Aristarchus no alternative but to use inequalities for connecting sides and angles.

Fourth, Greek geometry had not yet incorporated the division of the circle into 360°. Greeks expressed angles as arbitrary portions of the circle. Therefore, the expression of certain angles also becomes a bit complicated. For example, Aristarchus refers to 87° as a portion of a circle that is *less than a quadrant in one-thirtieth of the quadrant*, and to 2° as *one-fifteenth of a sign of the zodiac*.

Fifth, Aristarchus is unwilling to make certain simplifications that would significantly reduce the required step without resulting in differences to the final results. Thus, for example, Aristarchus takes practically a third of the treatise to solve a problem that could be reduced to a few lines if he accepted certain simplifications that would later be accepted by Ptolemy when he calculated the distances in the *Almagest*. We will discuss the reasons for Aristarchus's rejection later.

Taken one by one, none of these reasons is insurmountable. But all pushing in the same direction demand from the reader an almost-heroic degree of patience in understanding the details of the calculations. Nonetheless, we would not like these difficulties to prevent the reader from understanding the heart of Aristarchus's treatise, which is based, as we said, on clear intuitions that any reader can easily appreciate. In any case, reading together the translation and this analysis should be enough – or so we hope – for the treatise to be appreciated in its complexity and richness.

The general structure of *On Sizes*

The treatise contains 18 propositions[1] built on six hypotheses. The treatise begins with the enumeration of the six hypotheses and a brief description of the expected results. The first three hypotheses do not have numerical values and are intended to justify constructing the geometric diagrams representing the astronomical configurations. The first hypothesis states that the Moon receives its light from the Sun; the second, that the Earth can be considered a point and the center of the lunar orbit; the third, that, in a dichotomy (i.e., when exactly half a Moon is seen from the Earth), our eye is in the same plane of the circle dividing the dark and the illuminated parts of the Moon. The following three hypotheses establish numerical values. The first indicates that the lunar elongation in a dichotomy is 87°; the second, that, in a lunar eclipse, the size of the shadow of the Earth projected on the Moon is twice the size of the Moon; and the last one, that the apparent size of the Moon seen from the Earth is 2°.

---

1 Arabic manuscripts contain one more proposition at the end. It establishes the ratio between the Earth–Moon distance and the Moon's distance to the vertex of the cone containing the Sun and Earth. According to the proposition, the ratio is greater than 71/37 but smaller than 3. See Berggren and Sidoli (2007: 221). We do not incorporate it because it most likely does not belong to the original Greek text.

Note that these last three hypotheses already contain all the data necessary to apply the two basic ideas. To obtain the ratio between the distances, Aristarchus needs to know the angle between the Moon and the Sun, seen from the Earth in a dichotomy (hypothesis 4). To obtain the ratios of the Sun's and the Moon's sizes with the Earth's size, besides the already-known distances, Aristarchus needs to know the size of the Earth's shadow (hypothesis 5). As we already mentioned, the apparent size of the Moon (hypothesis 6) plays a subsidiary role that we will analyze later.

A resume anticipating the conclusions follows the enumeration of the hypothesis. In it, Aristarchus highlights as main results proposition 7 (the ratio between the distances), proposition 9 (the ratio between the sizes of the Moon and the Sun), and proposition 15 (the ratio between the sizes of the Earth and Sun).[2]

The 18 propositions follow this brief resume. The first aim of Aristarchus is to calculate the ratio between the Earth–Sun distance and the Earth–Moon distance. He does this in proposition 7. Propositions 1 to 6 are preparatory to arrive at that result. Once he obtained the ratio between the distances, it is effortless to calculate the ratio between the sizes (proposition 9). Proposition 8 establishes a premise that he needs. After he gets the ratio between the sizes, calculating the ratio between the volumes is almost trivial (proposition 10).

Once Aristarchus established the ratio between the sizes and the ratio between the distances, his next goal was to find the ratio between the Sun's and the Moon's sizes, not with each other, but with the Earth. In this case, he must take several steps to arrive at the result. Propositions 11 to 14 are preparatory for, in proposition 15, finally establishing the ratio between the diameters of the Sun and the Earth. Then, he easily calculates the ratio between their volumes (proposition 16). In proposition 17, he establishes the ratio between the diameters of the Moon and the Earth and, in proposition 18, the ratio between their volumes.

Therefore, the book has three clear objectives: (a) to arrive at the ratio between the Earth–Moon and Earth–Sun distances (proposition 7); (b) to arrive at the ratio between the sizes of the Sun and the Earth (proposition 15); and (c) to arrive at the ratio between the sizes of the Moon and the Earth (proposition 17). The first objective is obtained based on the first intuition (the right triangle in a dichotomy). The

---

2   It would be worth discussing why Aristarchus highlights these three results and not others. For example, it would seem natural to think that he should highlight propositions 7, 9, 15, and 17 as the main results. There are no reasons for excluding the ratio between the sizes of the Earth and the Moon (proposition 17). It is true that it is very easy to obtain that result from proposition 15, but the same can be said of the relationship between propositions 7 and 9, and nevertheless, the results of proposition 9 are made explicit in the text. In favor of the status quo, however, it can be said that the relationship between propositions 7 and 9 is not the same as the relationship between propositions 15 and 17. Once one has proposition 15 (and 7), proposition 17 can be obtained with a simple calculation. However, proposition 9 can only be obtained on the condition that proposition 8 is accepted, which introduces a certain novelty. Another reason to explain the presence of proposition 15 and the absence of proposition 17 in the resume could be that, for Aristarchus and his contemporaries, the most remarkable result of the treatise is the size of the Sun and not so much that of the Moon. Pappus partially modifies the presentation of the results. See Heath (1913: 412).

last two, on the second intuition (the projection of the Earth's shadow in a lunar eclipse).

In the analysis of the next section, we do not always follow Aristarchus's order: we introduce the propositions when the calculation requires them. In this way, the necessity of each step will be manifest. We divide the explanation into five parts. In the first, we analyze in some detail the six hypotheses. In the second, we take the necessary steps to arrive at proposition 7 (establishing the ratio between distances). In the third, we arrive at proposition 9 (the ratio between the sizes of the Sun and the Moon). Then we come to proposition 15 (the ratio between the sizes of the Sun and the Earth). Finally, we arrive at propositions 17 and 18 (the ratios between the sizes of the Moon and the Earth and between their volumes).

Analysis of the hypotheses

Aristarchus's treatise begins with the statement of six hypotheses. The first three allow the geometrical representation of astronomical configurations; the last three establish values that will allow for calculating the maximum and minimum limits of distances and sizes. Let us look briefly at each one of them.

First hypothesis: that the Moon receives its light from the Sun

This hypothesis could seem so obvious to contemporary eyes that it would not even deserve to be made explicit, but the truth is that it is not evident that the Moon receives its light from the Sun. More or less careful observations quickly show that the illuminated face of the Moon follows the Sun, but this does not demonstrate that the Moon is actually illuminated by the Sun. It could be explained – as Berosus, a Babylonian priest and astronomer of the fourth century BC, supposedly did – by stating that the Moon is a half-light, half-sky-color globe and that the illuminated part is always attracted to the Sun (Vitruvius 1999, IX, 2: 111–112). We have strong evidence that Anaxagoras was the first to explain the phases of the Moon correctly. In the *Cratylus*, Plato makes Socrates say, with some wonder, that Anaxagoras recently asserted that the Moon takes its light from the Sun (Cratylus 409a; Reeve 1997: 45–47). Later testimonies also attribute to Anaxagoras the correct explanation of eclipses: in solar eclipses, the Moon hides the Sun; in lunar eclipses, the Earth interposes between the Sun and the Moon, casting its shadow on the Moon. Furthermore, Anaxagoras stated that the Moon is closer to the Earth than the Sun and, as we have already mentioned, that the Sun is larger than the Peloponnese. He even affirmed that the Moon has valleys and craters and is of the same nature as the Earth (Hippolytus, *Refut.* I, 8; Marcovich 1986: 67–70).

Aristarchus lived two centuries after Anaxagoras. By Aristarchus's time, it was probably widespread knowledge that the Moon receives its light from the Sun. This hypothesis plays an essential role in the treatise. As we have already mentioned, knowing that the Moon gets its light from the Sun allows the illuminated portion of the Moon to give us information about the Sun's position.

Second hypothesis: that the Earth has a ratio of a point and a center with respect to the sphere of the Moon

Introducing this hypothesis, Aristarchus indicates that the lunar orbit is so large compared to the Earth that, for the purposes of calculations, we can consider the Earth to be a point located at the center of the orbit. This hypothesis aims to avoid the horizontal parallax of the Moon, identifying the whole Earth with its center.[3] Consequently, the eye of the observer will be located in the center of the Earth. In many diagrams, the same point will represent the observer's eye, the center of the Earth, and the center of the lunar orbit. Otherwise, Aristarchus would have to consider the difference between the observer and the center of the lunar orbit, which would greatly complicate diagrams and calculations.

Hypothesis 2 is a useful assumption, considering how much it simplifies the calculations. But it is not unproblematic. In the first place, it seems a too strong assertion. Before Aristarchus, Euclid, in his *Phenomena* (Berggren and Thomas 2006: 52–53), and after him, Ptolemy, in the *Almagest* (I, 6; Toomer 1998: 43), affirm that the Earth can be considered a point at the center of the sphere of the fixed stars, but the Moon would appear to be too close to the Earth to make that assumption without consequences.

In proposition 11, Aristarchus calculates that the lunar diameter is, at most, 30 times smaller than the Earth–Moon distance. And he also demonstrates in proposition 17 that the Earth's diameter is at least 108/43 times the lunar diameter (around 2.5 times greater). Thus, the lunar orbit has a radius, at most, 11.95 times greater than the Earth's radius. In other words, it is only 12 times bigger, at best. Consequently, assuming that the Earth should be considered a point is an exaggeration even for Aristarchus's parameters. Figure 3.1 represents the Earth and lunar orbit to scale.

But the problem with hypothesis 2 is not only that it is an exaggeration but also that it seems to be in contradiction with another hypothesis. Indeed, hypothesis 5 states that the projection of the Earth's shadow on the Moon is equal to two Moons, but if the Earth is considered a point, it makes no sense to talk about the size of its shadow (Berggren and Sidoli 2007: 216–217). Therefore, strictly speaking, propositions using both hypotheses simultaneously start with a contradiction. That is the case of propositions 13 and 14. The results of these propositions, in turn, are used to calculate the crucial values of proposition 15 (the ratio between the sizes of the Sun and the Earth) and everything coming later (including the ratio between the sizes of the Moon and the Earth). In other words, all the results of the second part of the treatise which use the Earth's shadow suffer from this contradiction.

Stated in this way, it would seem that much of Aristarchus's work starts from contradictory principles. If an axiomatic system starts from contradictory axioms, anything can be concluded (*ex falso quodlibet*). However, suppose we interpret hypothesis 2 as stating that observations made on the Earth's surface should be

---

3   For an explanation of the horizontal parallax, see p. 9 on Chapter 1.

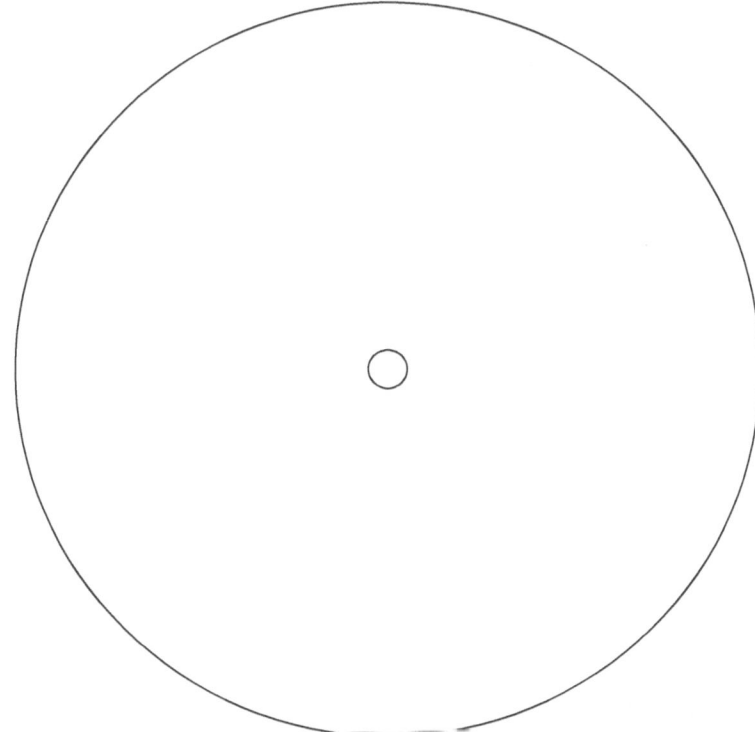

Figure 3.1 The Earth (the smaller circle) and the lunar orbit (the greater circle) to scale according to Aristarchus's parameters.

considered as made from its center – that is the spirit of the hypothesis, after all. In that case, then there is no contradiction: the Earth casts a shadow, but we must always consider that the observer's eye is at the center of the Earth.

As we mentioned when we described Ptolemy's method (pp. 10–13), precisely the denial of this hypothesis, and the consequent acceptance of the existence of lunar parallax, allowed him to calculate the Earth–Moon distance expressed in terrestrial radii.

Third hypothesis: that when the Moon appears to us halved, the great circle delimiting the dark and the bright parts of the Moon is inclined to our eye

The enunciation of the hypothesis seems not too clear. Still, a little later, Aristarchus himself explains its meaning by affirming that *being inclined to our eye* means that the great circle and our eye are in the same plane. Therefore, the

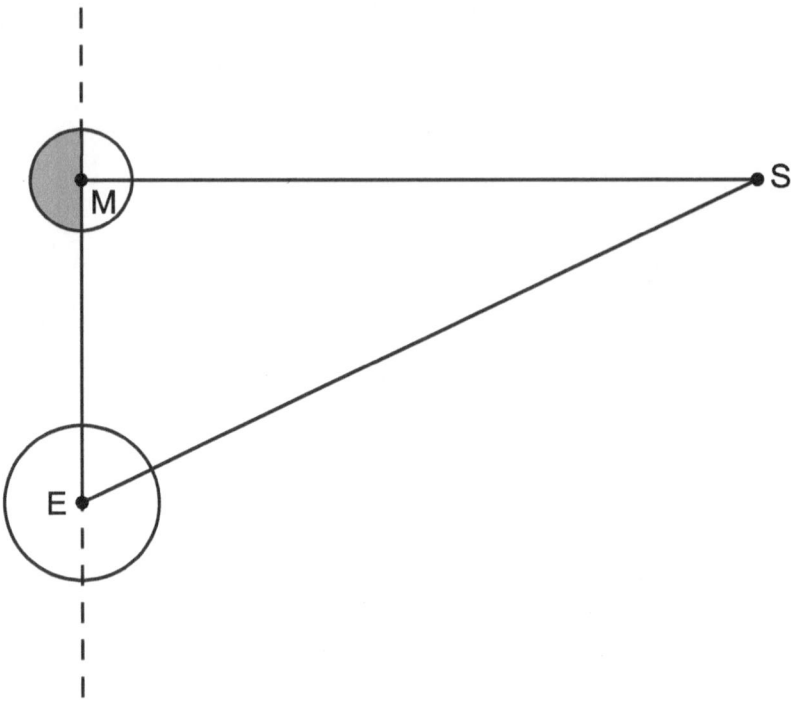

Figure 3.2 The plane that contains the circle dividing the dark and the illuminated parts of the Moon at a dichotomy also includes the center of the Earth.

hypothesis asserts that when the Moon appears to us half illuminated and half dark – that is, at the exact moment of a dichotomy – the plane that contains the circle dividing those parts also includes our eye. The center of the three bodies defines the plane of Figure 3.2. Therefore, the plane defined by the circle dividing the dark and the illuminated parts of the Moon is perpendicular to the plane of the figure. Line ME represents that plane. Hypothesis 3 asserts that this plane also contains the observer's eye. Since the previous hypothesis assumes that the observer's eye is at the center of the Earth, this hypothesis states that, in a dichotomy, the line that divides the dark and the illuminated parts of the Moon passes through the center of the Earth.

In the hypothesis, Aristarchus says that the circle dividing the dark and the illuminated parts of the Moon is a great circle. This assertion poses a problem because, strictly speaking, this is not the case. In Figure 3.3, the center of the Sun is *S*, that of the Moon is *M*, and that of the Earth, *E*. When the light source is larger than the illuminated sphere, it does not illuminate half a sphere, but a larger portion, as can be seen in the figure. The ellipse centered on *V* represents the circle

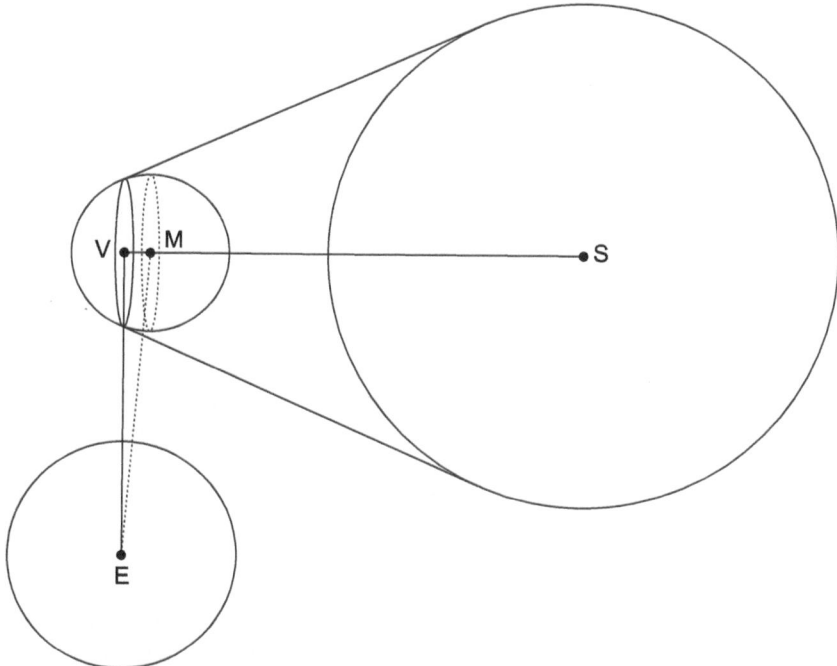

Figure 3.3 The circle dividing the dark and the illuminated parts of the Moon is not a great circle, because the Sun is greater than the Moon.

dividing the dark and the illuminated parts of the Moon, while the ellipse centered at *M* represents the great circle perpendicular to line SM. The figure shows that the circle that divides the dark and the illuminated parts of the Moon is not a great circle.

One might protest, saying that the objection is nothing more than an act of geometric pedantry. But the problem is that Aristarchus himself shows that the dividing circle is not maximum. In proposition 2, he demonstrates that since the Sun is greater than the Moon, it always illuminates more than half a sphere. Therefore, the circle dividing the illuminated and the dark parts of the Moon is never a great circle. The question, then, remains: why does Aristarchus affirm in the hypothesis what he will deny in the proposition? It is possible that the word μέγιστον (great) is an interpolation and was not in Aristarchus's original wording. It is also possible that Aristarchus did not want to go into details in the first formulation of the hypothesis (see note 5 of the translation, p. 47).

But whether it is a great circle or not, the relevant point of the hypothesis is to recognize that, in a dichotomy, the plane of the circle that divides the dark and the illuminated parts of the Moon contains the center of the Earth, in such a way that the line joining the center of that circle with the center of the Sun is perpendicular to the plane. In this way, the desired right angle is obtained.

Fourth hypothesis: that when the Moon appears to us halved, its distance from the Sun is one quadrant minus one-thirtieth part of a quadrant

We know that a quadrant measures 90°. The thirtieth part of a quadrant measures, therefore, 90°/30 = 3°. So a distance less than one quadrant in one-thirtieth of a quadrant is 87°. Aristarchus states that in a dichotomy, the Moon's elongation is 87°.

This hypothesis is the first one that introduces a numerical value. In principle, this value could be obtained by observation: one must wait for the exact moment of a dichotomy and measure the angle between the Moon, the eye, and the Sun. For example, you could place a small tube on one eye and turn it until you see the Sun, the same with another tube and the Moon, and then measure the angle between them. See Figure 3.4.

The problem arises with the practical difficulties when trying to measure due to many factors. In the first place, it is not easy to measure angles accurately when dealing with bodies as large in apparent size as the Sun and the Moon, since it is difficult to determine their centers. Secondly, the Sun is so bright that, in principle, it is difficult to observe it directly. This difficulty is not insurmountable, for one can use its shadow and not look directly at the Sun. But third and most important, it does

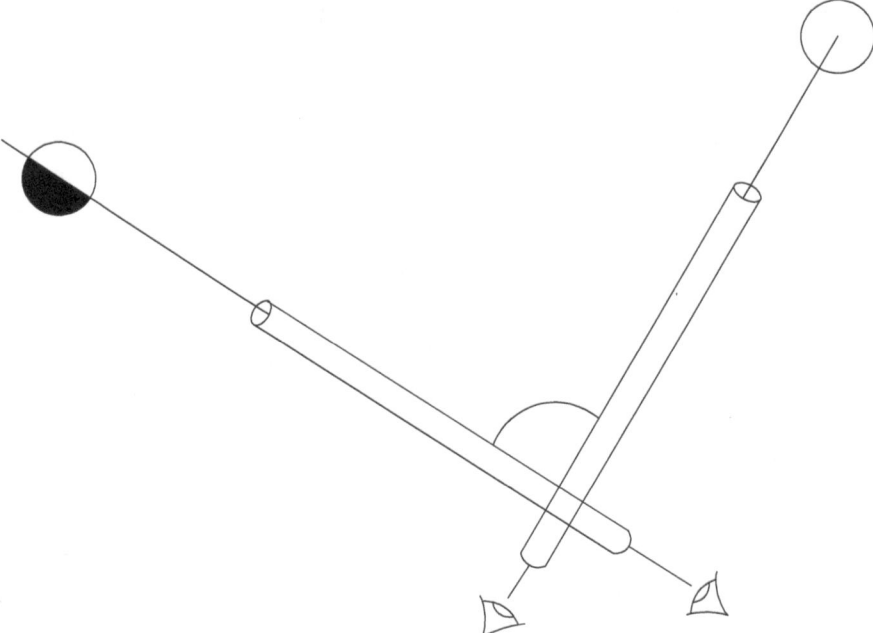

Figure 3.4 How to measure the lunar elongation.

not seem easy to determine the exact moment of a dichotomy. The synodic month – the time between a lunar phase and the return to the same phase – has approximately 30 days. Therefore, the Moon moves away angularly from the Sun at a speed of 360° every 30 days, implying 12° degrees per day or, approximately, half a degree per hour. Suppose one considers that the acceptable margin of error in elongation is 1° (which, anyway, would imply an enormous variation in the ratio between the distances). In that case, one should determine the moment of a dichotomy with an accuracy smaller than two hours, which is practically impossible, because the variation of the illuminated portion of the Moon is imperceptible in such a short time.

The correct value for the lunar elongation in a dichotomy is 89° 51′, separated from the quadrant not by 3° but by only 9 minutes. The ancients could not measure elongations with such precision. It would be impossible for them to distinguish that angle from a right angle. The difference from 87° to 89° 51′ is responsible for the error in Aristarchus's ratio between the distances. Aristarchus's ratio is much smaller than the currently accepted value. Indeed, the Sun is not 19 times but about 390 times farther than the Moon.[4]

But maybe Aristarchus did not measure that value but inferred it from a plausible reasoning. James Evans (1998: 72) proposed this possible explanation, reconstructing Aristarchus's possible reasoning as explained next.

We know that the time between two full moons is approximately 30 days. In Figure 3.5, the Sun is at *S*, the Earth is at *E*, and the Moon revolves around the Earth in the orbit ABCD. Point A represents the position of the Moon in a new moon; point C, the position of the Moon at full moon; B and D, the positions of the Moon in both dichotomies. Now, the Sun is not at an infinite distance, and the angle on the Moon between the Sun and Earth is one quadrant (hypothesis 3). Consequently, arcs BA and AD are smaller than arcs BC and CD. The angular distance traveled from the first to the second dichotomy (arc BC) is longer than the angular distance traveled from the second to the first (arc CD). Let us assume a constant lunar speed. Therefore, if the arc is smaller, so is the time. Less time will elapse from the new moon to the first quarter than from the first quarter to the full moon. We could assume that it takes 14 of the 30 days to complete arc BAD, while it takes the remaining 16 to complete arc DCB. But if this were the case, we should notice it, because we would perceive that there is at least one night of difference. But we do not perceive that difference. Therefore, the difference is smaller. Suppose, then, that the Moon completes arc BAD in 14 and a half days and DCB in the remaining 15 and a half days. Thus, it completes arc BA in 7 days and a

---

4   Note that, if Aristarchus had taken lunar parallax into account in this measurement and if, as can be deduced from its values, the Moon was at approximately 20 Earth radii, the difference between the angle measured on the surface of the Earth and the angle that would correspond from the center of the Earth would be $(2 \tan^{-1}(1/20) =) 2° 52′$: practically the difference between the real value and the value proposed by Aristarchus, although in the opposite direction. If what are measured from the Earth's surface were 87°, the corresponding angle from the center would be 84° 8′.

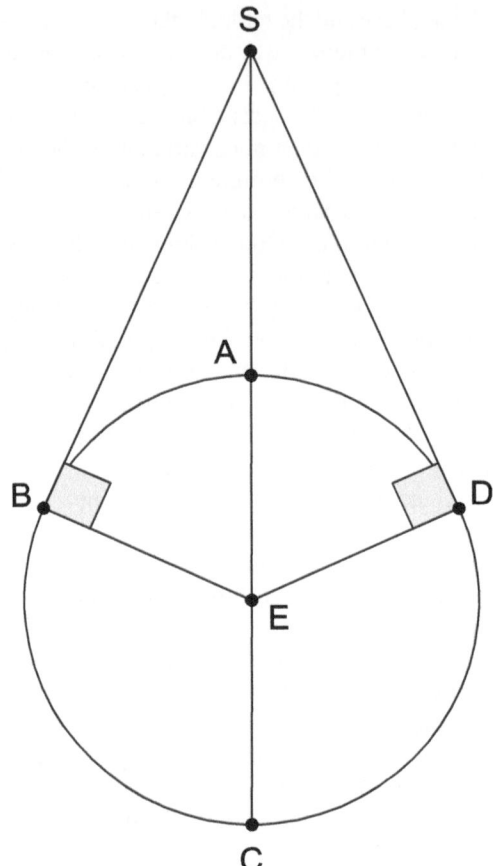

Figure 3.5 Calculation of the lunar elongation at dichotomy.

quarter. How long is arc BA? We know that the Moon completes one quadrant in 7 and a half days (because it completes four in 30 days). Therefore, BA is less than a quadrant for a quarter of a day. Now, in 30 days, the Moon covers four quadrants, so in one day, it covers 4/30 of a quadrant. In a quarter of a day, it completes, therefore, 1/30 of a quadrant. Consequently, when the Moon is in a dichotomy, its distance from the Sun is one quadrant minus one-thirtieth part of a quadrant.

Fifth hypothesis: that the width of the shadow (of the Earth) is two Moons

This hypothesis is the second one including numerical values. In it, Aristarchus argues that the size of the shadow of the Earth projected on the surface of the Moon in a lunar eclipse is equal to two Moons. The value is reasonably accurate, and in principle, it could be calculated from naked-eye observation, doing something similar to what Figure 3.6 shows. In a lunar eclipse photo, we have drawn the circle of the Moon in dotted lines. Then we draw a circle twice in size, trying to fit it with the arc delimiting the shadow of the Earth on the Moon. The figure shows that it works very well. Of course, the ancients could not take a picture of an eclipse. Still, they could do something similar with a drawing representing what they saw in a lunar eclipse as accurately as possible.

In Chapter 1, we have already mentioned much more sophisticated methods developed by Ptolemy (see p. 11), but it is not reasonable to expect that Aristarchus would employ anything of the kind. On the one hand, Ptolemy's method needs wholly developed solar and lunar models, not still available to Aristarchus. On the other, astronomers at the time of Aristarchus were still not too worried about obtaining such accurate values, as we will discuss later. However, there are other more straightforward methods to calculate the size of the Earth's shadow. Cleomedes (II, 1; Bowen and Tood 2004: 115), for example, argues that it can be shown that the size of the Earth's shadow is equal to two Moons from the following reasoning: the time it takes for the Moon since the eclipse begins until it is totally eclipsed is equal to the time that it remains totally eclipsed and also equal to the time that elapses from the beginning of the reappearance of the Moon until the

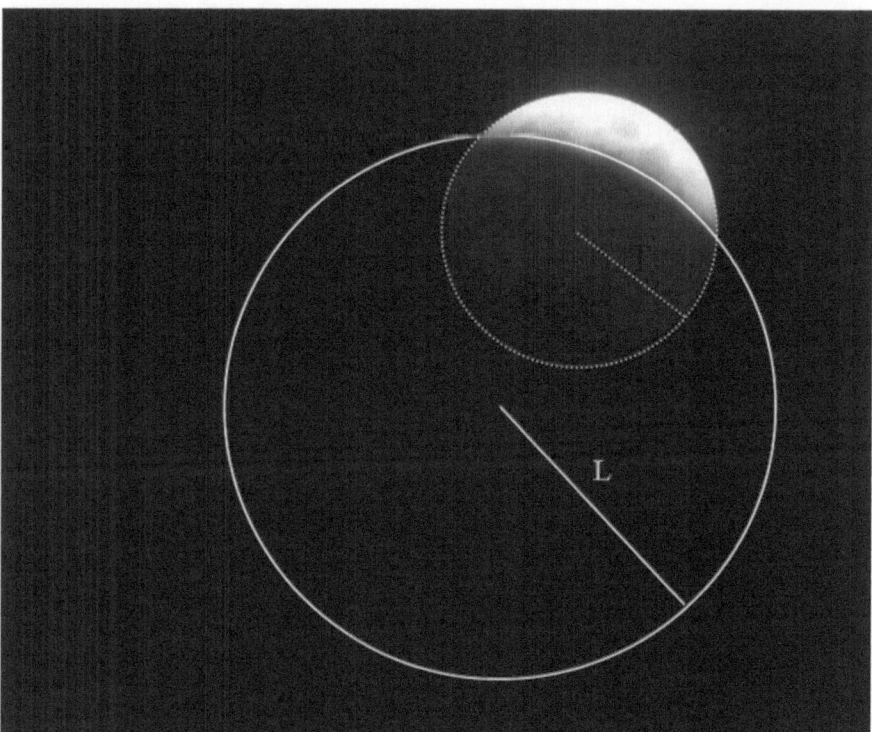

Figure 3.6 The apparent size of the shadow of the Earth is equal to two Moons.

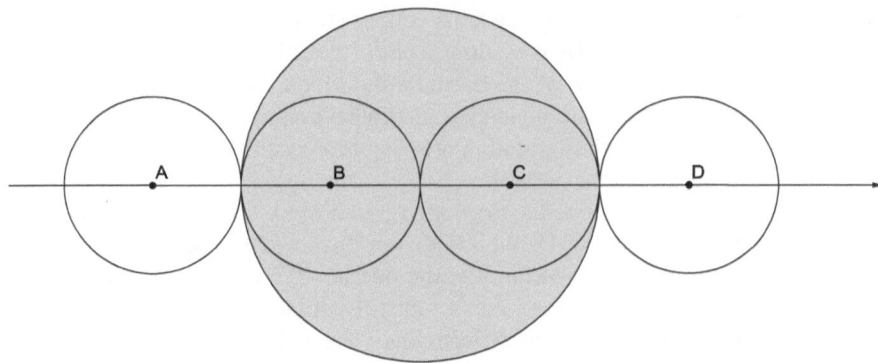

Figure 3.7 How to calculate the size of the apparent size of the shadow from the duration
of the different faces of a lunar eclipse.

eclipse ends. Thus, we have three equal intervals that can only be explained if the
diameter of the shadow is equal to two lunar diameters (cf. Neugebauer 1975:
654). In Figure 3.7, the bigger, shaded circle represents the shadow of the Earth,
which is equal to two Moons, represented by the smaller circles in four different
positions. The Moon begins to be eclipsed when it is in position A and begins to
be totally eclipsed when it is in position B. It will remain totally eclipsed until it
reaches position C, in which it will begin to be visible, but partially eclipsed, until
it comes to position D, in which the eclipse will end.

Therefore, it can be seen that the time between A and B is equal to the time
between B and C, and also between C and D, if and only if the Earth's shadow is
equal to precisely two Moons. So Cleomedes was right: from the analysis of this
phenomenon, one can conclude that the shadow of the Earth is equal to two Moons.
It is plausible that Aristarchus reasoned similarly.

Sixth hypothesis: that the Moon subtends one-fifteenth part of a sign of the zodiac

Aristarchus asserts that the Moon subtends one-fifteenth of a sign of the zodiac,
that is, that its apparent size is one-fifteenth of a zodiac sign. Twelve signs cover
the entire belt of the zodiac, so each one measures 30°; thus, the apparent size of
the Moon is 2°.

Hypothesis 6 is the most problematic. First of all, the apparent size of the Moon
is greatly exaggerated. One single naked-eye observation of the Moon would show
that it has an apparent size four times smaller: around half a degree. But the problem is not only that Aristarchus, without justification, uses such an inaccurate value
but also that, according to Archimedes's testimony, Aristarchus discovered
(εὑρηκότος) a much more accurate value. Indeed, in a passage from *Sand-Reckoner*
(Heath 1897: 223), Archimedes states that Aristarchus found that the apparent size

of the Sun is 720 parts of the circle of the zodiac, that is, half a degree. But because the apparent sizes of the Sun and the Moon are equal – something that Aristarchus shows in proposition 8 – also the apparent size of the Moon should be half a degree.

The most tempting explanation is to claim that it is merely an error in the transmission of the text. Manitius (Proclus 1909: 292, n.) suggests this hypothesis. He proposes that *one-fifteenth part* (πεντεκαιδέκατον μέρος) would be an alteration of *one-fiftieth part* (πεντηκοστόν μέρος) corresponding to 0.6°. But there is no way to make all the values found in the intermediary steps of calculations consistent with a value different to 2°.

Another explanation is to assert that Aristarchus wrote *On Sizes* before discovering that the value for the apparent size of the Moon was much smaller. This hypothesis is certainly reasonable, although it does not explain how Aristarchus may have been so wrong in his first observation. But perhaps it was not an observation, but a calculation, as in the case of hypothesis 4. A reasonable way to calculate it would be to measure the time that takes the Sun to rise, that is, from the first visibility of the Sun above the horizon to the moment in which the lower tip of the Sun leaves the horizon. The Sun makes a complete revolution (360°) in 24 hours. If we know how long it takes to rise and we divide this value by 24 hours and multiply it by 360°, we will obtain the apparent size of the Sun. We know that Macrobius (cited in Heath 1913: 311) describes an experiment of this type, which gives a similar result: 1° 40', since the estimated time is one-ninth of an hour. Aristarchus's result would imply that it takes 8 minutes for the Sun to rise.

Another hypothesis, suggested by Tannery (1893: 241, 1912: 375–376), is to assert that Aristarchus intentionally used an exaggerated value to highlight that he was not trying to find exact values but only to show the power of geometry. We cannot develop this hypothesis in detail here. It is enough to anticipate that the influence of the apparent size of the Moon is limited to two values: the maximum limit of the ratio between the sizes of the Sun and the Earth and the minimum value for the ratio between the sizes of the Earth and the Moon. Nevertheless, its influence is negligible. Therefore, by introducing such a high value and showing that the results do not vary, Aristarchus would want to note that this value could be chosen almost arbitrarily. Indeed, if the lunar diameter were actually 0, the maximum limit of the ratio between the diameters of the Sun and Earth would go from 7.16 to 7, and the lower limit of the ratio between the diameters of the Earth and the Moon would go from 2.51 to 2.57. If the value were half a degree, they would be 7.04 and 2.55, respectively. If instead the lunar diameter assumed the ridiculous value of 15° (a half sign of the zodiac), the values would be 8.51 and 2.11, respectively. But maybe Tannery's proposal is "too ingenious," as Heath (1913: 311) suggests. If Aristarchus wanted to show that the value could be exaggerated without consequences, why stop at 2° and not propose that it measured, for example, 15° or 30°? After all, 2° is not too far from the proposed values. Remember that Macrobius proposes 1° 40'.

But there is another possible explanation (Carman 2014). In the analysis of hypothesis 3, we mentioned that Aristarchus argues that the circle dividing the dark and the illuminated parts of the Moon is not a great circle. Therefore, there is an angular difference between the dividing circle and the great circle parallel to it, represented by the angle VEM of Figure 3.3. In proposition 4, Aristarchus shows that, although they are different, the angular difference between them is so small that it is imperceptible for an Earth-located observer. He calculates angle VEM and, obtaining that it is very small (1′ 36″), affirms that it is imperceptible. The value of that angle depends directly on the apparent size of the Moon, since it is a portion of that angle. Therefore, if the apparent size is greater, the angle between the two circles will be too. One can increase the size of the apparent size, but at some point, if the apparent size of the Moon is too large, the difference between the two circles will no longer be imperceptible. Following the style of Evans's proposal for the value of 87°, Aristarchus could have chosen the highest possible value for the apparent size of the Moon, which, however, still guarantees that the difference between the two circles is imperceptible. This value happens to be 2°. Of course, this is a highly speculative explanation, but it seems consistent with the spirit of the method presumably used in the other values by Aristarchus.

Observation and theory in obtaining the values of the hypotheses

As we have already mentioned, Aristarchus's treatise is particularly valuable because it is the first extant complete work applying geometry to the study of the sky to obtain determined values. For the first time in the history of science, starting from empirical values, geometry is used to obtain new values that cannot be directly measured, such as the sizes and distances of the Sun and the Moon. There-fore, it constitutes a unique piece in the birth of science. Under this perspective, it is convenient to reflect on the origin of the empirical values used as inputs in the treatise calculations. Let us remember that these values are stated in the last three hypotheses: the lunar elongation at the time of a dichotomy (87°), the ratio between the apparent size of the Moon's and the Earth's shadow (2/1), and the apparent size of the Moon (2°). Aristarchus postulates these values as hypotheses, without the slightest trace of how he may have obtained them. Considering the significant errors that some of them show, it is reasonable to assume that they do not come directly from observation. The methods that we proposed for obtaining each value, even if speculative, show an astonishing complexity.

We should probably leave aside from the analysis the value obtained for the shadow of the Earth, which, being approximately correct, could come from obser-vation. However, we have shown that the other two neither come from more or less crude observations nor have been chosen arbitrarily. They seem to arise from a complex interplay of theory and observation.

Let us start with the lunar elongation value. Presumably, the goal of Aris-tarchus's calculation is to show that the ratio between the distances (and, conse-quently, also the ratio between the sizes) is much greater than his contemporaries and predecessors commonly believed. Remember that the ratios listed by Archi-medes of authors before Aristarchus are always smaller than his value. Even Epi-curus affirmed that the Sun is not bigger than it appears, and Anaxagoras that it was greater than the Peloponnese. Aristarchus's value, therefore, would be shocking for the time in concluding a Sun much farther away and larger than expected. Accordingly, Aristarchus would have decided to test his hypothesis in the worst possible scenario. The smaller the lunar elongation in a dichotomy, the smaller the ratio between the distances. Aristarchus would have liked to show that the Sun is still much farther away than the Moon, even with the smallest possible elongation. Consequently, he would have used an elongation as small as possible. But the elongation cannot be smaller than 87°, since in that case, it would contradict a qualitative observation that is very accessible. We do not perceive any difference between the time from the first quarter to the last quarter and the time from the last quarter to the first quarter. But an elongation smaller than 87° would imply a perceivable difference between them (at least one night). Thus, we have two constraints. First, a theoretical consideration defines the wanted limit of the value: it must be the lowest possible elongation because we want to prove that the Sun is far away. Second, a simple observation helps to find the limit of the value: it cannot be smaller than 87°, because there is no perceivable difference in the time between the quarters of the Moon.

It is possible to find the same configuration in the explanation for the value of the apparent size of the Moon (2°). As we have already mentioned, even if negligible, the apparent lunar size plays a role in determining the upper limit of the ratio between the Sun–Earth sizes and the lower limit of the ratio between the Earth–Moon sizes. The larger the apparent lunar size, the greater the influence on these ratios. Presumably, Aristarchus would have wanted to show that his influence is negligible. In doing this, again, he would have assumed the worst possible scenario. He would therefore seek the greatest possible apparent size. And again, the limit is found using an observation. If the Moon had an apparent size greater than 2°, then the difference between the line dividing the dark and the illuminated parts of the Moon and a diameter of the Moon would be perceptible. But that difference is not perceived. Therefore, it cannot be bigger than 2°. Here again, we find two constraints. First, a theoretical consideration defines the wanted limit of the value: we want the largest possible apparent size to show that its influence on the calculation is not relevant. Second, a simple observation helps to find the limit of the value: it cannot be greater than 2° because, in that case, the unperceived difference between the two lines would be perceivable.

In synthesis, at least two of the three values – Carman 2014 gives a similar explanation for the third as well – would have arisen from a complex and not at all naive interplay between theory and observation. Let us now analyze, step by step, the path that leads from these departing values to the desired results.

Toward the ratio between the distances (propositions 1 to 7)

Conceptual introduction

As we have already pointed out, this first part of *On Sizes* aims to establish the relationship between the Earth–Sun and Earth–Moon distances. To do this,

Aristarchus builds a right triangle in which the Earth, the Moon, and the Sun are the vertices, the Moon being at the right angle (see Figure 3.2). As we have also mentioned, this diagram assumes that the line dividing the dark and the illuminated parts of the Moon passes through the Moon's center. However, strictly speaking, this assumption is not correct, because the illuminated part of the Moon is greater than half a sphere (see Figure 3.3). But the difference is so small that it can be ignored, as Aristarchus himself will demonstrate.

In *proposition 2*, Aristarchus demonstrates that, if a sphere illuminates a smaller sphere, the illuminated part is greater than half a sphere. To do this, he encloses both spheres in a cone and shows that the circle formed by the contact of the cone with the surface of the spheres is smaller than a great circle. Therefore, he dedicates *proposition 1* to showing that "[o]ne and the same cylinder envelopes two equal spheres, and one and the same cone [envelopes two] unequal [spheres, the cone] having its vertex in the same direction as the smaller sphere."

Aristarchus is now almost ready to argue that the difference between the great circle and the one dividing the dark and the illuminated parts is imperceptible. He will demonstrate that in *proposition 4*. But he still has to take a previous step. The size of the illuminated portion of the sphere depends not only on the size of the illuminating sphere but also on the distance between them. The closer both spheres are, the greater the illuminated portion will be, and therefore, the greater the difference between the dividing circle and a great circle, as can be seen in Figure 3.8.

The distance from the Moon to the Sun does not remain constant. Therefore, it will be necessary to establish that the difference is imperceptible in a dichotomy, where the Moon is at a right angle. In a dichotomy, the Moon–Sun distance is a little less than average (see the lunar position at points B or D in Figure 3.5 on p. 192).

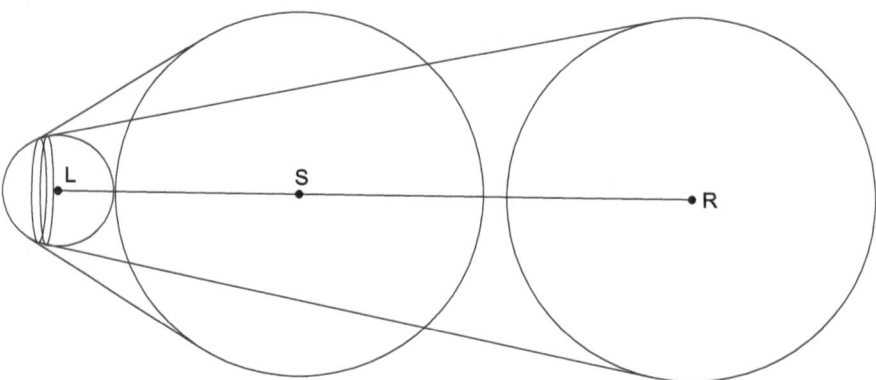

Figure 3.8 The circle that divides the illuminated and the dark parts depends on the size of the light source, but also on the distance between them. The sphere with center in L is illuminated by the greater sphere that is at two possible distances, in S and in R. In S it is closer, and thus the circle that divides the dark part from the light one is greater than the one formed when it is in R.

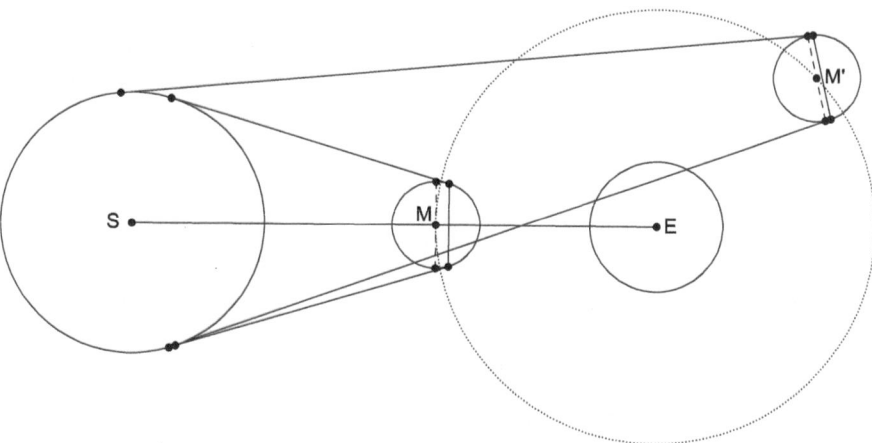

Figure 3.9 The center of the Sun is *S*, that of the Earth *E*, and that of the Moon *M* when it is in conjunction, and *M'* in any other position orbiting the Earth. The circle dividing the dark and the illuminated parts of the Moon is smaller in conjunction than in any other position (it is noticeable that its distance from the center of the sphere is greater, and therefore, the circle is smaller).

But Aristarchus will not demonstrate that the difference between the dividing circle and a great circle is negligible in a dichotomy, but in the worst possible situation, that is, when the Moon is as close as possible to the Sun. This happens in conjunction when the Sun, the Moon, and the Earth are in a straight line, in that order. Of course, if it works in the worst situation, it also works at a dichotomy. Thus, in *proposition 3*, Aristarchus will establish that conjunction is the worst possible scenario, that is, that in conjunction, the circle dividing the dark and the illuminated parts of the Moon is the minimum possible. As shown in Figure 3.9, in any other lunar position, the Moon–Sun distance will be greater, and therefore, the difference between the dividing circle and a great circle will be smaller.

Thus, in *proposition 1*, Aristarchus points out that "[o]ne and the same cylinder envelopes two equal spheres, and one and the same cone [envelopes two] unequal [spheres, the cone] having its vertex in the same direction as the smaller sphere." In *proposition 2*, he shows that "[i]f a sphere is illuminated by a sphere greater than it, more than a hemisphere [half a sphere] will be illuminated." In *proposition 3*, he demonstrates that the greatest difference between the illuminated and the dark parts occurs "when the cone enveloping the Sun and the Moon has its vertex towards our eye," that is, in conjunction. Finally, in *proposition 4*, Aristarchus shows that, even in this extreme situation, "[t]he circle delimiting the shaded and the bright parts on the Moon is perceptibly indistinguishable from the great circle on the Moon."

Once Aristarchus has justified that it is legitimate to assume that the light of the Sun divides into halves the Moon, he begins to construct the right triangle.

Hypothesis 3 allows him to affirm that, at the exact moment of a dichotomy, the circle dividing the dark and the illuminated parts of the Moon and our eye are in the same plane. Now that Aristarchus showed that this circle is indistinguishable from a great circle (proposition 4), he asserts in *proposition 5* that, when the Moon appears to us halved, our eye is in the plane containing the great circle that is parallel to the dividing circle. Consequently, the line joining our eye and the center of the Moon is perpendicular to the line joining the centers of the Moon and the Sun, forming a right angle.

Everything is ready for starting the long chain of calculations once Aristarchus introduces the value of the angle centered on the Earth (the lunar elongation in a dichotomy). This long calculation aims to obtain the maximum and minimum value of the ratio between the distances (*proposition 7*). However, previously in *proposition 6*, he does something that would seem unnecessary: he shows that the angle centered on the Earth – which hypothesis 4 already teaches us is 87° – is indeed less than a quadrant and that the Moon moves below the Sun, that is, that the Moon is closer to the Earth than the Sun. Both assertions are patent once you know that the angle centered on the Moon is right and the lunar elongation is 87°. However, it would seem that Aristarchus delays the introduction of the numerical value as long as possible and wants to show that it has to be smaller than one quadrant even before introducing the value. This delay in introducing the value is another interesting example of the interplay of theory and observation. It seems that Aristarchus's procedure was ruled by a methodological law asking to move as much as possible in a purely geometric level, avoiding the empirical contamination implied in the introduction of numerical values. The geometer/astronomer will introduce the value only when it is strictly necessary to obtain another value. But a qualitative property is never deduced from a numerical value if it can be inferred from already-accepted geometric properties. The numerical values seem to be treated as foreign bodies by geometry.

Now that we already analyzed the role that these first seven propositions play in the reasoning leading to the limits of the ratio of the lunar and solar distances, let us look at each of them separately. The first two propositions are purely geometrical, and we already described them sufficiently. Accordingly, we will start from proposition 3.

Proposition 3

In this proposition, Aristarchus wants to show that the circle that divides the dark and the illuminated parts of the Moon is the minimum possible when the Sun is as close as possible to the Moon. In Aristarchus's words, this happens "when the cone enveloping the Sun and the Moon has its vertex towards our eye," that is, in conjunction.

Proposition 3 is quite intuitive, as shown in Figure 3.10. The center of the Earth is point A, the center of the Sun is at point B, and the center of the Moon is in two different positions: in Γ when the Moon is on a straight line with the Sun and the Earth and between them, and in Δ when it is not. The lines perpendicular to the

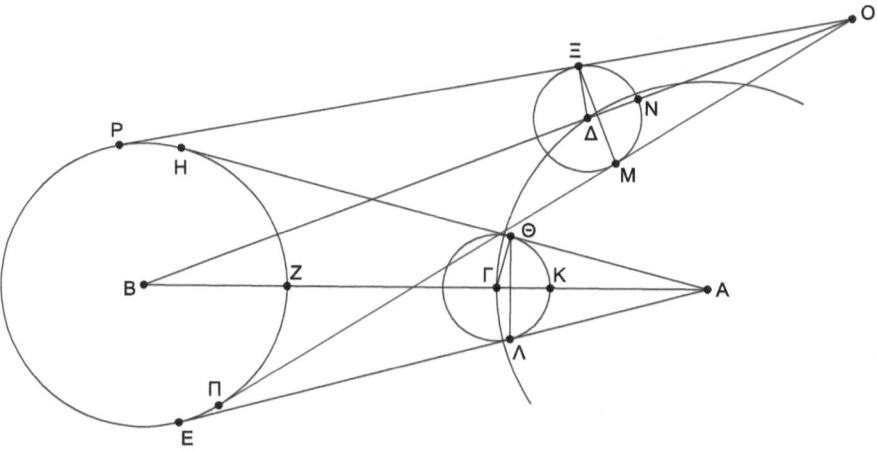

Figure 3.10 Diagram of proposition 3. The observer's eye is at *A*, the center of the Sun at
B, and the center of the Moon, when it is in conjunction, at *Γ* and, when it is
in not, at *Δ*.

axis of each cone join the points of the sphere of the Moon tangent to each of them.
Consequently, these are the lines dividing the dark and the illuminated parts of the
Moon in both configurations. The figure shows that line ΞM, corresponding to the
non-conjunction position, is greater than line ΘΛ, corresponding to conjunction,
since line ΞM is closer to the center of its circle (Δ) than line ΘΛ to its own (Γ).
Figure 3.11 shows that if a cone contains two equal spheres, the greater the distance between the spheres, the smaller the difference between the circle tangent to
the cone and the great circle.

Let us see how Aristarchus proceeds to demonstrate the proposition. Figure 3.11
shows that the cone, with a basis formed by circle GF, is higher than the cone with
a base formed by circle DE. In the plane of the figure, it means that OH is greater
than IJ. Because the tangent points of the sphere define the cone, the greater the
base, the higher the cone. And the other way around, the higher the cone, the
greater the base. Return now to the diagram of proposition 3, reproduced in Figure 3.10. Compare the cones with bases formed by the circles dividing the dark
and the illuminated parts of the Moon, that is, cones ΞMO and ΘΛA. Aristarchus
shows that the height of the cone with vertex at O is greater than the height of the
cone with vertex at A. By doing so, he demonstrates that the base of the first is
greater than that of the second. Now, since the bases represent the lines dividing
the dark and the illuminated parts of the Moon, Aristarchus showed that the dividing line is the smallest in the cone with vertex at A, representing the conjunction.
Therefore, its difference with a great circle is the maximum.

Let us remember that Aristarchus wanted to demonstrate this because he knew
that when we are at the exact moment of a dichotomy, the circle aligned with our
eye is not a great circle but the circle dividing the dark and the illuminated parts

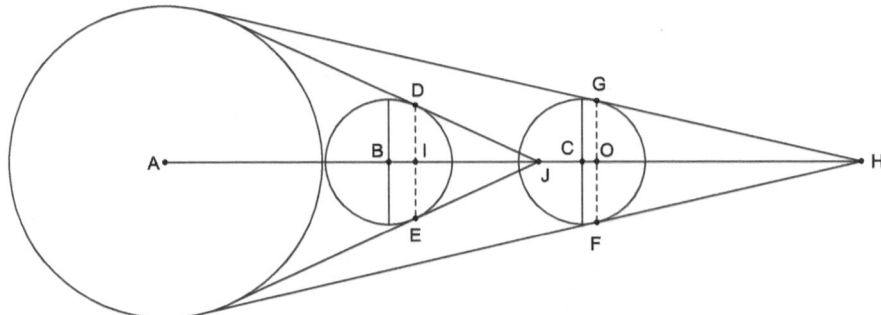

Figure 3.11 If the distance between two spheres contained by cones increases, the difference between the circle tangent to the cones and a great circle diminishes. The bold line perpendicular to axis AD represents the great circle of spheres centered at points B and C. The dashed line, the circle tangent to the cones.

of the Moon. He wants to show that, nevertheless, these two circles are indistinguishable from our eyes, even in the worst situation. A conjunction is the worst scenario because in it the difference between the two circles is maximum. In the following proposition, he will demonstrate that even in conjunction, they are indistinguishable.

Proposition 4

In proposition 4, Aristarchus wants to show that the circle dividing the dark and the illuminated parts of the Moon is indistinguishable from a great circle for an observer on the Earth. We know that the dividing circle varies in size. Accordingly, Aristarchus demonstrates that in the worst case (when the circle is the smallest and, therefore, it moves as far as possible from the great circle), it is indistinguishable. If so, it will be much more indistinguishable in the others.

Aristarchus's strategy to demonstrate that the great circle and the dividing circle are perceptually indistinguishable from the Earth consists of showing that the angle formed between the extremes of the two circles and the eye of an observer on Earth is so small that it is not perceived by the human eye. This angle represents the apparent size of the difference between the circles as seen from the Earth. Consequently, it depends on two variables: the real distance between the circles and the position of the observer.

As we already discussed, the real distance between the two circles depends on how close the Moon is to the Sun. In the previous proposition, Aristarchus has already shown that the worst situation occurs in conjunction when the Moon is as close as possible to the Sun.

But the apparent size of the difference also depends on the position of the observer with respect to these two circles. The distance from the observer is always the same because the Moon revolves around the Earth. But if the observer is located perpendicular to the planes of the circles, the angle formed by the circles' extremities will be the smallest (position A in Figure 3.12). If, on the other hand, it is located on the line that, starting from the center of the Moon, passes through the middle of the extremities of the two circles, it will be the greatest (position A' in Figure 3.12).

In Figure 3.12, the Moon is centered at B, and the observer is at A when the Moon is in conjunction, and at A' near a dichotomy. Line ΔΓ represents the circle dividing the dark and the illuminated parts of the Moon. Line EZ represents the great circle parallel to the dividing circle. Therefore, angles EAΓ and EA'Γ represent the apparent size of the difference between these two circles, seen from the two positions. The figure shows that the angle is much greater seen from A'. Thus, the worst situation occurs in a dichotomy.

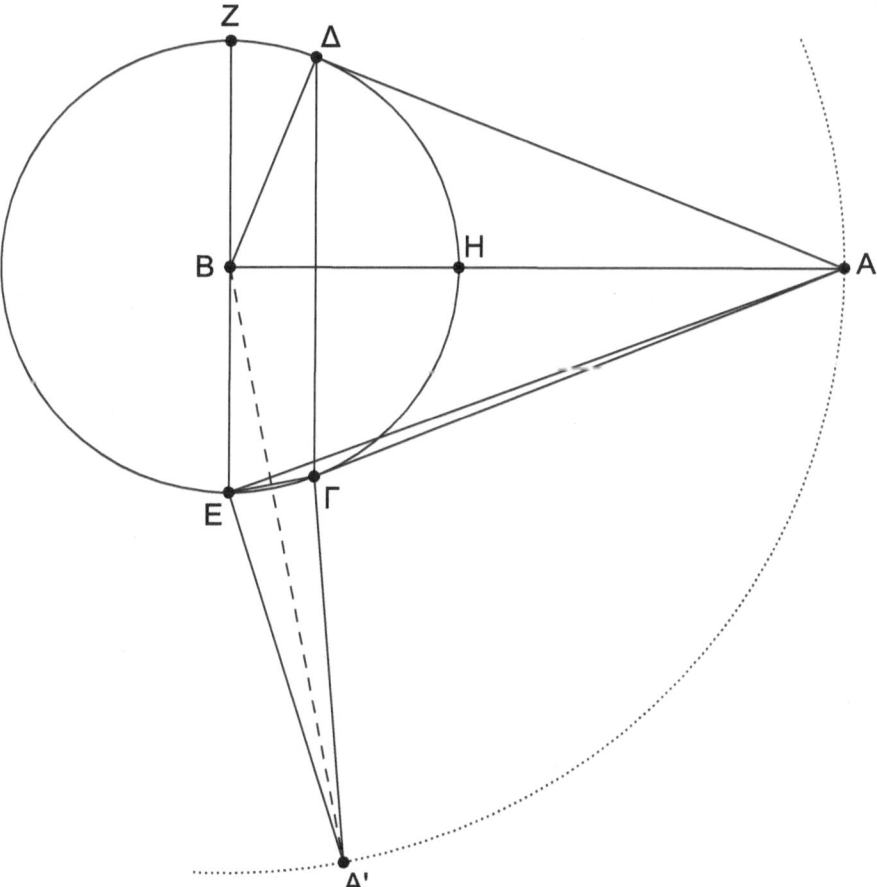

Figure 3.12 Simplification of the diagram corresponding to proposition 4. When the observer is located at A', the angular distance between the two circles, the great circle represented by line ZE, and the circle dividing the dark and the illuminated parts of the Moon (represented by ΔΓ) is maximum.

Since the maximum real difference between the two circles occurs in conjunction and the apparent size of the difference is maximum when they are seen close to quadrature, assuming an observer on the Earth, both variables cannot be in their worst situation simultaneously. In the demonstration, however, Aristarchus combines the worst situation of both variables, generating an impossible situation: the difference between the circles will be the greatest possible (the one that corresponds to conjunction), and the observer will be in the worst possible position (close to quadrature). To do this, Aristarchus asks to move arc EΓ to be in front of A, that is, arc ΘK, as Figure 3.13 shows. His aim is now to show that angle KAΘ (equal to EA′Γ) is imperceptible from A.

What remains is only a geometric procedure that allows Aristarchus to compare angle KAΘ – the one he wants to find out – with angle ΔAΓ – a known angle,

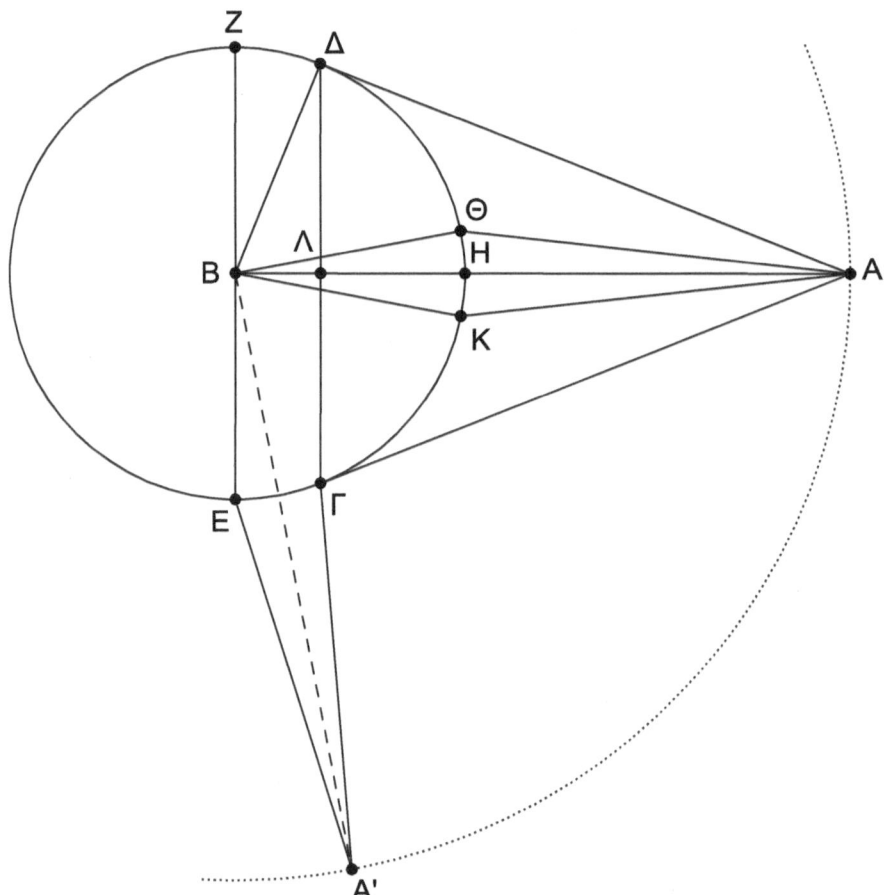

Figure 3.13 Diagram of figure corresponding to proposition 4. The Moon is centered at B, and the observer is at A. Aristarchus asks that arc EΓ, the one that would be seen from A′, be moved to be facing A. Thus, EΓ is equal to ΘK.

because it is the apparent size of the Moon, which, according to hypothesis 6, measures 2°. We will describe the geometrical steps at the end of this section. Applying geometry to that single datum, Aristarchus finds that angle KAΘ is, in his own words, smaller than 1/3960 parts of a quadrant. That is, it is smaller than 1 minute and almost 22 seconds of a degree. According to him, this angle is imperceptible to the human eye.

Then he takes one final step showing that if angle KAΘ is imperceptible to the human eye, angle EAΓ will be much more so (he refers to angle ZAΔ, but it is the same). That is to say much more imperceptible will be the angle seen from the Earth in conjunction. This last step seems to make no sense because this angle is smaller than the one he is interested in, which corresponds to an observer close to quadrature. Furthermore, strictly speaking, the difference between the two circles is imperceptible, not only because the angle is small, but also because, seen from A, the dividing circle hides the great circle. One can conjecture that those last paragraphs are not original, but that they belonged to some – not particularly smart – scholion that later, by mistake, became part of the main text. But we have not found textual evidence in any manuscript. If it is as we suppose, it must be a very early error.

Aristarchus has demonstrated that, seen from Earth, the difference between a great circle and the circle dividing the dark and the illuminated parts of the Moon is imperceptible to the human eye. There is no justification in the treatise for the assertion that a human eye cannot perceive an angle of 1′22″ (for a possible explanation, see Carman 2014). But it is worth noting that Aristarchus is right. The limit for the angular resolution perceived by the human eye is usually set between 4.38 minutes (Maar 1982) and 3.6 minutes (Wegman 1995: 290). Even the smallest value is greater than twice the value found by Aristarchus.

Let us go a bit deeper into the geometrical calculation to show how Aristarchus can find angle KAΘ. Refer again to Figure 3.13. By construction, KBΘ is equal to ZBΔ. Because lines ZE and ΔΓ are parallel, ZBΔ is equal to BΔΛ. Now, triangles BΔΛ and BAΔ are similar (because they are right triangles sharing one not-right angle: B). Consequently, BΔΛ is equal to BAΔ. Aristarchus knows angle BAΔ because it is half the apparent size of the Moon, that is, 1°. Therefore, KBΘ is equal to 1°.

KBΘ represents the difference between the circles seen from B, the center of the Moon, but Aristarchus wants to know that difference seen from A, the observer, that is, angle KAΘ. Of course, the ratio between KBΘ and KAΘ is equal to the ratio between their halves, HBΘ and HAΘ. Aristarchus assumes that BΘ/AΘ is greater HAΘ/HBΘ. Commandino provides an elegant proof of this inequality (see note E at p. 83 of the translation). Because Aristarchus already knows angle HBΘ, he needs to obtain a ratio equal to or greater than BΘ/AΘ. Let us see how he finds that ratio.

In right triangle BAΔ, Aristarchus assumes that BΔ/ΔA is smaller than the ratio between angle BAΔ and half a right angle. This inequality is again elegantly demonstrated by Commandino in note A of p. 79 of the translation. Consequently, BΔ/ΔA is smaller than 1/45. But BH is equal to BΔ and BA is greater than ΔA.

Therefore, also, BH/BA is smaller than 1/45. But BA is equal to BH plus HA, and BH is 1 when BA is 45; therefore, by separation (see appendix 1 at p. 295), we know that BH/HA is smaller than 1/44. Besides, Aristarchus knows that BΘ is equal to BH and that AΘ is greater than AH. Consequently, BΘ/AΘ is smaller than BH/AH. Consequently, also, BΘ/AΘ is smaller than 1/44. He has obtained the wanted ratio equal to or greater than BΘ/AΘ.

Because BΘ/AΘ is greater than HAΘ/HBΘ, also, 1/44 will be greater than HAΘ/HBΘ. But we already know that HAΘ/HBΘ = KAΘ/KBΘ. Consequently, 1/44 is also greater than KAΘ/KBΘ, meaning, that KAΘ is smaller than 44 times KBΘ. But KBΘ is 1°. Therefore, KAΘ is smaller than 1°/44 or 1′ 22″.

Proposition 5

Proposition 5 is very short. In hypothesis 3, Aristarchus had stated that when the Moon appears to us halved, the circle dividing the dark and the illuminated parts of the Moon is in the same plane of our eye. Thanks to proposition 4, he can now assume that both circles are in the same plane, since they are indistinguishable from our eye. Then, Aristarchus can affirm that, at the exact moment of a dichotomy, the greater circle, parallel to the dividing one, is in the same plane that our eye is in. In doing that, he is already in a position to build the so-desired right triangle.

Proposition 6

Proposition 6 demonstrates that (a) the Moon moves below the Sun and that (b) when the Moon is halved – in a dichotomy – it is less than one quadrant away from the Sun. As we have already mentioned, this is evident once a lunar elongation of 87° is assumed (hypothesis 4). But as we have seen, Aristarchus delays the introduction of values as long as possible. On this occasion, Aristarchus makes two assertions using *reductio ad absurdum*. On the one hand, that the Moon cannot be more than one quadrant (or even one quadrant) apart from the Sun. On the other, that it is also impossible for it to be farther away than the Sun (or even at the same distance). The reason is obvious: if the Sun, the Moon, and the Earth form a right triangle with the right angle on the Moon, neither of the other two angles can be greater than or equal to a right. Besides, the hypotenuse is the Earth–Sun distance, so the Sun will necessarily be farther away than the Moon (see again Figure 3.2 at p. 188).

Figure 3.14 is a more friendly and complete version of the diagram of proposition 6 but is essentially the same. The Earth is centered at *A*, and the Sun at *S*. The Moon assumes five different possible positions, numbered from 1 to 5. In each of them, we added two sections to the sphere of the Moon: the dotted-line section is always perpendicular to the line joining the Moon with the Sun and represents, therefore, the line that divides the dark and the illuminated parts of the Moon. The solid-line section represents the circle dividing the dark and the illuminated parts in a dichotomy, and therefore, it is in the same plane as *A*, the

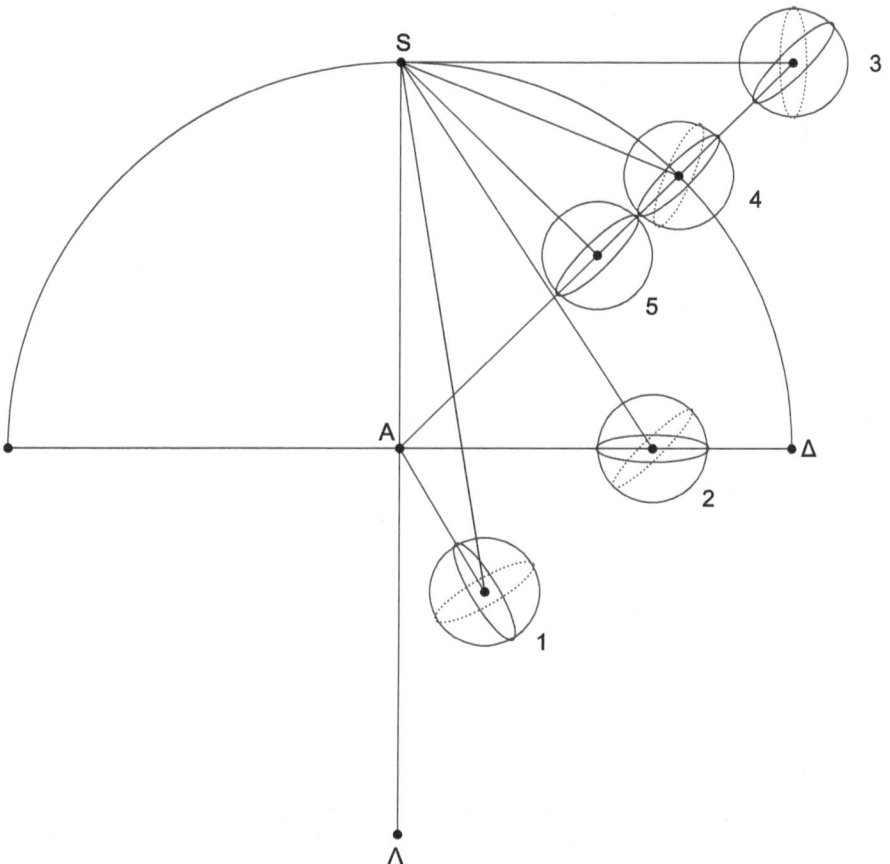

Figure 3.14 Diagram of proposition 6 partially modified. The Earth is at *A*, and the Sun at
S; the Moon takes five possible positions numbered from 1 to 5. Aristarchus
will prove that only position 5 is possible.

center of the Earth. We are wondering what the possible position of the Moon in
dichotomy is, where both sections should coincide. The figure shows that this is
the case in position 5, where the Moon is less than one quadrant from the Sun
and below it.

Therefore, the *reductio ad absurdum* has four steps, corresponding to positions
1 to 4 in Figure 3.14. In the first, Aristarchus demonstrates that the lunar elongation
cannot be greater than one quadrant; in the second, that the angle cannot be precisely a quadrant either. It follows, therefore, that the Moon is less than one quadrant apart from the Sun. Assuming that this is the case, he shows, in the third step,
that the Moon cannot be farther from the Earth than the Sun and, in the fourth, that
it cannot even be at the same distance. Consequently, the Moon is closer to the
Earth than the Sun (position 5).

The strategy, in all the steps, is the same. Aristarchus knows that the Sun, the Moon, and the Earth form a right triangle with the right angle on the Moon in a dichotomy. Therefore, he will show that if the Moon is not below the Sun or closer to it than one quadrant, it is impossible to build the right triangle. This is so because the angle centered on the Earth should also be obtuse (first case) or at least right (in the second) or because one of the sides must be greater than the hypotenuse (third case) or at least equal to it (fourth case). In the diagram, Aristarchus somehow represents the impossibility of each situation in doubling the circle that divides the dark and the illuminated parts of the Moon. Each circle will fulfill one of the two constraints. The dotted line will always be perpendicular to the Sun because the Sun is the source of light. The bold line will always be in the plane of our eye because we know we are dealing with a dichotomy. Only in position 5 both circles coincide.

Strictly speaking, Aristarchus is not demonstrating that the Moon always moves below the Sun, but only when it is in a dichotomy, which is the case he has analyzed. But if we add that the Moon and the Sun rotate in circular orbits around the Earth,[5] their distances from the Earth never change. Assuming that, the proof working on quadrature can be extrapolated to every possible lunar position. From the way Aristarchus formulates the proposition, it is clear that he states that he has demonstrated that the Moon always moves below the Sun so that we can assume the implicit premise of the circularity of the orbits.

Proposition 7

Introduction

In some way, it seems that proposition 6 is there to justify the geometric construction of proposition 7, in which Aristarchus will place the center of the Moon in the first quadrant and closer to the Earth than the Sun. In it, Aristarchus will obtain the maximum and minimum limit of the ratio between the solar and lunar distances. Here, for the first time, he will introduce the value proposed in hypothesis 4: the lunar elongation at the exact moment of a dichotomy is 87°. This value is the only datum he needs to obtain the limits. As we have said, with trigonometry at hand, it is an almost-trivial operation: to get the cosine of 87°. But trigonometry had not yet been developed. Aristarchus faces the need to bypass trigonometry. He has one angle of a triangle, and he needs to obtain the ratio between the two sides of that angle. Aristarchus knows some proto-trigonometric inequalities between sides and angles or arcs that he will use to calculate the two limits.

5　Or the Earth around the Sun. Aristarchus is not assuming geocentricism here. It is only necessary that the Earth–Sun distance remain fixed, which happens both in geocentricism and heliocentrism, as long as the orbits are circular.

Proposition 7a minimum limit[6]

Figure 3.15 shows two right triangles of the same height, ABC and ABD. Aristarchus knows that there is a proportion between the ratio of the angles formed by the equal leg and the hypotenuse (angles α and β) and the ratio between the unequal legs (BD and BC): if the greater angle is in the numerator, the ratio between the angles is smaller than the ratio between the sides (see geometric theorem 3 in Appendix 1 on p. 294):

$$\frac{\alpha}{\beta} < \frac{BD}{BC}$$

Strictly, the ratio between the sides is equal to the ratio between the tangents of the angles. But because the ratio between the angles is always smaller than the ratio between the tangents, the ratio between the sides will be greater than the ratio between the angles.

To apply this proto-trigonometric inequality, Aristarchus draws a diagram in which side BD represents the solar distance, and side BC, the lunar distance. He

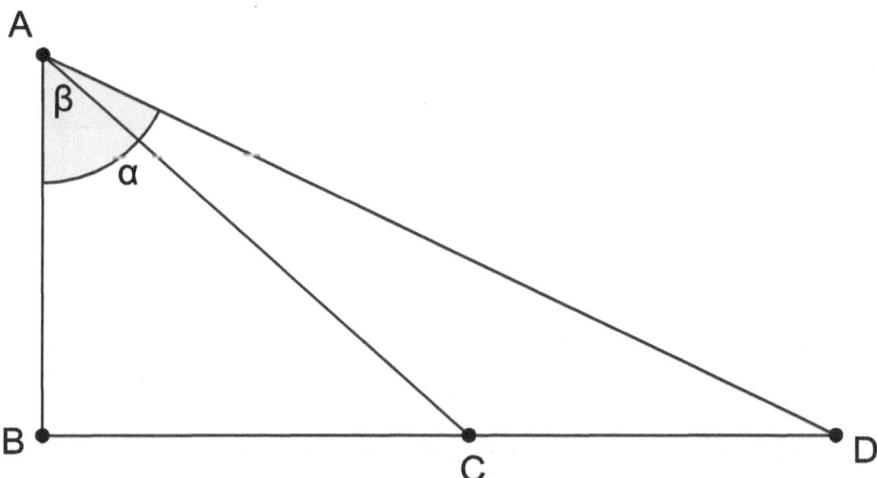

Figure 3.15 Geometric theorem 3: α/β < BD/BC.

6   When, in the same proposition, Aristarchus calculates more than one value, we will distinguish the calculations that lead to the different results using letters. Thus: 7a will be the calculation that concludes in the lower limit for the ratio between the solar and lunar distances, and 7b the calculation concluding in the upper limit. In general, if more than one value is calculated in the same proposition, they are upper and lower limits of the same ratio, as in the case of propositions 9, 15, 16, and 17. But proposition 13 calculates the upper and lower limits of two ratios and the lower of a third. Consequently, we will distinguish 13a, 13b, 13c, 13d, and 13e. We introduced this subdivisions for didactic reasons and have no philological foundation in Aristarchus's treatise.

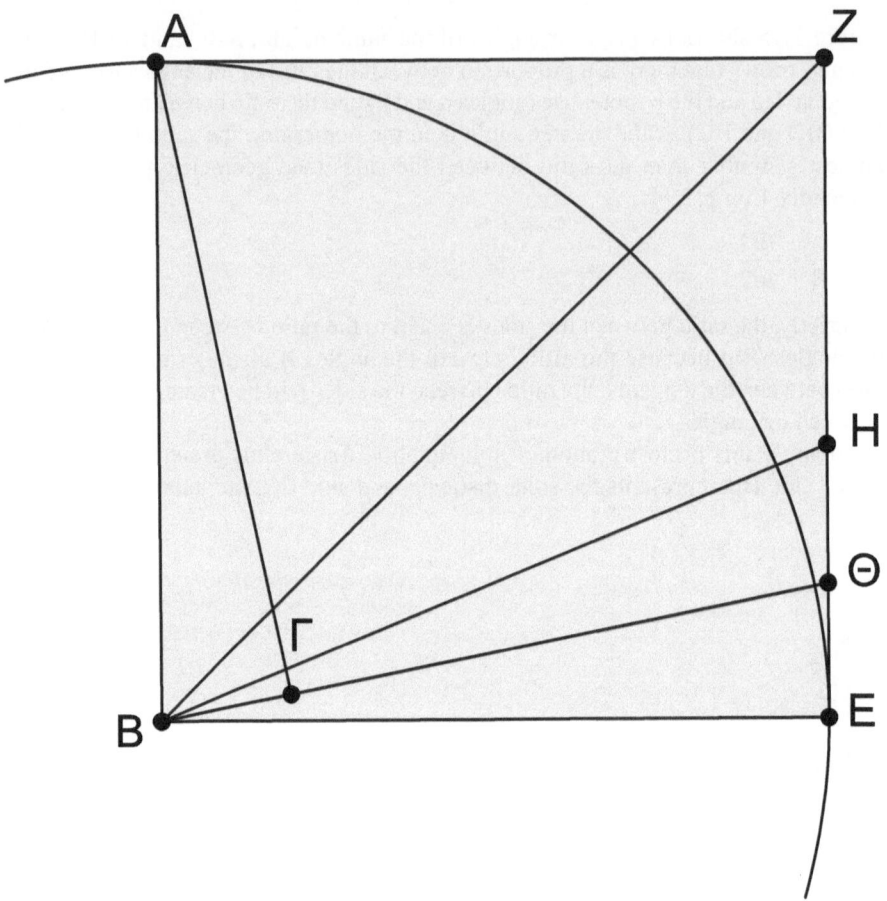

Figure 3.16 Partial reproduction of the diagram of proposition 7. *A* is the Sun; *B* the Earth, and *Γ* the Moon. In the right triangle *ABΓ*, the angle in *Γ* is right.

can obtain angles α and β from the lunar elongation (87°). Finally, he applies the proto-trigonometric inequality to obtain the minimum limit of the ratio between the distances: the ratio between the distances will be at least as large as the ratio between the angles.

The idea behind the procedure is very simple. Aristarchus inscribes the Earth–Moon–Sun triangle in a square whose sides are equal to the solar distance. In Figure 3.16, B is the Earth, A the Sun, and Γ the Moon. Therefore, the Moon is closer to the Earth than the Sun and, at an angular distance, smaller than one quadrant, as requested by the previous proposition. Of course, ZE is equal to AB. Therefore, ZE also represents the solar distance. Now, since AB and ZE are parallel, the angles ABΘ and BΘE are equal, and so are the angles AΓB and BEΘ, which are right. Therefore, the triangles EBΘ and ΓAB are similar. But since the sides

AB and BE are also equal, the triangles are not only similar but also equal. Thus, line ΘE is equal to line BΓ. But line BΓ represents the lunar distance. So line ΘE represents the lunar distance too. Aristarchus has succeeded in placing the two segments that he wishes to compare on the same line: ZE represents the solar distance, and ΘE the lunar distance. He knows, therefore, that the ratio between the sides is greater than that between angles ZBE and ΘBE. And both angles are known. On the one hand, since ABEZ is a square, angle ZBE is 45°. On the other, angle ΘBE is equal to ΓAB, that is, 3°. Therefore, the ratio between the sides should be greater than 45/3, that is, 15.

But Aristarchus is not satisfied with this ratio and seeks a more precise one, applying the theorem not to sides ZE and ΘE but to sides HE and ΘE. Since the ratio between the angles is smaller, the approximation will be better. Angle HBE is, by construction, half ZBE, so Aristarchus already knows it. It remains to establish the ratio between ZH and ZE. Aristarchus obtains that through several steps that we will follow in detail.

In the first place, applying the proto-trigonometric inequality to angles HBE and ΘBE, and sides HE and ΘE, he obtains that:

$$\frac{HE}{\Theta E} > \frac{HBE}{\Theta BE} \tag{1}$$

But HBE is half ZBE, and consequently, it is 22.5°, and ΘBE is 3°; hence:

$$\frac{HE}{\Theta E} > \frac{22.5}{3} = \frac{15}{2} \tag{2}$$

Second, Aristarchus knows (Euclid, *Elements* VI, 3) that if a line bisecting an angle of a triangle cuts the opposite side of that triangle, the ratio of the segments of that side is equal to the ratio of the remaining sides of the triangle. Consequently, because line BH bisects angle ZBE:

$$\frac{ZB}{BE} = \frac{ZH}{HE} \tag{3}$$

And of course, the equation still holds if every term is squared:

$$\frac{ZB^2}{BE^2} = \frac{ZH^2}{HE^2} \tag{4}$$

Third, because triangle ZBE is a right triangle, Aristarchus can apply the theorem of Pythagoras. Because the triangle is also isosceles, he obtains that:

$$ZB^2 = 2 \cdot BE^2 \tag{5}$$

And consequently, replacing ZB² by 2BE² in (4), he obtains:

$$\frac{2 \cdot BE^2}{BE^2} = 2 = \frac{ZH^2}{HE^2} \tag{6}$$

That is, that:

$$ZH^2 = 2 \cdot HE^2 \tag{7}$$

This means that ZH/HE = √2, an irrational number. But fortunately, Aristarchus is working not with equations but with inequalities. So he asserts that it is also true that:

$$7^2 < 2 \cdot 5^2 \tag{8}$$

And therefore, from (7) and (8), that:

$$\frac{7^2}{2 \cdot 5^2} < \frac{ZH^2}{2 \cdot HE^2} \tag{9}$$

And extracting the square root of all the terms and dividing by two the denominators:

$$\frac{7}{5} < \frac{ZH}{HE} \tag{9}$$

But ZE = ZH + HE, consequently (by composition):[7]

$$\frac{7+5}{5} < \frac{ZH + HE}{HE} \rightarrow \frac{12}{5} = \frac{36}{15} < \frac{ZE}{HE} \tag{10}$$

But from (2) and (10), by equality of terms:[8]

$$\frac{HE}{\Theta E} > \frac{15}{2} \text{ and } \frac{ZE}{HE} > \frac{36}{15} \rightarrow \frac{ZE}{\Theta E} > \frac{36}{2} = 18$$

That is, the ratio between ZE and ΘE must be greater than 18. But ZE represents the solar distance, and ΘE the lunar distance. This second step allows him, then, to move the minimum limit from 15 to 18. The Sun is at least 18 times farther than the Moon, but how far can it be?

Proposition 7b maximum limit

To obtain the upper limit, Aristarchus will use another proto-trigonometric inequality establishing a proportion between arcs and sides. According to this theorem (see geometric theorem 4 in Appendix 1 on p. 294), the ratio between the arcs is greater than the ratio between their chords if the greater arc is in the numerator:

---

7 For an explanation of this operation, see Appendix 1 on p. 295.
8 For an explanation of this operation, see Appendix 1 on p. 296.

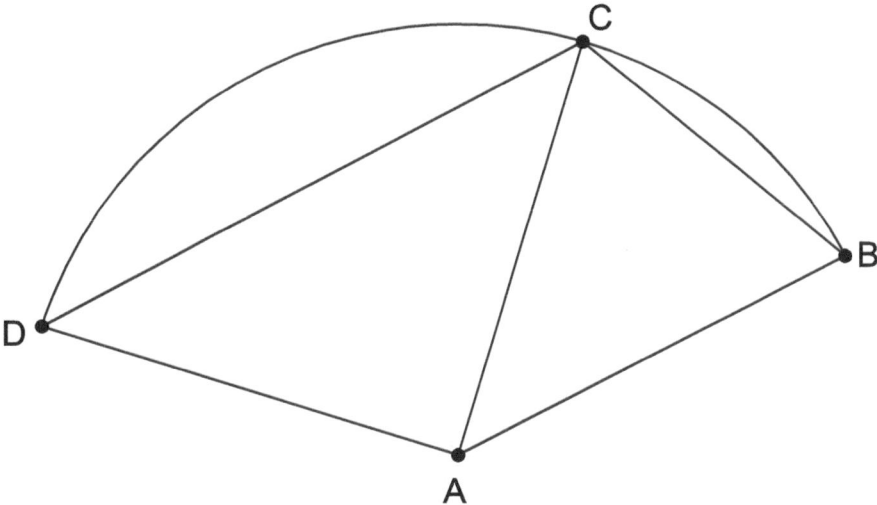

Figure 3.17 Geometric theorem 4: arc DC/arc CB > chord DC/chord CB.

$$\frac{arc\,DC}{arc\,CB} > \frac{chord\,DC}{chord\,CB}$$

In Figure 3.17, arc DC is greater than arc CB. Therefore, the ratio between the arcs (DC/CB) is greater than the ratio between their chords (lines DC and CB).

Aristarchus must, therefore, compare an arc the chord of which corresponds to the solar distance, with another arc the chord of which corresponds to the lunar distance. The advantage of this theorem is that the ratio between the chords is smaller than the ratio between the arcs, allowing Aristarchus to establish a maximum limit for the distances.

To achieve this configuration, Aristarchus draws a circle on Figure 3.16 with the solar distance as diameter and passing through points B and K. See Figure 3.18. Point B represents the Earth, and point K is at the same distance as Γ (the Moon) from B, but on the line BA. Thus, he already obtains the arc corresponding to the lunar distance inscribed in the circle: arc BK. Then he inscribes a hexagon in the circle with one of its vertices at the Earth, point B. The sides of the hexagon are equal to the radius of the inscribing circle. Since the diameter of the circle is the solar distance, one side of the hexagon represents half the solar distance. Consequently, while the chord of arc BK represents the lunar distance, the chord of arc BΛ, one side of the hexagon, represents half the solar distance. Aristarchus successfully identifies the chords corresponding to the distances and can now apply the proto-trigonometric inequality: the ratio between the arcs is greater than the ratio between the chords. On the one hand, from the Sun, arc KB is equal to 3°, but the Sun is at the circumference of the circle. Therefore, the same arc is

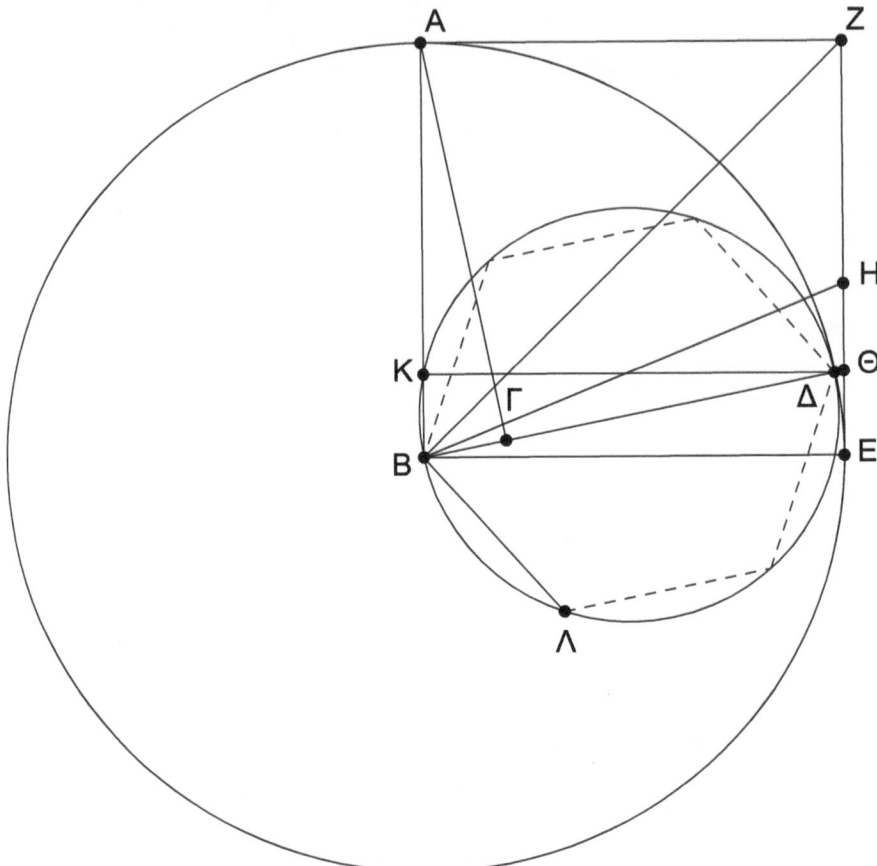

Figure 3.18 Diagram of proposition 7: *A* is the Sun, *B* is the Earth, and *Γ* the Moon.

double when measured from the center of the circle: 6°.[9] On the other hand, the arc that corresponds to one side of the hexagon is one-sixth of a circle, that is, 60°. The ratio between them, therefore, is 1 to 10. Applying the proto-trigonometric inequality, Aristarchus knows that the ratio between one side of the hexagon and the lunar distance is smaller than 10. But one side of the hexagon is half the solar distance. Consequently, the ratio between the solar and lunar distances is smaller than 20. The Sun is farther than 18 times but closer than 20 times the Moon.

9 Euclid in *Elements*, III, 20 proves that the angle of a chord measured from the center of the circle doubles the angle measured from the circumference of that circle.

Conclusion

Aristarchus carries out two different calculations to obtain the two limits of the ratio between the distances. Starting from a ratio between angles (or arcs), he must arrive at a ratio between sides. He takes advantage of two different proto-trigono-metric inequalities for obtaining the two limits. The first one, establishing that in triangles sharing one side, the ratio between the sides is greater than the ratio between the angles (geometric theorem 3), allows him to get the minimum limit. The second one, establishing that in a circle the ratio between the arcs is smaller than the ratio between their chords (geometric theorem 4), allows him to get the maximum limit. The first part of the proof consists, therefore, in constructing a suitable triangle to apply geometric theorem 3; the second part, in building a circle, with chords representing the distances to apply the geometric theorem 4.

Conclusion to this section

Aristarchus has proved that the Sun is farther than 18 times but closer than 20 times from the Earth than the Moon. To do so, he first showed that when a sphere is illuminated by a greater sphere, the illuminated part is greater than half a sphere (proposition 2). The difference between the illuminated and the dark parts of the Moon generated a problem for him because his geometric construction needed that exactly half of the Moon was illuminated in a dichotomy. Therefore, in proposition 4, he showed that seen from Earth, the difference between half a sphere and the illuminated part cannot be distinguished, even in the worst possible situation. The worst configuration was, according to proposition 3, a conjunction. Therefore, he was justified in supposing that, in a dichotomy, the line that joins the eye with the Moon's center also divides the dark from the illuminated part (proposition 5). Then Aristarchus showed that, in this configuration, seen from Earth, the Moon must be below the Sun and closer than one quadrant from it (proposition 6). Finally, in proposition 7, introducing the fundamental empirical value for this part of the treatise – that the angular distance between the Sun and the Moon at the exact moment of a dichotomy is 87° – he calculated the limits of the ratio between the distances: the Sun is between 18 and 20 times farther than the Moon.

Towards the ratio between the sizes of the Sun and the Moon

Conceptual introduction

Once Aristarchus calculated the ratio between the distances, knowing the apparent sizes of the bodies, he can also obtain the ratio between the real sizes. And since they are spheres, knowing the ratio between real sizes, Aristarchus can also get the ratio between the volumes.

The apparent size depends on two factors: the real size of the object and the distance between the object and the observer. For this reason, when a body moves away, it seems to shrink, even if its actual size does not change. Therefore,

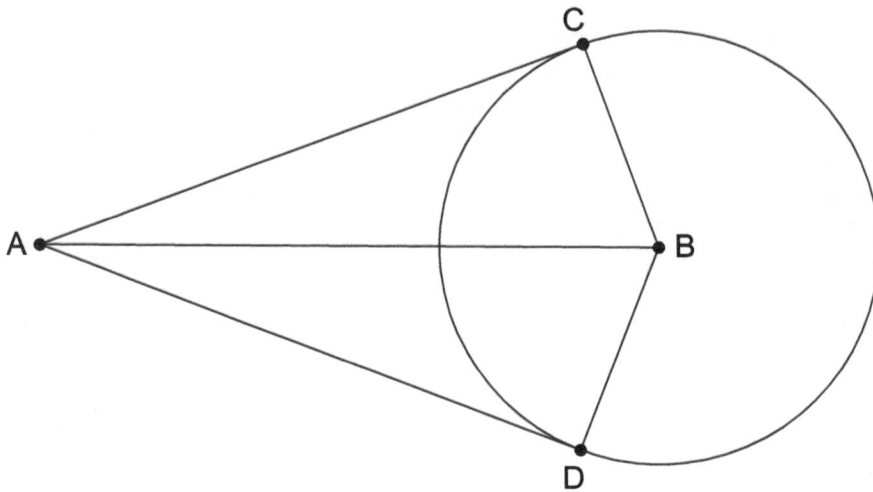

Figure 3.19 Apparent and real size. The sphere centered at *B* is seen by an observer at *A*
with the apparent size CAD. The real size depends on the apparent size (CAD)
and the distance between the object and the observer (AB).

knowing its apparent size and distance is enough to obtain its real size. Again, we
are facing a straightforward trigonometry problem. In Figure 3.19, the radius *BC*
of the sphere centered at *B* is equal to the sine of half the apparent size (angle CAB)
multiplied by the distance (AB).

It is important to recall that Aristarchus is not calculating absolute values but
ratios. He already knows the ratio between the distances. He still needs to obtain
the ratio between the apparent sizes. Aristarchus finds that ratio in proposition 8.[10]
In proposition 9, he calculates the ratio between their real sizes as the ratio between
their diameters. Finally, in proposition 10, he obtains the ratio between their
volumes.

Proposition 8

Indeed, in proposition 8, he demonstrates that every time "the Sun is wholly
eclipsed, one and the same cone envelopes the Sun and the Moon, [the cone] having its vertex towards our eye," that is, the apparent sizes of the Sun and the Moon
are the same. The demonstration is straightforward and much more empirical than
the others. Precisely because of this, and because Aristarchus silently assumes this

---

10   As Neugebauer (1975: 635) clarifies, Aristarchus assumes this proposition from the beginning of
the treatise.

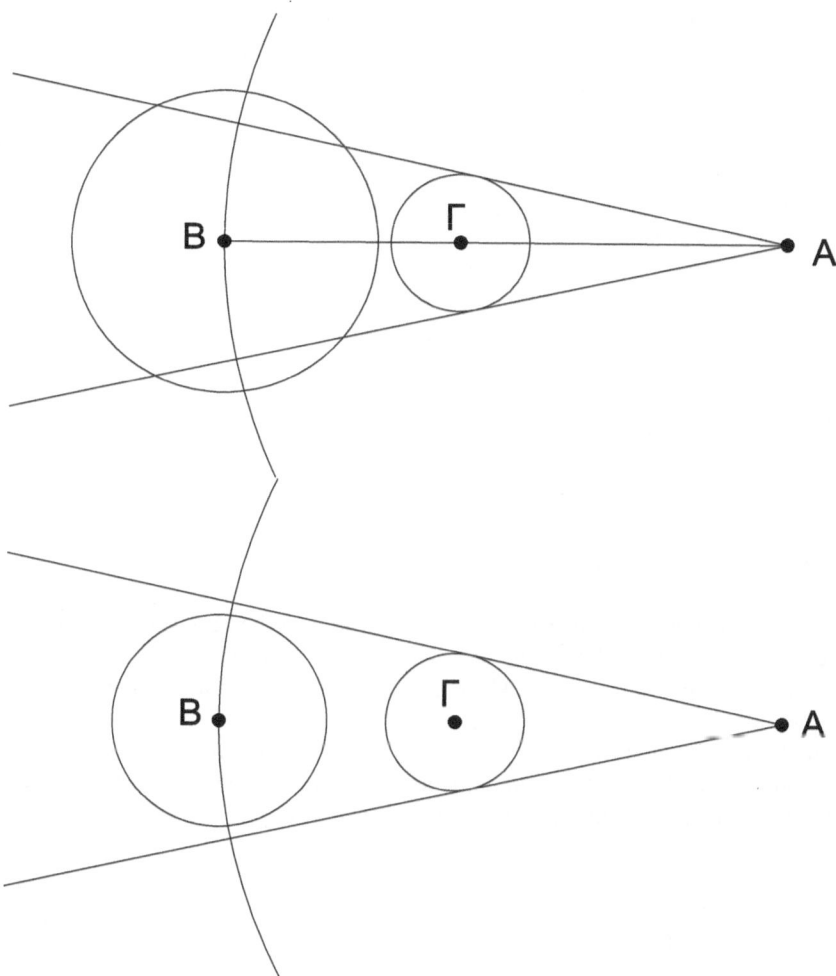

Figure 3.20 The observer is at *A*, *B* represents the center of the Sun, and *Γ* the center of the
Moon. The apparent size of the Sun cannot be greater than that of the Moon
because, if not, there would not be total solar eclipses (figure above). And
because the Sun is totally occulted by the Moon just one instant, the apparent
size of the Sun cannot be smaller than that of the Moon (figure below). Therefore, the apparent sizes of the Sun and the Moon are equal.

11 This excludes the possibility of annular solar eclipses (those in which the apparent size of the Sun
is slightly larger than that of the Moon). Annular eclipses are very rare, and there is not enough
evidence that ancient astronomers have recorded any of them. For a discussion about the frequency
of annular eclipses in Greece and the historical evidence of their possible observations, see Carman
(2015: 99).

proposition from the very beginning of the treatise, it is valid to ask why he does not include this proposition as one of the hypotheses.

Aristarchus analyzes what happens in a total eclipse of the Sun. On the one hand, he states that the entire Sun is eclipsed – showing that the apparent size of the Sun is not greater than that of the Moon. On the other, he affirms that the Sun does not remain eclipsed for some time but begins to reappear as soon as it disappears – showing that the apparent size of the Sun is not smaller than that of the Moon. If it is neither greater nor smaller, the apparent size of the Sun will be equal to that of the Moon.[11]

In Figure 3.20, A is the center of the Earth, Γ the center of the Moon, and B the center of the Sun. In the upper diagram, the apparent size of the Sun is greater than that of the Moon; in the lower one, it is smaller.

Proposition 9

Because the apparent sizes of the Sun and the Moon are the same, the ratio between the real sizes is equal to the ratio between the distances. Since the Sun is 18 to 20 times farther away than the Moon, it will be 18 to 20 times larger. Aristarchus invests proposition 9 in showing this.

The geometric proof is very simple. In Figure 3.21, A represents the eye of the observer, Γ the center of the Moon, and B the center of the Sun. If the Sun and the Moon are inscribed in the same cone with the vertex towards the observer, the triangle having the radius of the Moon as its base (ΓZA) is similar to the triangle having the radius of the Sun as its base (BKA). Thus, the ratio between its sides is equal. Therefore, the ratio between the distances (AB/AΓ) and the ratio between their radii (BK/ΓZ) are proportional. Consequently, the diameter of the Sun will be between 18 and 20 times greater than that of the Moon.

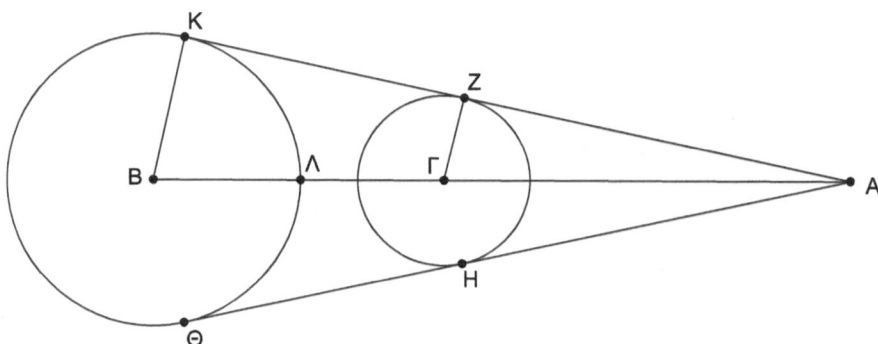

Figure 3.21 Figure of proposition 9. The observer is at *A*, *Γ* is the center of the Moon, and *B* the center of the Sun. The Moon and the Sun have the same apparent size and, therefore, are enclosed in the same cone, with its vertex towards the observer's eye.

Proposition 10

If you know the maximum and minimum limits of the ratio between diameters, calculating the ratio between volumes of spheres is almost trivial, cubing the maximum and minimum limits of the diameters. That is what Aristarchus does in proposition 10.

Of course, this does not establish the volume of the bodies but only the ratio between them. In Aristarchus's times, the formula for a sphere's volume ($4/3\pi r^3$) had not yet been found. It will be discovered by Archimedes one generation later. But being that $4/3\pi$ is constant, the ratio between the volumes is equal to the ratio between the cube of their radii (r^3), which are the only variables. Even though Aristarchus does not know the formula, he certainly knows that the volume depends on the cube of the radius. Aristarchus will establish that the volume of the Sun is bigger than 5,832 ($= 18^3$) times but smaller than 8,000 ($= 20^3$) times the volume of the Moon.

The conclusion to this section

This proposition concludes the second part of the treatise, in which Aristarchus set out to calculate the ratio between the sizes and volumes of the Sun and the Moon. In the remaining propositions, Aristarchus intends to obtain the ratio between sizes and volumes, but not comparing the Sun with the Moon but both with the Earth.

Towards the ratio between the sizes of the Sun and the Earth

Conceptual introduction

As we already explained, Aristarchus bases the calculation of this part on a new intuition: one can obtain the size of an object by knowing the size of its shadow, the size of the light source, and the distances between the light source, the object, and the screen on which the shadow is projected. Aristarchus realizes that this situation occurs during a lunar eclipse when the shadow of the Earth is projected on the Moon. In Chapter 1, we have already shown how Aristarchus could have straightforwardly obtained the requested value. However, the calculation he offers is much more complicated, mainly because Aristarchus is unwilling to make certain approximations that would significantly simplify it.

Let us briefly recall the simplified version of calculating the ratio between the sizes of the Earth, the Sun, and the Moon that we have already presented on p. 7. In Figure 3.22, A is the center of the Sun, B is the center of the Earth, and Γ is the center of the Moon. NΞ represents the shadow of the Earth projected on the Moon. Aristarchus wants to obtain the value of BΛ, the radius of the Earth. We know that BΛ = BE + EΛ, that BE is equal to two Moons, and that ΠΩ/ΛE = ΩΞ/EΞ. But ΩΞ/EΞ = 20/1 and ΠΩ = 17, so ΛE = 17/20 = 0.85. Consequently, BΛ = (BE + EΛ =) 2.85. That is, the Earth is 2.85 times greater in size than the Moon. Since the Sun is about 19 times bigger than the Moon, it is (19/2.85 =) 6.66 times bigger than the Earth.

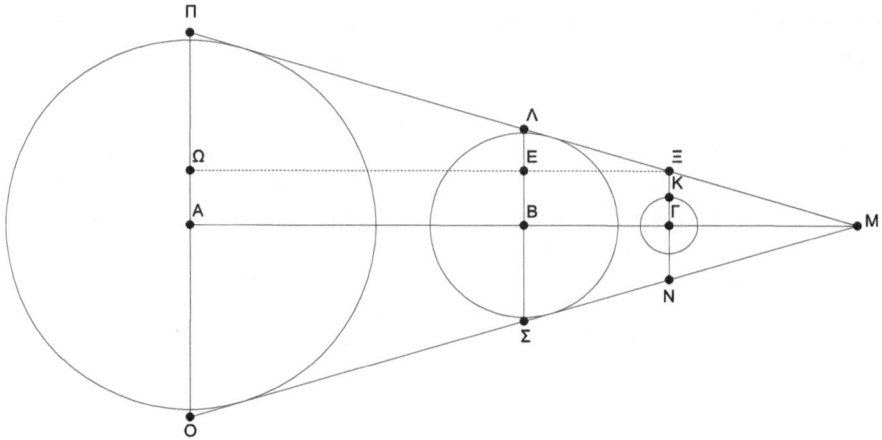

Figure 3.22 Simplified scheme of a lunar eclipse. *A* is the center of the Sun, *B* is the center of the Earth, and *Γ* is the center of the Moon.

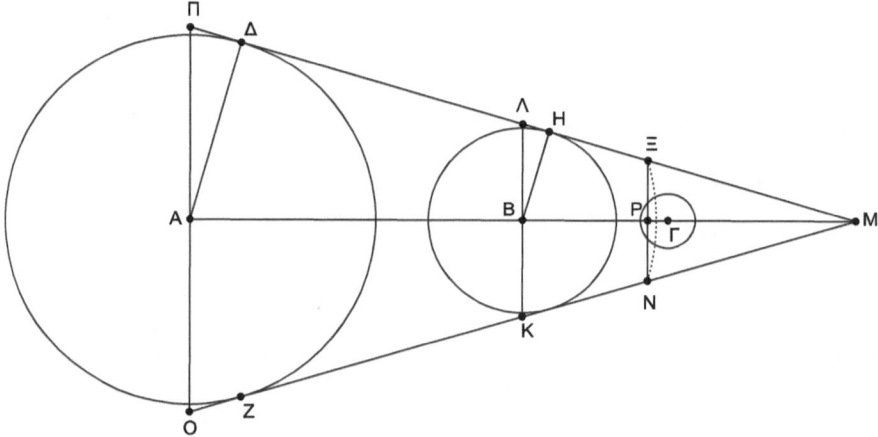

Figure 3.23 Diagram of proposition 15. *A* is the center of the Sun, *B* is the center of the Earth, and *Γ* is the center of the Moon.

Unwanted simplifications

Compared with the calculation that Aristarchus offers in proposition 15, the calculation exposed in the last paragraph has three simplifications. Figure 3.23 is the diagram of proposition 15 and represents the configuration without these simplifications. The first simplification is related to the fact that Aristarchus considers that line ΠΟ is greater than the diameter of the Sun. The second is linked to how he interprets that "the width of the shadow of the Earth is two moons." For reasons that we will discuss shortly, the line on which the shadow is projected (ΞN) is a little closer to the Earth than the center of the Moon (Γ), implying that the distance between the Earth and the shadow (BP) is not equal to the Earth–Moon distance (BΓ). Finally, the third simplification relates to the fact that, according to Aristarchus, ΞN is not exactly two lunar diameters.

The three simplifications that our calculation assumed, therefore, consist of assuming (a) that the radius of the Sun was AΠ when it is AΔ; (b) that P is in Γ, that is, that the line where the shadow of the Earth is projected goes through the center of the Moon; and (c) that line ΞN is equal to two lunar diameters. Resuming, we assumed that AΠ = AΔ, that BP = BΓ, and that ΞN = two lunar diameters when, in fact, AΠ > AΔ, BP < BΓ, and ΞN < two lunar diameters.

The reason for the first simplification: the radius of the base of the cone is not equal to the radius of the sphere

Refer again to Figure 3.21. We performed our calculation using sides ΠA, ΛB, and ΞΓ. The three sides are parallel to each other. The advantage of using these parallel sides is that they allowed us to work with the similarity of triangles ΠΩΞ and ΛEΞ. But ΠA and ΛB are slightly larger than the radii of the Sun and the Earth, respectively. We can still perform the calculation using ΠA. Nevertheless, in that case, we have to obtain the ratio between the radii of the base of the cone and that of the Sun (ΠA and ΔA in Figure 3.22, respectively).

The reason for the second and third simplifications: the circle dividing the visible and the invisible parts of the Moon

The reason for the second simplification is related to how Aristarchus interprets that "the width of the shadow of the Earth is two moons." We already mentioned that because the Sun is greater in size than the Moon, it illuminates more than half a lunar sphere. But from the same kind of diagram, we can infer that also the eye at the center of the Earth sees less than half a lunar sphere. Indeed, in Figure 3.24, the portion of the Moon that the eye sees is not the portion delimited by the great circle, represented by line ZE, but the portion delimited by the circle represented by line ΔΓ. Although Aristarchus did not explicitly mention it, he took this into account by asserting in proposition 4 that the angle representing the apparent size of the Moon is the angle ΔAΓ and not ZAE (see p. 204).

We are dealing with two different circles at the same time, each of them dividing the sphere of the Moon with its own criterion, which we should not confuse. The circle that divides the dark and the illuminated parts of the Moon depends exclusively on the position of the Moon and the Sun. The circle is always perpendicular to the line joining the centers of the Sun and the Moon. The position of the Earth plays no role at all in it. Furthermore, as the Moon–Sun distance varies, the size of that circle varies. This circle is the smallest in conjunction when it is as distant

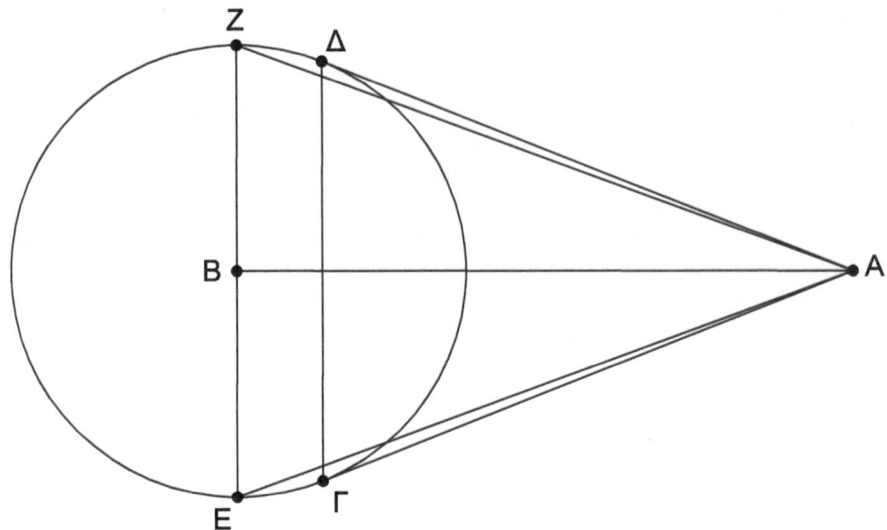

Figure 3.24 The eye, located at *A*, sees less than half a sphere of the Moon, centered at *B*.
The portion of the circle seen from *A* is the portion at the right of line *ΔΓ*.

as possible from a great circle, as Aristarchus demonstrated in proposition 3. On
the other hand, the circle that divides the visible and the invisible parts of the Moon
from the Earth depends exclusively on the position of the Moon and the Earth. The
Sun plays no role at all in it. This circle does not change in size, because the Earth–
Moon distance is invariable (since for Aristarchus the Moon revolves in a circular
orbit around the Earth). But it does change position, being always perpendicular
to the line joining the center of the Moon with the center of the Earth. For example,
in a dichotomy, the sizes of the two circles are different, and their orientations are
perpendicular to each other.

In Figure 3.25, the eye is at E, the Moon is centered at B, and the Sun at I. The
cone that contains the Sun and the Moon is partially drawn by the lines tangent to
the Moon at H and Θ. Therefore, circle HΘ divides the dark and the illuminated
parts of the Moon. The figure shows that this circle is smaller than a great circle
(not drawn in the figure, but passing through B). And ΔΓ is the circle that divides
the visible and the invisible parts of the Moon. In this configuration, it is smaller
than and perpendicular to HΘ.

Even if both circles are different in size and orientation, there is one particular
configuration in which the two coincide. In conjunction, the cone that contains the
Sun and has a vertex in our eye also contains the Moon. This is so because the Sun
and the Moon have the same apparent size seen from the Earth. Figure 3.26 represents that configuration. The center of the Sun is E, the center of the Moon B,
and A is the observer's eye located at the center of the Earth. The portion of the

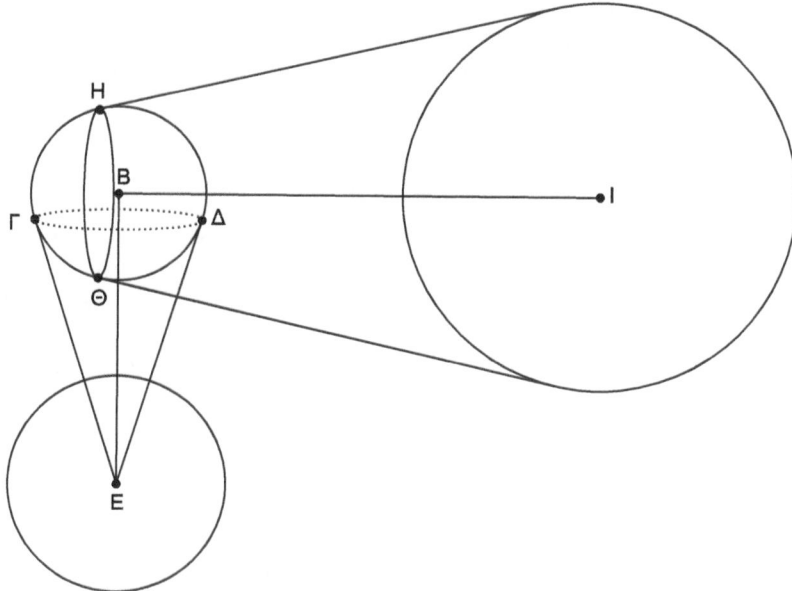

Figure 3.25 The figure represents a dichotomy. The center of the Earth is *E*, the center of the Moon is *B*, and the center of the Sun *I*. The circle dividing the dark and the illuminated parts of the Moon (HΘ) and the circle dividing the visible and invisible parts from the Earth (ΓΔ) do not coincide in size, and they are perpendicular to each other.

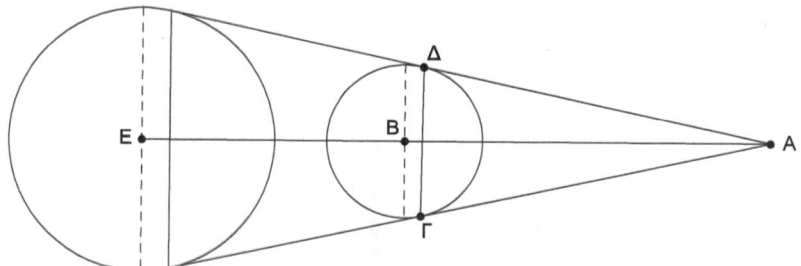

Figure 3.26 The figure represents a lunar eclipse. The center of the Sun is *E*, the center of the Moon is *B*, and the eye of the observer is at *A*, the center of the Earth. The dark portion of the Moon is the same portion visible from the Earth.

Moon at the right of line ΓΔ is, at the same time, the part visible from Earth and the dark part of the Moon.

From proposition 11 onwards, Aristarchus no longer refers to the circle dividing the dark and the illuminated parts of the Moon (which is only relevant in dichotomy and, therefore, in the first propositions) but to the circle that divides the visible and invisible parts. Curiously, however, he continues to call it the circle that divides the dark and the illuminated parts (cf. Carman 2014). To avoid confusion, we will call it the circle that divides the visible and the invisible parts of the Moon or, shorter, the *visibility circle*.

We must now ask ourselves how Aristarchus interprets that "the width of the shadow of the Earth is two moons," as hypothesis 5 asserts. Of course, the most straightforward interpretation would be to hold that the diameter of the shadow is equal to two lunar diameters, as Figure 3.27 shows. Centuries later, Ptolemy interprets it in that way (V, 15; Toomer 1998: 257–258).

But this is not Aristarchus's interpretation. Since we do not see half a lunar sphere but the portion delimited by the visibility circle, Aristarchus asserts that the Earth shadow is equal to two visibility circles of the Moon. In this case, then, the shadow will be smaller than in the previous case, as seen in Figure 3.28.

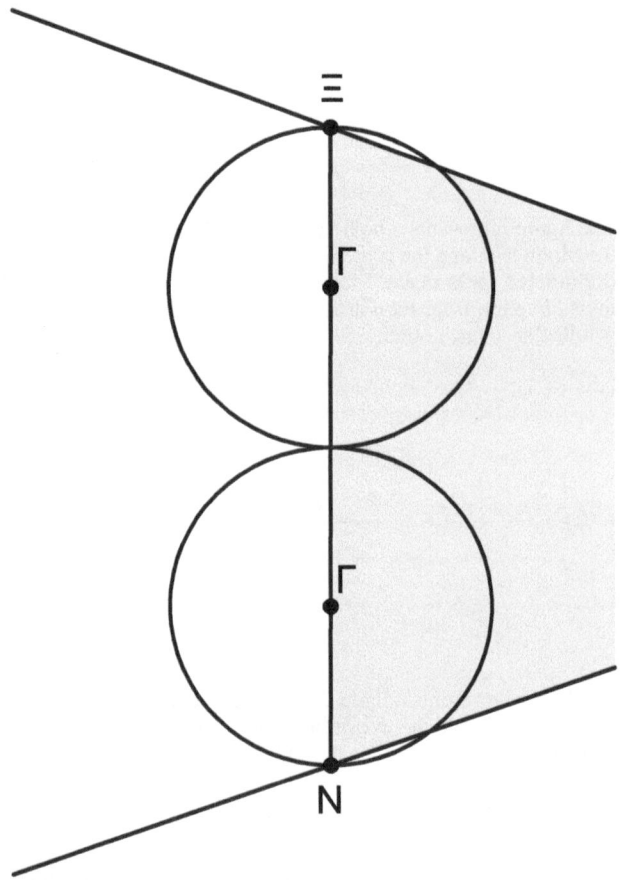

Figure 3.27 The diameter of the Earth shadow is equal to two lunar diameters. The center of the Moon is Γ, and line ΞN is the shadow line, equal to two lunar diameters.

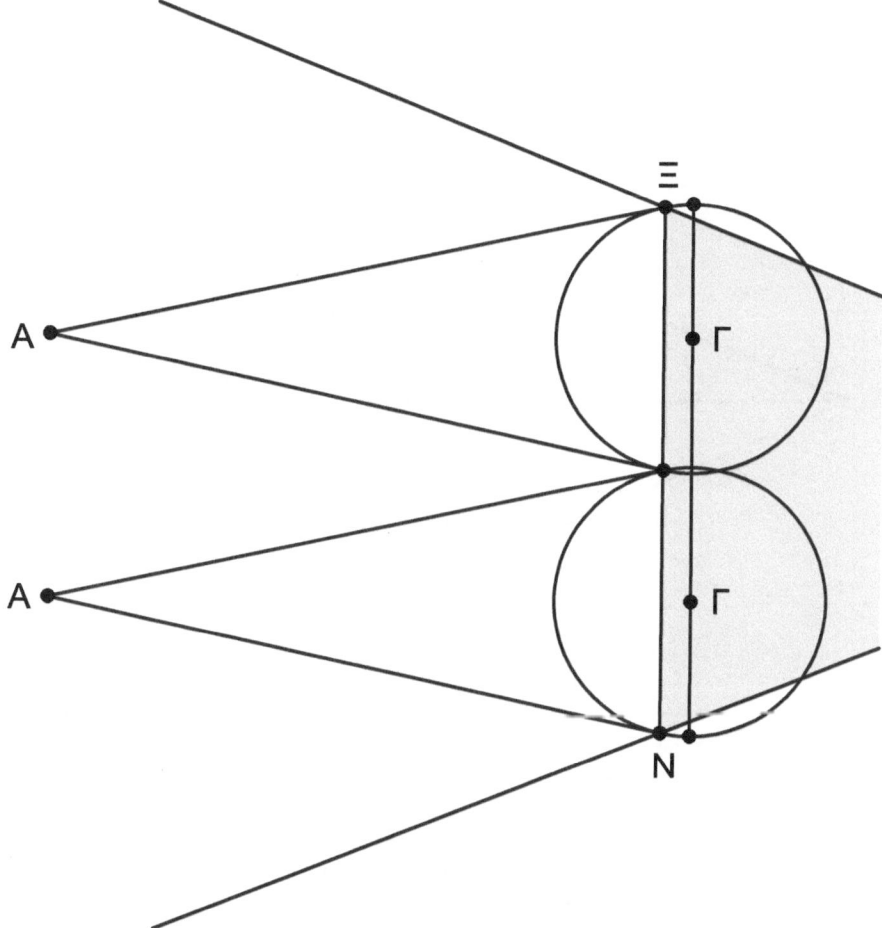

Figure 3.28 The diameter of the Earth shadow is equal to two circles of visibility. The center of the Moon is *Γ*, and line *ΞN* is the shadow line, smaller than two lunar diameters. The observer is duplicated at both positions of *A*.

However, Figure 3.28 is not entirely correct, since the center of the Earth (point A) is duplicated in two different places. In Figure 3.29, which is the one that Aristarchus finally adopts, we have unified the two points *A*. In this diagram, as can be seen, the size of the shadow is even smaller.

Accordingly, when in hypothesis 5 Aristarchus states that the width of the shadow of the Earth is equal to two Moons, he is referring to apparent sizes measured in angles. The apparent size of the width of the shadow (ΞN) is equal to the apparent size of two Moons. Strictly speaking, Aristarchus is right: if we see that the shadow of the Earth is equal to two Moons, what we actually observe is that

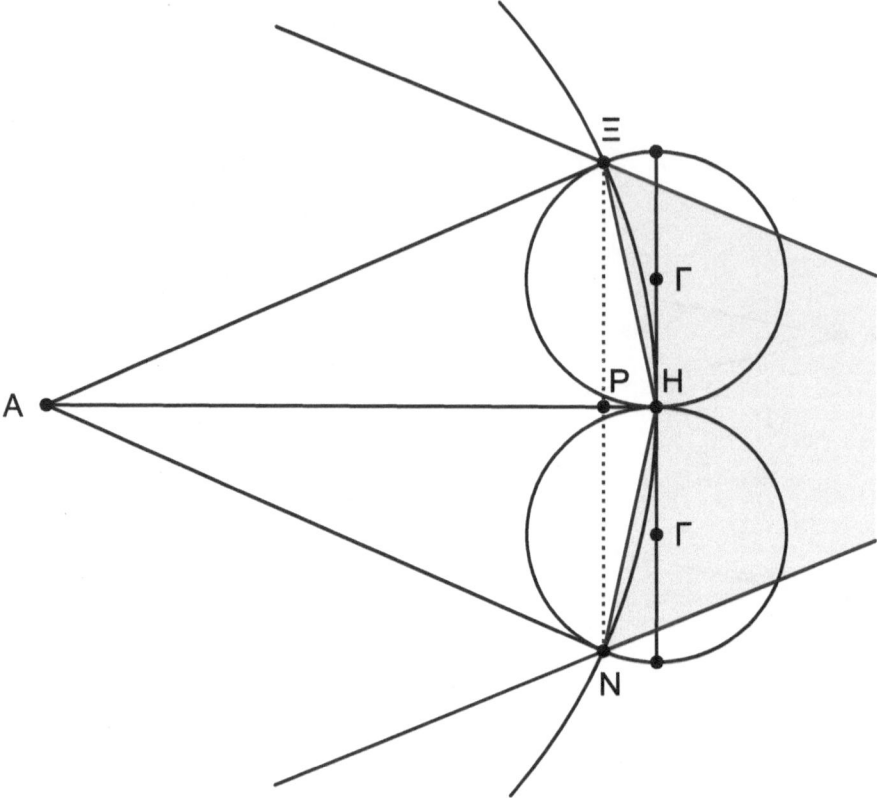

Figure 3.29 The width of the Earth shadow is equal to two Moons, according to Aristarchus's interpretation. The center of the Moon is *Γ*, and lines *ΞH* and *HN* define the shadow line. The observer is at *A*.

the size of the shadow is equal to two apparent sizes of the Moon. A great subtlety that, however, could be ignored without any relevant consequence on the resulting values. So much so that Otto Neugebauer claims that taking this into account implies "pure mathematical pedantry" (1975: 643) from Aristarchus's side. We believe, however, that Neugebauer's reproach involves judging Aristarchus with the current epistemological criteria. Aristarchus is still at the birth of empirical science, and a lot of water will flow under the bridge of epistemology from his time to Ptolemy's. The latter will have no problem making this simplification, aware that it has no relevant consequences on the results. Undoubtedly, Aristarchus knew that ignoring that detail would not significantly affect the results. Still, his epistemic values seem to be others: he is not aiming to obtain an accurate empirical value but to show an impeccable method from a geometric point of view. The decision of not making the simplification will result in greatly complicating the calculations. As we have already said, practically a third of the treatise is devoted only to that. Aristarchus faces these difficulties with courage and, with patience, solves them. We beg the reader for similar courage and patience to face what follows.

The dotted line ΞN (Figure 3.29) plays a significant role in the calculations of proposition 13. Aristarchus names it "[t]he straight line subtending the arc cut off in the shadow of the Earth from the circle on which travel the extremities of the diameter of the [circle] delimiting the shaded and bright parts on the Moon."

If we replace "the shaded and bright parts" with "the visible and invisible parts," we see that the sentence refers to that line. The diameters of the circle dividing the visible and invisible parts of the Moon are ΞH and HN; its extremities are, there-fore, Ξ, H, and N. The arc of the circle in which these extremities move is an arc centered at A passing through these three points (drawn in Figure 3.29). The points where this arc meets the Earth shadow are the intersecting points between the arc and the cone containing the Sun and the Earth. They are, therefore, again, points Ξ and N. Consequently, the line extended under that arc is the dotted line ΞPN. From now on, we will call this line *the shadow line*.

P is the midpoint between *Ξ* and *N* and the point where the shadow line inter-sects the axis of the cone. The distance from the center of the Earth to *P* is slightly smaller than the distance from the center of the Earth to the center of the Moon (AH). Not considering both distances to be identical is the third simplification that Aristarchus is unwilling to make. Accordingly, Aristarchus must calculate the dis-tance from the center of the Earth to the point P.

Propositions 11 to 14. Aristarchus's calculations to avoid simplifications

Let us summarize the three differences between our first scheme and Aristarchus's. First of all, the radius of the base of the cone (ΠΟ) does not coincide with the diameter of the Sun, being slightly larger. Second, even though the Moon is the screen on which the Earth's shadow is projected, the point at which the shadow is measured (P) does not coincide with the center of the Moon (Γ). Third, line ΞN, which measures the size of the shadow, is a little less than two lunar diameters.

We already know that Aristarchus calculates the upper and lower limits of the ratios because he cannot find the ratio directly. But for the sake of clarity, let us put aside the lower and upper limits for a moment and take a general look at the ratios themselves. To obtain the ratio between the Sun and Earth diameters (15a and 15b), Aristarchus needs to know two ratios between diameters and one ratio between distances. The ratios between diameters are the ratio between the shadow line and the solar diameter (13c and 13d) and the ratio between the shadow line and the base of the cone (13e). The ratio between distances is the one between the Earth–Moon distance and the Moon–shadow line distance (14), allowing Aris-tarchus to go from the Earth–Sun and Earth–Moon distances (7a and 7b) to dis-tances to the shadow line.

Moreover, to calculate those ratios, Aristarchus needs to determine others. First, to calculate the ratio between the shadow line and the solar diameter (13c and 13d), he needs the ratio between the lunar diameter and the shadow line (13a and 13b). Then, knowing the ratio between the solar and lunar diameters (9a and 9b), he obtains the ratio with the solar diameter. Second, to calculate the ratio between the shadow line and the base of the cone (13e), in addition to the already-obtained ratio between the solar diameter and the shadow line (13d), Aristarchus has to know the ratio between the solar diameter and the base of the cone. In turn, Aristarchus obtains this last ratio from the one between the circle that divides the dark and the illuminated parts of the Moon and the lunar diameter (12). Third, to obtain the ratio between the Earth–Moon and Moon–shadow line distances (14), in addition to the already-obtained ratio between the shadow line and the diameter of the Moon (13a), Aristarchus needs to know the ratio between the Earth–Moon distance and the lunar diameter (11).

The following table represents these complicated relationships of logical implication.

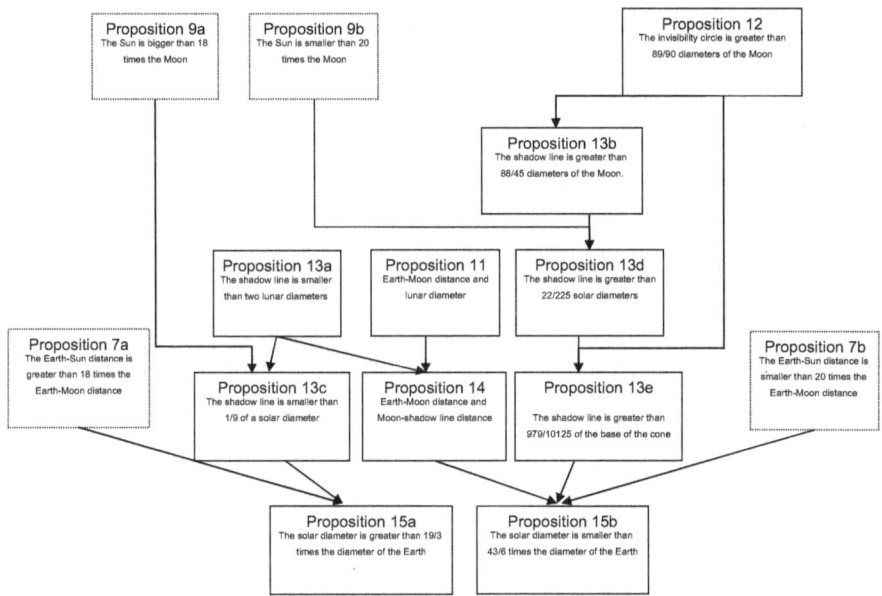

We have described the relationships of logical implication between the different parts of propositions 11 to 14. As expected, directly or indirectly, all lead to propositions 15a and 15b. Now we will show the details of the calculations of each of the propositions, from 11 to 14, in the order in which they appear in Aristarchus. We will conclude the development with the presentation of proposition 15.

Proposition 11

In proposition 11, Aristarchus wants to calculate the upper and lower limits of the ratio between the Earth–Moon distance and the diameter of the Moon. In this calculation, the Sun does not play any role. The only necessary datum is the

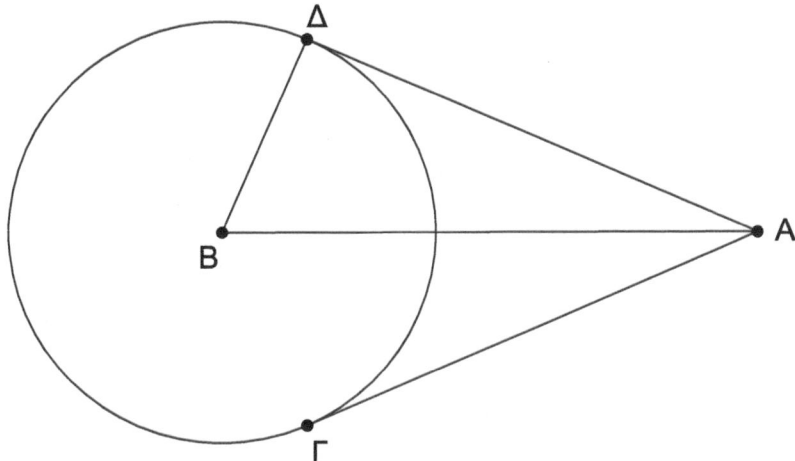

Figure 3.30 Calculation of the ratio between the Earth–Moon distance (AB) and the lunar
radius (BΔ) using trigonometry.

apparent size of the Moon (established in hypothesis 6). With trigonometry at
hand, the solution would be straightforward. Refer now to Figure 3.30. In the right
triangle AΔB, we know the angle at *A* (it is half the apparent size of the Moon,
determined by hypothesis 6), line AB is the Earth–Moon distance, and line BΔ is
the radius of the Moon. Therefore, the required ratio is the ratio between the hypotenuse and the opposite leg, which can be obtained by the sine of the angle. We
already mentioned that the angle is half the apparent size of the Moon (i.e., 1°).
Therefore, the sine function of 1° is 0.017. But the sine expresses the ratio between
the distance and the radius of the Moon. We must double it to obtain the ratio with
the diameter: 0.034. The lower limit obtained by Aristarchus is 0.033 (1/30), and
the upper one 0.044 (2/45). The correct value is between them, but much closer to
the lower limit.

Once again, the lack of trigonometry is the cause of the complications in the
calculation. In what follows, we describe Aristarchus's analysis. However, let us
recall that the only necessary datum to find this ratio is the apparent size of the
Moon, established in hypothesis 6.

First, Aristarchus calculates the *upper limit*, finding that the ratio between the
diameter and the distance is smaller than 2/45.

In Figure 3.31, angle ΔAΓ is the apparent size of the Moon (2° according to
hypothesis 6). Therefore, half of this angle, BAΓ is 1°. Aristarchus established in
proposition 4 that side BΓ is smaller than 1/45 of line AΓ.[12] Now, he wants to obtain

12 Aristarchus demonstrated it using lines BΔ and ΔA and not BΓ and ΓA, but the correspondence
    between them is obvious. It is an application of geometric theorem 1. See the proof in Appendix
    1, p. 292.

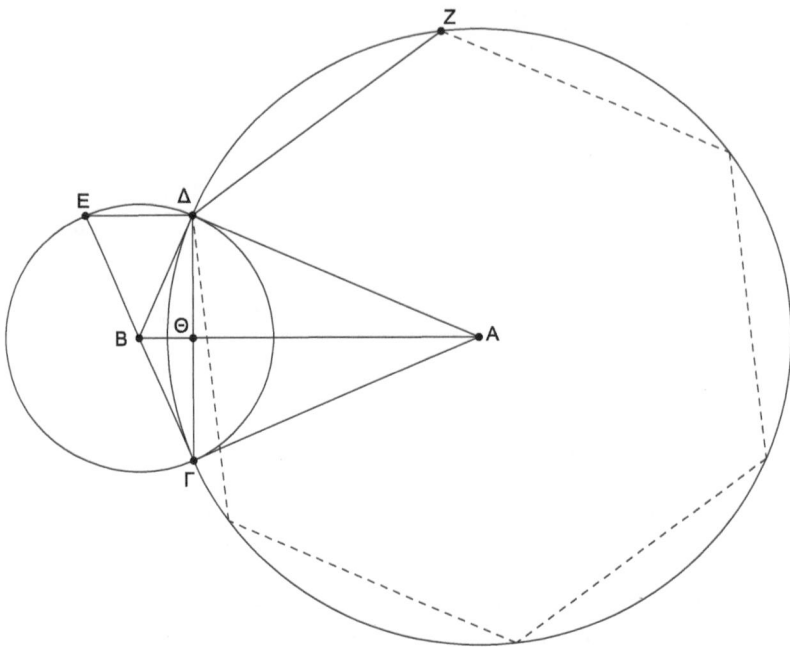

Figure 3.31 Diagram of proposition 11. The center of the Earth is *A*, the center of the Moon
 B. Angle ΔАΓ represents the apparent size of the Moon.

the inequality with AB (the Earth–Moon distance), not with AΓ. But since AB is
slightly larger than AΓ (it is the hypotenuse in triangle BΓA), the same inequality
still holds: BΓ is (even more) smaller than 1/45 parts of AB. Since BΓ is the radius
of the Moon and he wants to establish the ratio with the diameter, he doubles the
ratio. Accordingly – and this is the result – the diameter of the Moon is smaller
than 2/45 parts of the distance from the center of the Moon to our eye.

After the upper limit, Aristarchus calculates the *lower limit*.

First, he establishes that the triangles ΓEΔ and AΓB are similar. Consequently,
their sides are proportional. In particular, he is interested in the proportion AΓ/AB
= ΓΔ/ΓE. By exchanging the positions of *AΓ* and *ΓE* in the equation, he can also
affirm that ΓE/AB = ΓΔ/AΓ. Note that the first term of this equation is our goal: the
ratio between the Earth–Moon distance (AB) and the lunar diameter (EΓ). The
second term, on the other hand, establishes a ratio between the visibility circle (ΓΔ)
and the distance from the Earth to one extreme of this circle (AΓ).

To find out this last ratio, Aristarchus follows a strategy similar to the one used
in proposition 7b. There, he built a circle of a given radius and inscribed a hexagon
in it, which allowed him to establish the ratio between arcs and sides. Here,

Aristarchus knows arc ΓΔ (the apparent size of the Moon, 2°). It remains to find out the arc corresponding to line AΓ. So like in proposition 7b, Aristarchus draws a circle with center at *A* (our eye) and radius *A*Γ and inscribes a hexagon in that circle. Remember that the side of the hexagon inscribed in a circle is equal to the radius of that circle. Accordingly, AΓ (which is the radius of the circle) equals ΔZ (one side of the hexagon).

But Aristarchus also knows that arc ΓΔ measures 2°, while arc ΔZ measures 60°. Consequently, the ratio between the arcs is 1/30. The ratio between their chords, therefore, will be smaller than 1/30. Consequently, Aristarchus obtains that ΓΔ/AΓ > 1/30.

But he already established that ΓE/AB = ΓΔ/AΓ. Therefore, also, ΓE/AB > 1/30. To summarize, the ratio between the Earth–Moon distance and the lunar diameter is greater than 1/30 parts and smaller than 2/45 parts.

Proposition 12

In proposition 12, Aristarchus calculates the upper and lower limits of the ratio between the diameter of the Moon and the visibility circle. Let us remember that in proposition 4, he demonstrated that the difference between them is undetectable by our eye. But that demonstration was not intended to identify the two circles in the remaining calculations (as Ptolemy does, for example, in his calculation) but only to justify the geometric configuration in a dichotomy used in proposition 7. Therefore, in proposition 12, he calculates the difference between them.

The upper limit is obvious: the visibility line is smaller than one lunar diameter. The lower limit requires a calculation. To obtain the lower limit, like in propositions 7b and 11b, Aristarchus inscribes both lines in a circle and compares their arcs. In this case, however, he does not have to build *ad hoc* circles, since both lines are inscribed on the circle representing the sphere of the Moon. So Aristarchus's objective is to compare arc HZ with arc ΔEΓ in Figure 3.32.

Refer to Figure 3.32. Line ΔΓ represents the visibility line, and line HZ is the diameter of the Moon. We already know that the angle ΔAΓ represents the apparent size of the Moon. Therefore, angle BAΓ, half of it, measures 1°. Aristarchus also knows that angle BAΓ is equal to angle ΓBZ. He demonstrated it in proposition 11, showing that triangles ΓEΔ and AΓB are similar. Point E in Figure 3.31 is no longer part of the proposition diagram, but we wanted to include it, labeling [E*]. Therefore, angle ΓBZ also measures 1°. Since arc ZE measures one quadrant (90°), arc ΓZ measures 1/90 of arc ZE. So arc EΓ measures 89/90 parts of arc ZE. Since arc ΔΓ doubles EΓ and arc HZ doubles arc ZE, also arc ΔΓ measures 89/90 parts of arc HZ. Once again, Aristarchus applies *geometric theorem 4*. In this case, because the numerator is smaller than the denominator, the ratio between the arcs is smaller than the ratio between their chords. Therefore, the visibility line (ΔΓ) measures more than 89/90 parts of the diameter of the Moon (HZ). To summarize, the visibility line is less than 1 but greater than 89/90 parts of the lunar diameter.

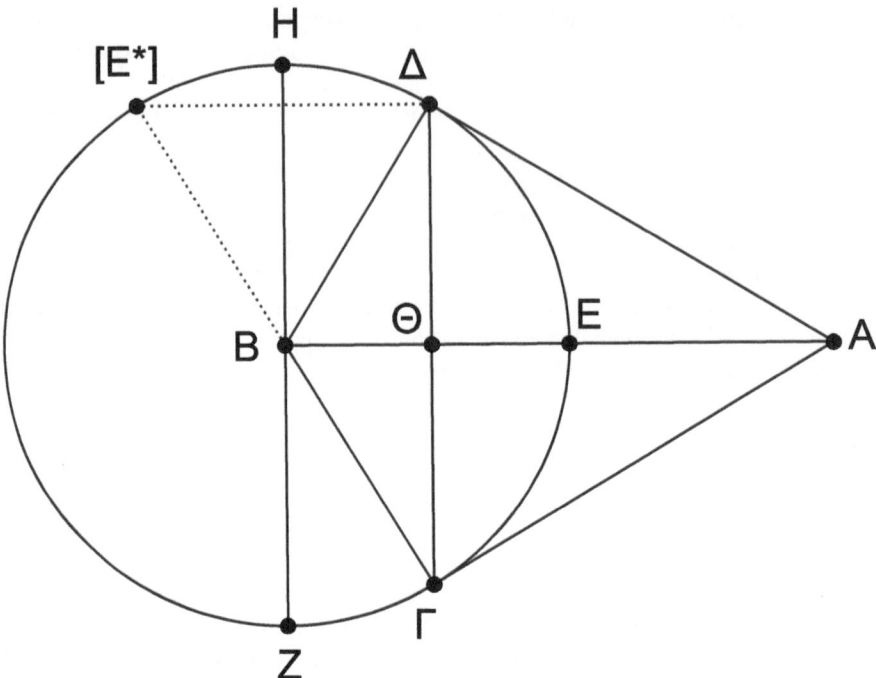

Figure 3.32 Diagram of proposition 12. The observer is at *A*, the center of the Earth. *B* is the center of the Moon. Angle *ΔAΓ* represents the apparent size of the Moon.

Proposition 13

In proposition 13, Aristarchus collects all the calculations related to the size of the shadow line. He establishes the ratio between the shadow line and (1) the diameter of the Moon, (2) the diameter of the Sun, and (3) the base of the cone (which, as we have seen, is a little greater than the diameter of the Sun).

He finds the upper and lower limits of the first two and only the lower limit of the last one. Accordingly, he calculates the upper limit (13a) and the lower limit (13b) of the ratio between the shadow line and the diameter of the Moon, the upper limit (13c) and the lower limit (13d) of the ratio between the shadow line and the diameter of the Sun, and finally, the lower limit of the ratio between the shadow line and the base of the cone (13e).

As we have already mentioned, the two values he needs for proposition 15 are 13c and 13e. But to get them, he has to calculate the other three. Let us analyze in detail, then, each calculation of proposition 13.

*Proposition 13a: the upper limit of the ratio between the shadow line
and the lunar diameter*

Proposition 13a holds that the shadow line is smaller than two lunar diameters.
Recall that in hypothesis 5, Aristarchus asserted that the width of the shadow was
equal to (and not smaller than) two Moons. However, once we understand how
Aristarchus interprets hypothesis 5, there is no contradiction. The way to prove
that the shadow line is smaller than two lunar diameters is straightforward.

In Figure 3.33, the isosceles triangle ΞNΛ has the shadow line as the base (ΞN)
and the two diameters of the visibility circle (ΞΛ and ΛN) as the equal sides.
Because the base of the triangle is smaller than the sum of the two other sides, the
shadow line is smaller than twice the diameter of the visibility circle. But we also

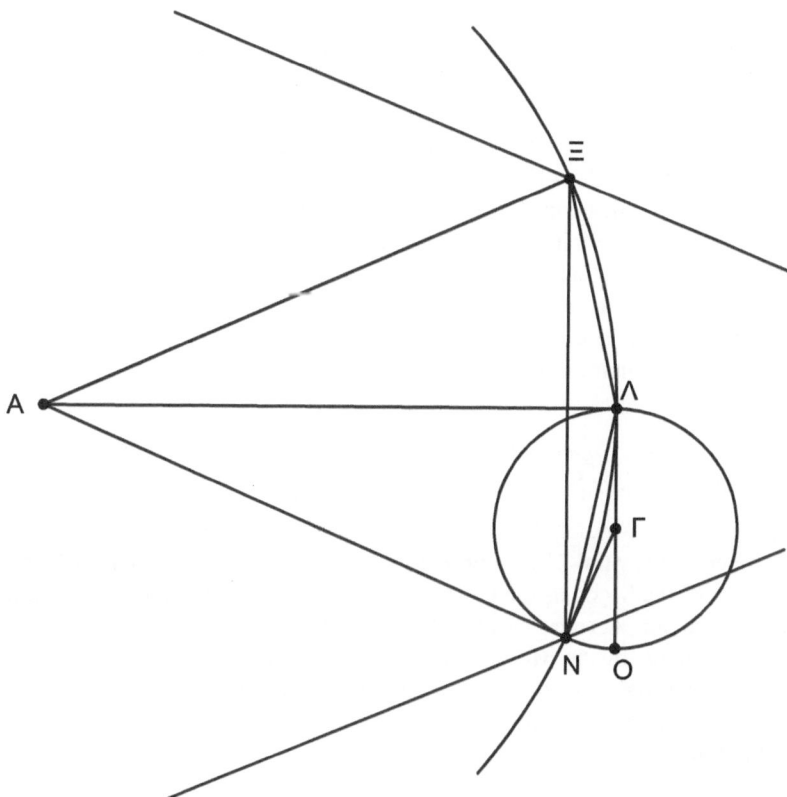

Figure 3.33 Comparison between the lunar diameter (ΛO) and the shadow line (ΞN). The
center of the Moon is Γ.

know that the diameter of the visibility circle is also smaller than one lunar diameter (ΛΟ). Therefore, the shadow line must be smaller than two lunar diameters.

Proposition 13b: the upper limit of the ratio between the shadow line and a lunar diameter

In proposition 13b, Aristarchus shows that the shadow line – although smaller than two lunar diameters – is greater than 88/45 parts of the lunar diameter (i.e., 1.9555). He already knows the ratio between the visibility line (NΛ) and the diameter of the Moon (ΛΟ). In proposition 12, he has previously demonstrated that the ratio is greater than 89/90. But Aristarchus looks now for the ratio between the radius of the Moon (ΛΓ) and the shadow line (NΞ). ΛΟ represents the diameter of the Moon, and ΛΓ the radius of the Moon. Consequently, ΛΟ is twice ΛΓ. It remains for him to establish the ratio between the visibility line (NΛ) and the shadow line (NΞ) to be able to use the visibility line as a mean term between the lunar radius and the shadow line.

Aristarchus also knows that triangles NΞΛ and NΛΓ are similar. Therefore, the ratio between the shadow line and the visibility line is equal to the ratio between the visibility line and the radius of the Moon. But he already established this ratio in proposition 12. Consequently, with the corresponding calculations, Aristarchus arrives at NΞ/ΛΟ > 7921/4050 (= 1.9558).

Now, this ratio will be multiplied by 20 in proposition 13d, and the result, in turn, multiplied by 89/90 in proposition 13e. To avoid handling such large values, Aristarchus highlights that, since 7921/4050 is greater than 88/45 (= 1.9555), it is also true that NΞ/ΛΟ is greater than 88/45.[13] In summary, Aristarchus establishes that the shadow line is smaller than two (proposition 13a) but greater than 88/45 (proposition 13b) lunar diameters.

Proposition 13c: the upper limit of the ratio between the shadow line and the diameter of the Sun

In proposition 13c, Aristarchus asserts that the shadow line is smaller than 1/9 of the diameter of the Sun. The demonstration is straightforward. Aristarchus uses the diameter of the Moon as the mean term for comparing the shadow line and the diameter of the Sun. Aristarchus already knows that the shadow line is smaller than twice the diameter of the Moon (proposition 13a), and that the diameter of the Moon is smaller than 1/18 times the diameter of the Sun (proposition 9a). Therefore, from both, it follows that the shadow line is smaller than 1/9 times the diameter of the Sun.

---

13 To establish 7921/4050 > 88/45, Aristarchus probably followed an algorithmic procedure known as antyphairetic calculation. Another use of this procedure can be seen in proposition 15, when he states 71755875/61735500 > 43/37. See Fowler (1999: 52–53).

Proposition 13d: the lower limit of the ratio between the shadow line
and the diameter of the Sun

In proposition 13d, Aristarchus shows that the shadow line is greater than 22/225 parts of the diameter of the Sun. Again, he uses the diameter of the Moon as a mean term. He knows that the ratio between the shadow line and the diameter of the Moon is greater than 88/45 (proposition 13b). He also knows that the ratio between the diameter of the Sun and the diameter of the Moon is greater than 1/20 (proposition 9b). Aristarchus concludes, therefore, that the ratio between the diameter of the Sun and the shadow line is 88/900, which, divided by 4, is 22/225.

Proposition 13e: the lower limit of the ratio between the shadow line
and the base of the cone

In proposition 13e, Aristarchus asserts that the shadow line is greater than 979/10125 parts of the base of the cone containing the Sun and the Earth (line ΟΠ of Figure 3.34). He already knows the ratio between the shadow line and the diameter of the Sun (proposition 13d). Consequently, he first looks for the ratio between the base of the cone and the diameter of the Sun, and then he uses the diameter of the Sun as a mean term.

In Figure 3.34, there are two cones. The cone containing the Sun and the Earth, that is, the cone produced in a lunar eclipse, is represented by triangle ΠΜΟ. And the cone containing the Sun and the Moon, that is, the cone produced in a solar eclipse, is represented by triangle ΣΒΤ. Aristarchus wants to establish the ratio with the base of the first cone. From the figure, it is easy to see that the base of the cone depends on the distances between the spheres, but also on the ratio between

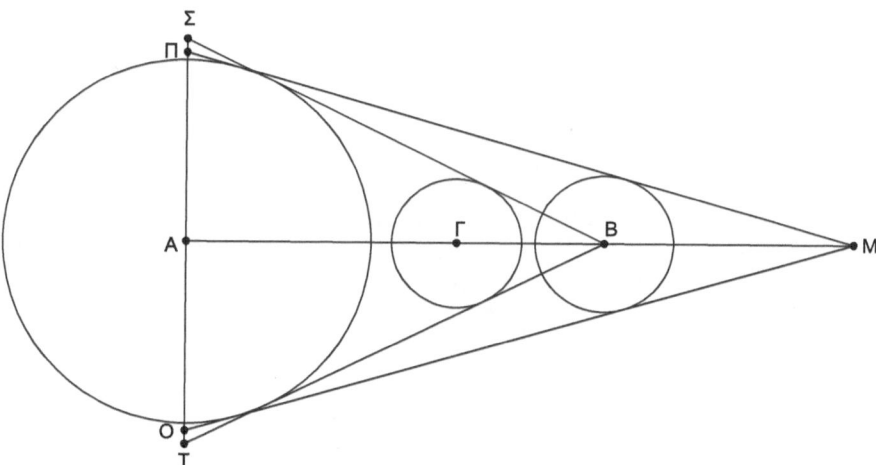

Figure 3.34 The base of the cone ΣΤ containing the Sun and the Moon (ΣΒΤ) is greater than the base of the cone (ΠΟ) containing the Sun and the Earth (ΠΜΟ).

their sizes. Consequently, to obtain the size of the base of the cone, Aristarchus has to know the ratio between the sizes of the Sun and Earth, which is what he is look-ing for. However, because the base of cone ΣBT is greater than the base of cone ΠMO and he needs a value equal to or greater than the base of the cone, Aris-tarchus calculates the ratio between the base of cone ΣBT and the diameter of the Sun, and then he will say that this ratio is even greater in the case of cone ΠMO.

In Figure 3.35, A is the center of the sphere with radius AY. Line YΦ is the diameter of the circle of the sphere tangent to the cone. Line ΣT is the diameter of the base of the cone passing through the center of the sphere. Aristarchus wants to find the ratio between the base of the cone (ΣT) and the diameter of the sphere. Since AΣ is half ΣT and AY is the radius of the sphere, the required ratio is equal to AΣ/AY. Now, since XY and AΣ are parallel, AY/XY is equal to AΣ/AY. Accord-ingly, knowing the ratio AY/YX allows Aristarchus to obtain ratio AΣ/AY.

Suppose we assume that in Figure 3.35 the circle centered in A is the Sun, and the observer at the center of the Earth is at B. In that case, the circle defined by radius YΦ is the circle that divides the part of the Sun that illuminates the Earth and the part that does not illuminate it. This circle also divides the part of the Sun visible from the Earth and the invisible part. Accordingly, line YΦ is the visibility

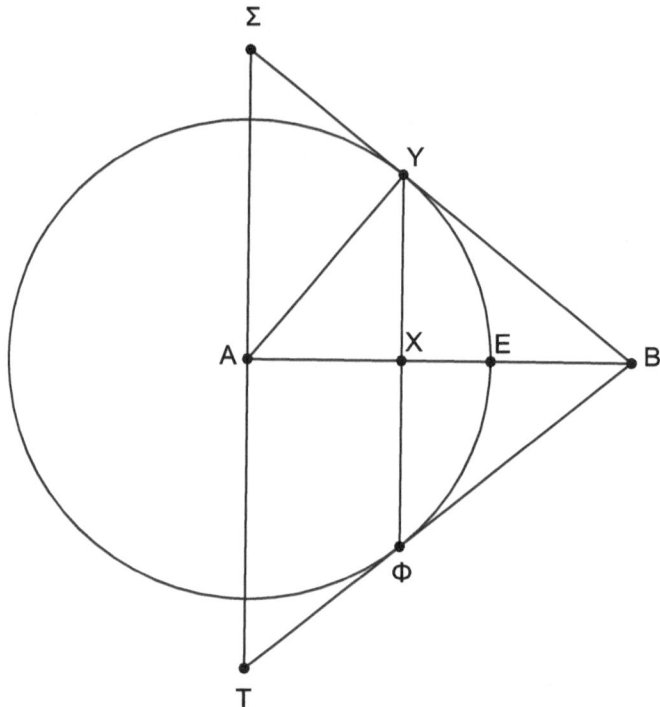

Figure 3.35 The observer is at *B*, and *A* is the center of the sphere. The ratio YX/AY is equal to the ratio AY/AΣ.

line of the Sun. Therefore, we can think of ratio AY/YX as the ratio between the radius of the Sun and half the visibility line of the Sun. But because the Sun and the Moon belong to the same cone with vertex at our eye (see again Figure 3.34), the ratio between the radius and the visibility line is the same in both spheres. Aristarchus already knows the ratio in the sphere of the Moon. In proposition 12, he obtained that the ratio between the dividing line and the diameter of the Moon is 89 to 90. Therefore, in the case of the sphere of the Sun, the same ratio holds. In this way, he obtains the ratio AY/YX for the Sun. We already showed that this ratio is equal to the ratio AΣ/AY, which, in turn, is equal to the ratio between the base of cone ΣBT and the diameter of the Sun. Remember that Aristarchus is not looking for the ratio with the base of this cone, but with cone ΠMO containing the Sun and the Earth. However, since the base of the latter is smaller than the base of cone ΣBT, the inequality still holds.

Consequently, Aristarchus asserts that the ratio between the diameter of the Sun and the base of the cone containing both the Sun and the Earth (ΠMO) is greater than 89/90. By proposition 13d, he also knows that the ratio between the shadow line and the diameter of the Sun is greater than 22/225. Therefore, the ratio between the shadow line and the base of the cone is greater than (22 · 89) / (225 · 90) = 1958/20250 or, equivalently, when divided by 2: 979/10125.[14]

In summary, Aristarchus established the ratio between the shadow line and the diameters of the Sun and the Moon (in both cases, the upper and lower limits) and the lower limit of the ratio between this line and the base of the cone containing the Sun and the Earth.

Proposition 14

In proposition 14, Aristarchus wants to establish the upper limit of the ratio between the Earth–Moon distance and the distance from the center of the Moon to the shadow line. Calculating this ratio requires several geometric steps, but the entire

---

14   Aristarchus applies the operation called *equality of terms*. See Appendix 1, p. 296. While *DS* is a common term, *225* and *89* are not.

$$\frac{N\Xi}{DS} > \frac{22}{225} \quad \text{and} \quad \frac{DS}{\Pi P} > \frac{89}{90}$$

To make them equal, Aristarchus must multiply the first ratio by 89 and the second by 225.

$$\frac{N\Xi}{DS} > \frac{22 \cdot 89}{225 \cdot 89} \quad \text{and} \quad \frac{DS}{\Pi P} > \frac{89 \cdot 225}{90 \cdot 225}$$

And now he can apply the *equality of terms* operation:

$$\frac{N\Xi}{DS} > \frac{22 \cdot 89}{225 \cdot 89} \quad \text{and} \quad \frac{DS}{\Pi P} > \frac{89 \cdot 225}{90 \cdot 225} \Rightarrow \frac{N\Xi}{\Pi P} > \frac{22 \cdot 89}{90 \cdot 225} = \frac{1958}{20250} = \frac{979}{10125}$$

Aristarchus makes both steps at the same time. For a similar explanation by Commandino, see note X in p. 147.

calculation depends only on two values: (1) the ratio between the shadow line and the lunar diameter (smaller than two, according to proposition 13a) and (2) the ratio between the Earth–Moon distance and the lunar diameter (smaller than 2/45, according to proposition 11). Aristarchus finds that the Earth–Moon distance is greater than 675 times the distance from the center of the Moon to the shadow line.

The calculation is extremely long, but unlike others, proto-trigonometric operations do not appear here. They are not required, because Aristarchus always operates with lines – distances or diameters. Each step is easy to understand, but it is difficult to see the place that each of them takes in the global demonstration. We will not analyze every geometric step in detail, but we will present a table with path of the reasoning behind them. In each box, we introduce the ratio reached at that step; the arrow with the description indicates the operation carried out to go from the previous to the following ratio. For understanding the logic of the reasoning, it could be convenient to start with the result in the lower box and go up. We follow the down-top direction in the subsequent explanation.

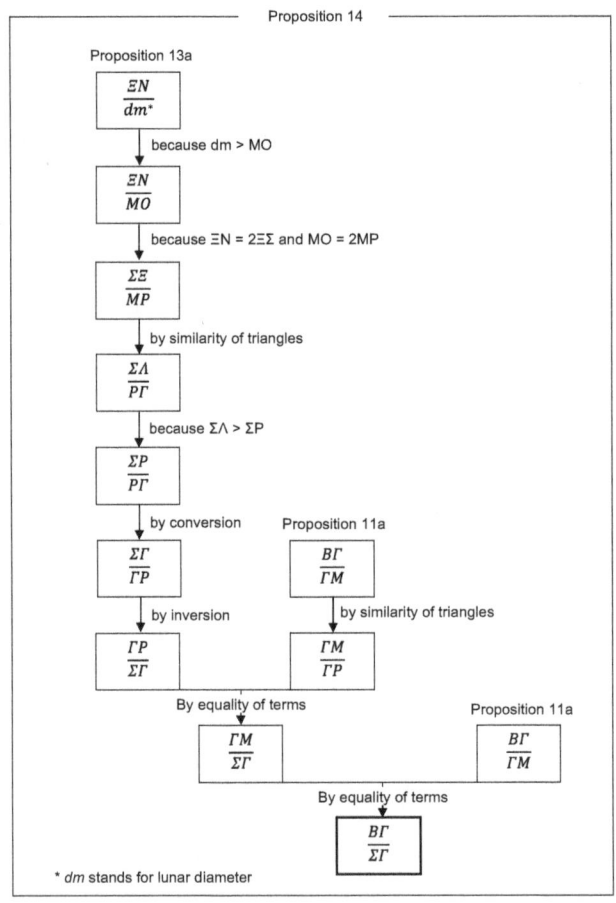

In Figure 3.36, the center of the Earth is B, Γ is the center of the Moon, and the shadow line is ΣΝ. While line ΒΓ represents the Earth–Moon distance, line ΣΓ represents the distance from the center of the Moon to the shadow line.

Aristarchus wants to establish the ratio between the Earth–Moon distance (ΒΓ) and the distance from the center of the Moon to the shadow line (ΣΓ). Since from proposition 11a he knows the ratio between the Earth–Moon distance (ΒΓ) and a lunar radius (ΓΜ), he wants to establish the ratio between a lunar radius (ΓΜ) and the distance from the center of the Moon to the shadow line (ΣΓ). Then, by the equality of terms, using the lunar radius as the mean term, he will obtain the required ratio between ΒΓ and ΣΓ.

Aristarchus obtains the ratio ΓΜ/ΣΓ by combining two other ratios. For obtaining one of them, proposition 11a becomes functional again. For obtaining the other, he follows a much longer path that will end in proposition 13a. Indeed, ratio ΓΜ/ΣΓ can be obtained from ratios ΓΜ/ΓΡ and ΓΡ/ΣΓ, applying the equality of terms and using ΓΡ as the mean term.

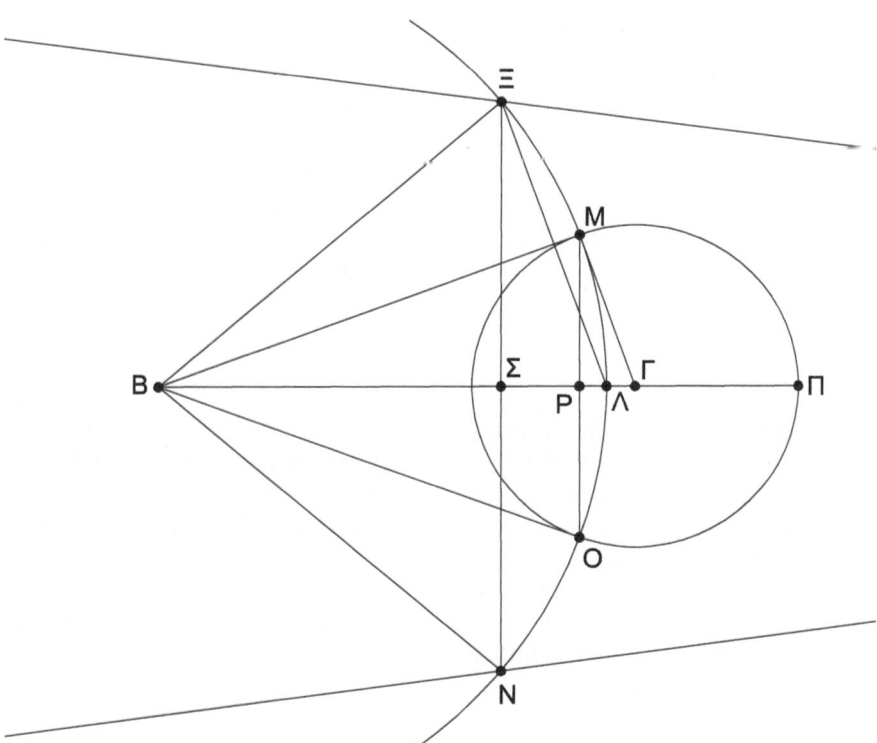

Figure 3.36 Detail of the diagram of proposition 14. The center of the Earth is *B*, and the center of the Moon is *Γ*. *ΞΝ* is the shadow line.

Aristarchus easily obtains the ratio ΓM/ΓP from the similarity between triangles BMΓ and MPΓ. This similarity of triangles allows Aristarchus to establish the proportion ΓM/ΓP = BΓ/ΓM. But BΓ/ΓM is the ratio between the Earth–Moon distance and a lunar diameter, already established in proposition 13a.

The path for obtaining ΓP/ΣΓ is longer. In the first place, Aristarchus inverts the terms of the ratio, obtaining ΣΓ/ΓP. Now, since ΣΓ is the sum of ΣP and PΓ, by conversion of ΣΓ/ΓP, he can obtain ΣP/ΓP. Also, since ΣP is smaller than ΣΛ, the same inequality holds for ΣΛ/ΓP. But by the similarity of triangles ΛΞΣ and MPΓ, Aristarchus knows that ΣΛ/ΓP = ΣΞ/MP. The second ratio of the proportion is not difficult to obtain. Since the ratio is between half the shadow line and half the visibility line, it is equal to the ratio between NΞ/MO. Although Aristarchus does not know the ratio between these two lines, from proposition 13a he does know the ratio between the shadow line and a lunar diameter. He also knows that MO is a little smaller than a lunar diameter. Therefore, Aristarchus obtains the ratio NΞ/MO from the ratio between the shadow line and a lunar diameter.

The road is long, and it is not difficult to get lost. However, once the logic of the argument is understood, it is clear that there are no unnecessary steps. Without the slightest deviation, each step leads towards the final aim: to calculate the ratio between BΓ and ΣΓ.

In summary, in proposition 14, Aristarchus establishes that the ratio between the Earth–Moon distance and the distance from the center of the Moon to the shadow line is greater than 675 to 1. To obtain that result, he used the values calculated in proposition 13a (the shadow line is smaller than two lunar diameters) and in proposition 11a (the ratio between the lunar diameter and the Earth–Moon distance is smaller than 2/45). The obtained value is necessary to calculate the ratio between the sizes of the Sun and the Moon. This calculation is finally performed in proposition 15.

Proposition 15

In the previous section, we analyzed how to obtain all the necessary values to calculate the ratio between the diameters of the Sun and the Earth. In this one, we follow in more detail the particular way in which Aristarchus calculates the limits of that ratio. Like in proposition 14, the calculation does not use proto-trigonometric operations but is extremely long. Because of that, while each step is easy to understand, it is hard to see the role that each of them plays in the general plan. Accordingly, we also offer here a table similar to that of proposition 14. On the left, we enumerate the steps taken to obtain the lower limit (the ratio between the diameters of the Sun and Earth is greater than 19/3), and on the right, the steps to obtain the upper limit (the ratio between the diameters is smaller than 43/6). The almost-perfect symmetry between the two calculations is easily appreciable in the table.

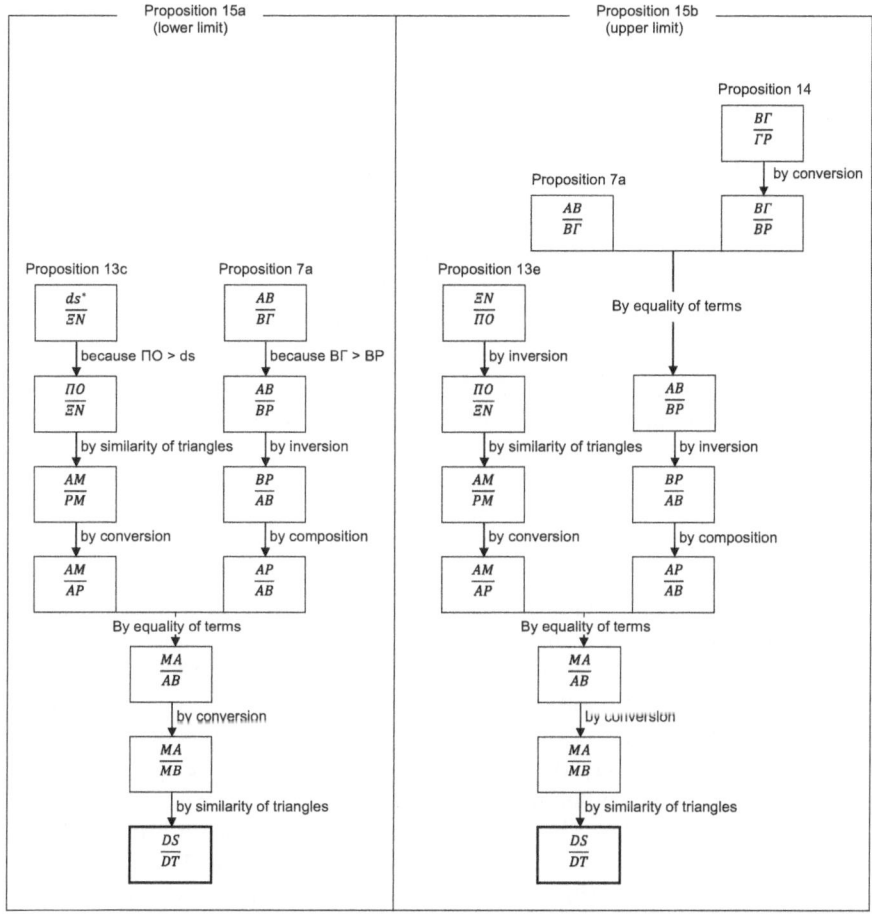

As before, it is probably easier to understand the logic of the calculation if we start from the bottom up. See Figure 3.37. Aristarchus wants to obtain the ratio between the radii of the Sun (AΔ) and the Earth (BH). Because of the similarity of triangles BHM and AΔM, the diameters are proportional to the distance from the center of each sphere to the vertex. That is, AΔ/BH = MA/MB. Accordingly, he obtains the ratio between the distances of both bodies to the vertex of the cone (MA/MB). Aristarchus finds this ratio by conversion from the ratio between the distances from the Sun to the vertex and from the Earth to the Sun (MA/AB). Therefore, this new ratio becomes Aristarchus's goal.

To obtain this last ratio, Aristarchus applies the equality of terms using the distance from the Sun to the shadow (AP) as the mean term. Accordingly, he needs to know the ratio of both terms with the distance from the Sun to the shadow: (1) the ratio between the distances from the Sun to the vertex and to the shadow (AM/

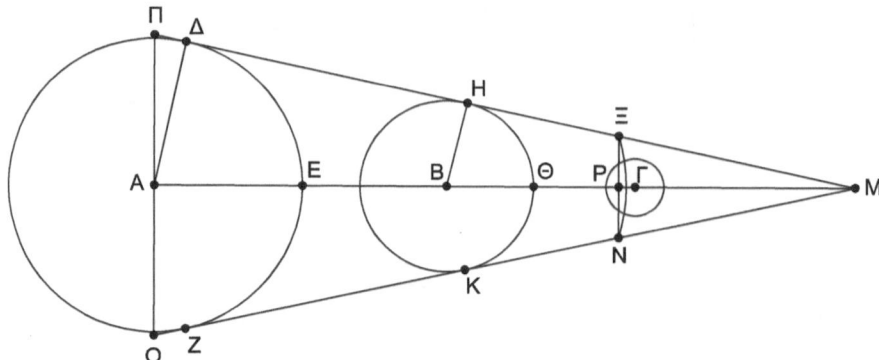

Figure 3.37 Diagram of proposition 15. The center of the Sun is *A*, the center of the Earth is *B*, and the center of the Moon is *Γ*. The shadow line is *ΞN*, crossing the axis of the cone in *P*.

AP) and (2) the ratio between the Sun–shadow and Earth–Sun distances (AP/AB). Aristarchus obtains each of these two ratios in different ways.

On the one hand, (1) Aristarchus obtains the ratio between the distances from the Sun to the vertex and to the shadow (AM/AP) by conversion from the ratio between the distances from the Sun to the vertex and from the shadow to the vertex (AM/PM). Because of the similarity of triangles AΠM and PΞM, this ratio is equal to AΠ/PΞ. But AΠ is half the base of the cone containing both the Sun and the Earth (ΠO), and PΞ is half the shadow line (ΞN). Consequently, AΠ/PΞ is equal to ΠO/ΞN. Aristarchus follows different paths for obtaining the (a) lower and (b) upper limit of this ratio. As before, the path aiming to get the lower limit is on the left of the table; the path seeking to obtain the upper limit is on the right. For obtaining the lower limit (a), he needs a ratio equal to or smaller than ΠO/ΞN. Because the solar diameter is smaller than ΠO, the ratio between the solar diameter and the shadow line (ΞN) is smaller than ΠO/ΞN. Aristarchus already obtained this ratio in proposition 13c. For obtaining the upper limit (b), Aristarchus needs a ratio equal to or greater than ΠO/ΞN. He already calculated this ratio in proposition 13e.

On the other hand, (2) the ratio between the distances from the Sun to the shadow and the Earth (AP/AB) can be obtained by composition from the ratio of the distances from the Earth to the shadow and the Sun (BP/AB). Aristarchus obtains this ratio inverting AB/BP. When calculating the lower (c) and upper (d) limits, he follows different paths to calculate the ratio AB/BP. For obtaining the lower limit (c), he needs a ratio equal to or smaller than AB/BP. Now, since the shadow line is closer to the Earth than the Moon, the ratio between the distances from the Earth to the Sun and the Moon (AB/BΓ) is smaller than the ratio between the distances from the Earth to the Sun and the shadow (AB/BP). Aristarchus already calculated the ratio between the Earth–Sun and Earth–Moon distances (AB/BΓ) in proposition 7a. In the upper-limit calculation (d), for obtaining the ratio AB/BP, he applies the equality of terms using the Earth–Moon distance (BΓ) as the mean term. Accordingly, he needs the ratio between the distances from the Earth to the Sun and the Moon (AB/BΓ) and the ratio between the distances from the Earth to the Moon and the shadow (BΓ/BP). Aristarchus already calculated the first ratio in proposition 7a. He obtains the second ratio by conversion of the ratio between the Earth–Moon distance and the Moon–shadow distance, already obtained in proposition 14.

Through these long paths, Aristarchus obtains that the Sun is at least 19/3 (= 6.33) times and, at most, 43/6 (= 7.16) times bigger than the Earth. Recall that, according to our preliminary calculation, assuming the simplifications that Aristarchus is unwilling to accept, the result was 6.66. The average between the limits of Aristarchus is 6.75. Once again, the road is long, but we will see that we are going straight to the target if we follow it carefully.

Proposition 16

Once Aristarchus established the upper and lower limits of the ratio between the diameters of the Sun and the Earth in proposition 15, obtaining the ratio between the volumes is almost trivial, cubing the diameters. In proposition 15 he found that 43/6 is the upper and 19/3 the lower limit; in proposition 16 he obtains the ratio between the volumes: $43^3/6^3 = 79507/216$ and $19^3/3^3 = 6859/27$.

Conclusion to this section

We have followed a very long journey to obtain the ratio between the sizes of the Sun and Earth. As we said more than once, the intuition behind it is straightforward. It is possible to calculate the size of an object knowing, on the one hand, the sizes of the shadow and the light source and, on the other, the distances between the light source, the illuminated object, and the screen on which the shadow is projected. The calculation, even without trigonometry, should not have been so tortuous: what is largely responsible for the complication is the resistance of Aristarchus to identify point P with the center of the Moon (Γ) and the solar diameter with the base of the cone.

Towards the ratio between the sizes of the Moon and the Earth

Introduction

In proposition 17, Aristarchus calculates the ratio between the diameters of the Moon and the Earth. Then, in proposition 18, he calculates the ratio between their volumes.

Proposition 17

In proposition 17, Aristarchus aims to establish the ratio between the diameters of the Moon and the Earth. He already found the ratio between the diameters of the

Sun and the Moon (proposition 9) and the ratio between the diameters of the Sun and the Earth (proposition 15). Accordingly, he can obtain the ratio between the diameters of the Moon and the Earth by simply combining both results using the Sun as a mean term.

We know that the Sun is, at most, 43/6 times bigger than the Earth and at least 18 times bigger than the Moon. Therefore, the Earth will be at least $(18 \cdot 6/43) =$ 108/43 times bigger than the Moon. In the same way, we know that the Sun is at least 19/3 times bigger than the Earth and, at most, 20 times bigger than the Moon. Therefore, the Earth will be at most $(20 \cdot 3/19) = 60/19$ times bigger than the Moon. Consequently, Aristarchus concludes that the ratio between the diameters of the Earth and Moon is greater than 108/43 (= 2.51) and smaller than 60/19 (3.16). The Earth is between two and a half and three and a quarter times greater in size than the Moon. The correct value is slightly higher than the upper limit found by Aristarchus but still smaller than 4.

Proposition 18

Then again, Aristarchus only cubes the ratio between the distances to get the ratio between the volumes. In proposition 18, Aristarchus obtains: $108^3/43^3 =$ 1259712/79507 and $60^3/19^3 = 216000/6859$.

Conclusion

Let us briefly review the results here. On the one hand, the Sun is between 18 and 20 times farther away than the Moon. Therefore, it is also between 18 and 20 times bigger than the Moon. On the other hand, the Earth is between 108/43 and 60/19 (i.e., between 2.51 and 3.16) times bigger than the Moon; the Sun, in turn, is between 19/3 and 43/6 (i.e., between 6.33 and 7.16) times bigger than the Earth. Resuming, the Sun is approximately seven times bigger than the Earth, and the Moon, a third of it.

Although it is usual to present Aristarchus's values for distances in terrestrial radii (Heath 1913: 350; van Helden 1986: 8; Evans 1998: 70–71; Neugebauer 1975: 639), he does not do that. It is true that, even without trigonometry, it is not difficult to obtain them from Aristarchus's results. The calculation is straightforward, using the results of proposition 11. In this proposition, Aristarchus found that the ratio between the lunar diameter and the Earth–Moon distance is smaller than 2/45 but greater than 1/30. That is, the Earth–Moon distance is between 22.5 and 30 lunar diameters or, if you like, between 45 and 60 lunar radii. Besides, the ratio between the lunar and terrestrial diameters is between 108/43 and 60/19 (proposition 17). Consequently, the Earth–Moon distance will be between 14.25 tr and 23.88 tr. It is enough to multiply these distances by the ratio between the Earth–Moon and Earth–Sun distances to obtain the Earth–Sun distance expressed in terrestrial radii: 256.5 tr and 477.78 tr.

Why Aristarchus did not express the distances in terrestrial radii is not clear. Van Helden (1986: 9) proposes two possible explanations. In the first place, he notices that Aristarchus would have obtained more reasonable distances using the value of the apparent size of the Moon of half a degree, which we know from Archimedes that Aristarchus proposed (see p. 194). However, the standards of consistency would not have allowed him to introduce a different value for the apparent size of the Moon in the middle of the treatise. Consequently, he decided not to express the distances in terrestrial radii whatsoever. In the second place, van Helden suggests that Aristarchus was not concerned at all in absolute distances. We tend to agree with this last proposal. It is important to highlight that what van Helden calls "absolute distances" are distances expressed in terrestrial radii. The question, therefore, is this: what advantage would Aristarchus get by expressing distances in Earth radii rather than simply using ratios of distances? Before some consensus about the value of the Earth's circumference in stades has been reached, the terrestrial radius does not have much meaning, since it does not yet allow Greeks to imagine its magnitude. Accordingly, we believe it was simply not of interest to Aristarchus to express distances in terrestrial radii. The situation will change later. For later authors who know the size of the terrestrial radius expressed in stades, it is interesting for them to express distances in terrestrial radii and in stades. To criticize Aristarchus for not having distances described in terrestrial radii is an unjustified anachronism. The expression of Neugebauer (1975: 636) when he says that "despite the title of the treatise, the distances, however, are not calculated by Aristarchus" cannot but be taken as an exaggeration.

4   Scholia

General introduction

In some manuscripts of Aristarchus's *On Sizes*, the text is accompanied by scholia. In total, there are about 140 scholia. Around 100 have been edited and translated into Latin (1810: 88–199) and into French (1823: 41–88) by Fortia. Noack (1992: 363–380) edited, translated into German, and commented briefly about 40 more scholia. Usually, they are very short. She also renumbered the scholia published by Fortia. We follow Noack's numbering and add Fortia's number in parentheses.

The contribution of the scholia to the understanding of *On Sizes* is very uneven. We have decided to publish only the relevant ones. The Greek text is accompanied by the diagrams like they are found in the manuscripts, a modern interpretation of them, an English translation, and a short commentary.[1]

Among those that are relevant, most consist of original geometric proofs of Aristarchus's assumptions. Others compare his results with those of later authors, such as Hipparchus or Ptolemy.

There are several reasons that we have excluded the other scholia. First of all, to avoid redundant information. Commandino published in his translation of *On Sizes* most of the scholia that are cross-references or those referring to the propositions of Euclid's *Elements*, *Optics*, *Catroptics*, or Theodosius's *Sphaerics* applied in the demonstrations (which are undoubtedly helpful to deepen the study of *On Sizes*). The scholia translated by Commandino are already included in our translation.

Others help to understand the mathematics of *On Sizes* for non-educated readers. Several other scholia consist of making explicit premises implicit in Aristarchus's reasoning but which are so obvious that they are not worth editing. Thus, scholion 15 affirms that "And, again, the three angles of a triangle measure two right angles" (Καὶ πάλιν ἔσονται αἱ τρεῖς τῶν τριγώνων γωνίαι μείζους δύο ὀρθῶν), and scholion 53 clarifies: "Because, 60 is ten times 6" (Τὸ μὲν γὰρ ξ' τοῦ ς' ἐστὶτὰ [δὲ] δεκαπλάσια). Other scholia make explicit mathematical operations that complement Aristarchus's results. For instance, in proposition 11, Aristarchus concludes that the ratio between the lunar diameter and the Earth–Moon distance is between

---

1 To know the manuscripts in which each scholion is copied, see Noack (1992: 72–79).

DOI: 10.4324/9781003184553-4

2/45 and 1/30. Scholion 65(38) calculates the interval between these two distances. The difference between the two is equal to 40/3600, or as expressed in the scholia, 40″. Still others describe operations implicit in Aristarchus's *On Sizes*. Scholion 84(53), for example, clarifies how to compare two reasons with three terms. Others explicate implicit operations. In some cases, the operations are elementary. For instance, scholion 25(14) demonstrates that an angle equal to 1/180 of a whole circle is the same as an angle equal to 1/45 of a quadrant. Sometimes the scholia develop the proto-trigonometric relations that Aristarchus only mentions. Thus, for example, in proposition 15, Aristarchus affirms that, by equality of terms, from PA/AB < 19/18 and MA/AP < 9/8, it is concluded that MA/AB < 171/144. To make the operation explicit, scholion 116(82) multiplies the first ratio by 8, obtaining 19/18 = 152/144, and the second by 9, obtaining 9/8 = 171/152. In this way, the mean term *PA* has the same value (152) in both ratios and can be established that MA/AB < 171/144. It is representative of these scholia that they begin with the expression Συνάγεται οὖν οὕτως ("This has been calculated thus").

Scholion 3(5)

The scholion consists of comparing the values given by Aristarchus in his six hypotheses of *On Sizes* with those of Ptolemy's *Almagest*.

The scholiast accepts the first and third hypotheses as true. The first hypothesis – that the Moon receives its light from the Sun – is justified from lunar eclipses and the fact that the illuminated side of the Moon always faces the Sun. In lunar eclipses, the Earth is aligned with the Sun and the Moon, blocking the light coming from the Sun and, consequently, darkening the Moon.

To justify the third hypothesis – that in a dichotomy, the circle dividing the dark and the illuminated parts of the Moon is in the same plane as our eye – the scholiast refers to Optics 22 (Heiberg and Menge 1895: 32–33), where Euclid demonstrates that an arc of a circle placed on the same plane as the eye appears to be a straight line.

Further, the scholiast considers two hypotheses false, the second and the sixth. In the former, Aristarchus affirms that the Earth has a ratio of center and point to the lunar orbit. The scholiast accepts that the Earth is at the center of the Lunar orbit but criticizes that it can be considered a point. He refers to the daily parallax to justify his falsehood.

The sixth hypothesis, asserting that the apparent lunar size is 2°, is considered false by referring to the *Almagest*, while emphasizing that the lunar distance varies considerably and, with it, also the apparent size of the Moon. In conclusion, it is necessary to consider the distance when the apparent size is measured.

Finally, the scholiast affirms that the fourth and fifth hypotheses are close to the true. The fifth affirms that the width of the Earth shadow is equal to two Moons. Here he compares this value with that of Ptolemy's *Almagest* (2.6). And the result is that both values are not too different.

About the fourth hypothesis, however, the scholiast says nothing. In this hypothesis Aristarchus postulates that in a dichotomy, the Moon has an elongation of 87°. As we have already mentioned in the Introduction, this is not true (not even close to the true) in Ptolemy's model. Ptolemy's ratio between the solar and lunar distances is within the limits set by Aristarchus's values only when the Moon is at its maximum distance from the Earth. This happens, according to Ptolemy, at syzygies. Nevertheless, Aristarchus's method takes the angle in a dichotomy, and in the Ptolemaic model, during a dichotomy, the Moon reaches its minimum distance from the Earth. In this configuration, the ratio between the distances is 32.93 and the elongation, 88.26°. This is not within Aristarchus's limits. See note 5 in Carman (2009).

Scholion 5(4)

The scholion consists of an enumeration of the distances and absolute radii of the Sun and the Moon, as reported in the *Almagest* (V, 13–16, Toomer 1998: 247–257). It contains several inaccuracies, presumably due to transcription errors. In the first mention of the lunar distance at apogee, the given value is 64 and 1/60 tr when it is 64 and 10/60 or 1/6 tr; in the second time, it is correct. The solar distance is right: 1,210 tr. It also includes the distance from the Earth center to the vertex of the shadow cone. While Ptolemy's value is 268 (*Almagest* V, 15, Toomer 1998: 257), in the scholion it is written 68. Presumably, the σ for 200 before ξη (68) has been lost. The absolute radii of the Sun and the Moon expressed in terrestrial radii are correct. And of course, since the apparent sizes are equal, the absolute size is proportional to the distances, as the scholion highlights at the end.

Scholion 13(8)

In the second part of the first proposition, Aristarchus demonstrates that two unequal spheres are contained by the same cone, whose vertex is on the side of the smaller sphere. The strategy followed to prove it is straightforward. It consists of showing that the segment starting at the vertex of the cone that is tangent to the smaller sphere (line KZ in the diagram of the proposition and the scholion) and the segment that, starting from point Z, is also tangent to the greatest sphere (ΓZ in both diagrams) belong to a straight line (ΓZK). If those two segments did not form a single straight line, both spheres would not be contained by the same cone.

See Figure 4.1. Lines *KZ* and *ΓZ* are not part of the same straight line: in this case, there are two cones, one with a vertex at K, and the other at K'. Of course,

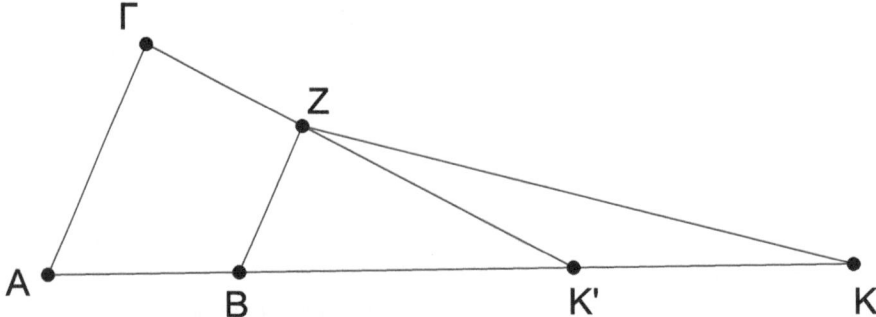

Figure 4.1 Lines ΓZ y ZK do not belong to the same straight line. Thus, there are two cones, one with vertex in *K*, and another with vertex in *K'*.

in the figure, the only cone that contains both spheres (with radii AΓ and BZ) is the one with vertex at K′; the other only contains the sphere with center at B, but this inaccuracy of the diagram is unavoidable simply because you cannot draw the impossible.

Scholion 13 offers a more detailed demonstration that line ΓZK cannot but be a straight line. See Figure 4.2. The scholiast asks that lines ΓZ and ZK, joined in Z, be crossed in Z by another line, ΔZB, and shows that the angles formed by ΔZ and both ΓZ and ZK amount to two right angles, that is, angles ΓZB and KZB add up to 180°. Therefore, both segments belong to the same straight line. The scholion has a peculiarity: it attributes letters to angles. Thus, it calls M angle ΓZB; Θ, angle KZB; E, angle ZΔΓ; and Λ, angle ΓZΔ. We have placed them in lowercase to avoid confusion with the letters designating points, but they are uppercase in the manuscripts. The proof is clear, is accurate, and does not skip steps.

Scholion 14(9)

Close to the end of proposition 1, Aristarchus asks to draw lines that, perpendicular to the axis of the cone, reach the tangent point, that is, the point where the surface of the cone touches the spheres. These lines do not divide the spheres into halves but into unequal portions, being the smaller portion, the one closest to the vertex. Scholion 14(9) demonstrates that these lines neither cross the center of the sphere nor fall in the larger sections. It proves that by a *reductio ad absurdum*. See Figure 1b of proposition 1 on p. 51. The scholiast first shows that if the line crossed the center, then the triangles formed for the larger sphere (AΓK) or the smaller sphere (BZK) would have two right angles. The scholion does not specify why they would be right angles, but it is indeed so. Angles centered on *A* and *B* would be right by construction because the lines are requested to be perpendicular to the axis. Angles centered on *Γ* and *Z* would also be right because the angle from the tangent point to the center of the sphere is also always right. Then, the scholiast shows that if the lines fell on the

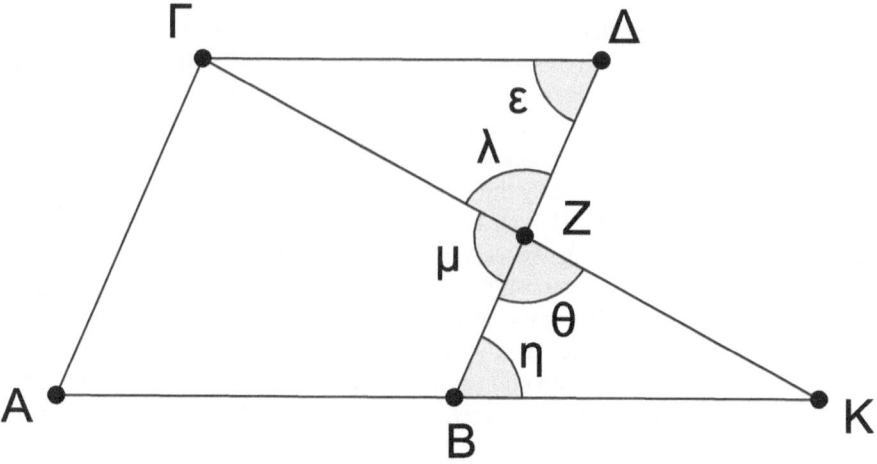

Figure 4.2 Diagram of scholion 13(8).

larger section (that is, they crossed the axis between Ξ and A for the larger sphere or between H and B for the smaller sphere), on the one hand, the angle in the axis of the cone would again be right (by construction), but on the other, the angle centered at the tangent points, Γ or Z, would be obtuse. But this is also impossible.

Scholion 19(11)

Scholion 19 demonstrates that the ratio of the radii of the spheres contained by a single cone is equal to the distance from the center of these spheres to the vertex of the cone. This is stated but not proved at the end of scholion 5(4). A similar but somewhat more-summarized proof is offered in note D of Commandino.

Refer to the figure of proposition 3 on p. 65. The demonstration indicates that triangles BHA and $\Gamma\Theta A$ are similar, and therefore, so are the ratios between their sides. In each triangle, one of the sides is the radius and the other is the distance to the vertex. Thus, the ratios between the distance to the vertex and the radius of each triangle are the same. By alternation, then, also, the ratio between both radii is equal to the ratio between both distances to the vertex.

Scholion 21(13)

In the third proposition, Aristarchus demonstrates that the circle that divides the dark and the illuminated parts of the Moon is the smallest at conjunction when the Sun, the Moon, and the Earth are on the same straight line.

Refer to the figure of proposition 3. In doing that, Aristarchus has to demonstrate that line $\Theta\Lambda$ is smaller than line ΞM. He only states that in both cases, the spheres (representing the Moon) are equal and that the distance from the vertex of the cones to the center of the respective spheres is smaller in the case of $\Theta\Lambda$. He then states that "according to the lemma," line $\Theta\Lambda$ is smaller than line ΞM. The lemma does not appear in Aristarchus's text. We can assume, therefore, that this explanation is an interpolation. In note G, Commandino says that he has not found the lemma but that it can be found in Euclid's *Optics*. The scholiast offers a demonstration. Admittedly, the proof is unnecessarily long. Not in the sense that it contains extra steps, but rather that it makes a painstaking effort to demonstrate that the cone in the figure that follows is the same in every respect as the smaller cone belonging to the figure earlier. But this could have been avoided simply by using only the previous figure and starting with the two cones already superimposed on the same axis. Nizze (1856: 8, n.b) offers a brief and ingenious proof that Heath (1913: 363, n. 1) reproduces. We present here essentially the same proof, but a bit more explained.

See Figure 4.3. Triangle EAH, with a right angle at A, is cut by the line $A\Sigma$, perpendicular to the base, EH, forming two triangles: $EA\Sigma$ and $AH\Sigma$. Euclid, in *Elements*, VI, 8 (Heath 2002, 2: 209–210), demonstrates that if a right triangle is drawn perpendicular to the base that intersects the right triangle, the triangles formed are similar to each other and to the original triangle. Thus, we know that triangles $EA\Sigma$ and EHA are similar. Therefore, the following proportion between the sides is given: $EH/EA = EA/E\Sigma$. Now, Euclid also demonstrates (*Elements*, VI, 17; Heath 2002, 2: 228–229) that if

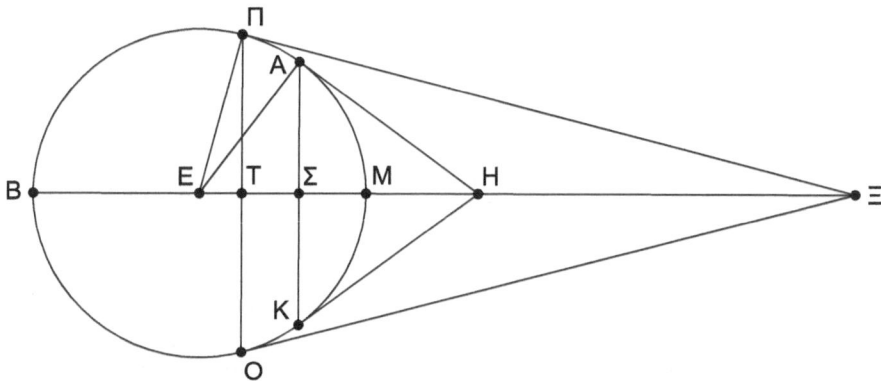

Figure 4.3 Diagram of scholion 21(13) with line EΠ added.

A/B = B/C, then A·C = B². Therefore, we know that EA² = EH·EΣ. The same reasoning, when applied to triangle EΠΞ, implies that EΠ² = EΞ·ET. We know that EA and EΠ are equal since they are both radii of the same circle. Therefore, we also know that EH·EΣ = EΞ·ET. But by construction, EH < EΞ. Therefore, it is also true that ET < EΣ. But ET and EΣ measure the distance from the center to ΠO and AK, respectively, and the one farthest from the center will be the smaller. Thus, AK < ΠO, *Q.E.D.*

Scholion 32(16)

This scholion offers a general demonstration of what we have called *geometric theorem 4* used several times in *On Sizes* (see Appendix 1). The theorem holds that, when the major arc is on the numerator, the ratio between the arcs is greater than the ratio between the chords. The scholion demonstrates an equivalent relationship: in triangle ABΓ, where line AB is smaller than line AΓ, line AB has with line AΓ a ratio greater than the ratio between arc AB and arc AΓ. See diagram of scholion 32(16) on p. 269. The proof is clear and complete. Like Aristarchus, the scholiast uses operations with ratios very fluently, occasionally performing three operations in the same step. The scholion appears when Aristarchus uses this theorem in proposition 4, when he affirms that the ratio between lines BΘ and ΘA is greater than the ratio between angles BAΘ and ABΘ (see the diagram of the proposition 4). On note E, Commandino offers another proof, somewhat briefer.

Scholion 37(18)

This scholion offers a demonstration similar to Pappus's proof in the second part of his commentary. Pappus and the scholiast demonstrate that the difference between the great circle and the circle that divides the illuminated and dark parts of the Moon is much more imperceptible if we observe them in syzygy than if we do it in quadrature. This statement, however, is neatly justified in this scholion in a much simpler way than that offered by Pappus's.

Scholion 45(23)

Scholion 45 appears at the end of proposition 7, in which Aristarchus demonstrates that the Earth–Sun distance is between 18 and 20 times greater than Earth–Moon distance. The scholiast affirms that the results obtained are in agreement with the values of Ptolemy's *Almagest*. Ptolemy first obtains the distances and, based on them and knowing the apparent size of both, calculates the absolute sizes (which have the same ratio as the distances because the Sun and the Moon have the same apparent size). The scholiast, however, proceeds the other way around. He highlights the values of the real sizes of the *Almagest*, states that the ratio between them falls within the limits established by Aristarchus, and then demonstrates that, since they have the same apparent size, the same ratio has to hold also between the distances. It is true, as we have already mentioned, that Ptolemy's values are within the limits of Aristarchus. Indeed, 5 30′ tr / 0 17′ 33″ tr is 18 48′. The scholiast expresses the fractions of the terrestrial radii in the sexagesimal system, a prevalent custom among ancient Greeks (used, for example, systematically in the *Almagest*). Thus, 5 30′ means 5.5 (like 5 hours 30 minutes means 5.5 hours), and 0 17′ 33″ is $0 + 17/60 + 33/3600 = 0.2925$. However, let us remember that this ratio is valid only at the maximum distance in Ptolemy's model. Aristarchus, however, calculates the distances using the quadrature, in which the Moon is at its minimum distance.

Scholion 47(30)

In the development of proposition 7, Aristarchus affirms that, as the square of ZB is to the squared of BE, so is the square of ZH to the square of HE (see the diagram of the proposition). Line BH bisects angle ZBE. The scholion consists of describing the enunciation of Euclid's *Elements* VI, 3 (Heath 2002, 2: 195–197) – that Aristarchus is clearly applying here – but in terms of the letters of the diagram of proposition 7. Commandino's note D develops essentially the same argument, a little more condensed but referencing explicitly to Euclid's *Elements*.

Scholion 50(27)

In the development of proposition 7, Aristarchus asks to draw a circle around triangle ΔKB. See diagram of proposition 7 on p. 99. Aristarchus affirms that since the angle at K is right, line ΔB is a diameter of the circle just drawn. The scholiast assumes that every triangle drawn in a semicircle has a right angle at the vertex that does not divide the semicircle and proves the converted proposition. That is, that every triangle having a right angle divides the circle in which it is inscribed in half with the hypotenuse. Accordingly, since triangle ΔKB is inscribed in the circle and the angle at K is right, side ΔB, the hypotenuse of the triangle, halves the circle and, therefore, will be its diameter. In Figure 4.4, if angle ABΓ is right,

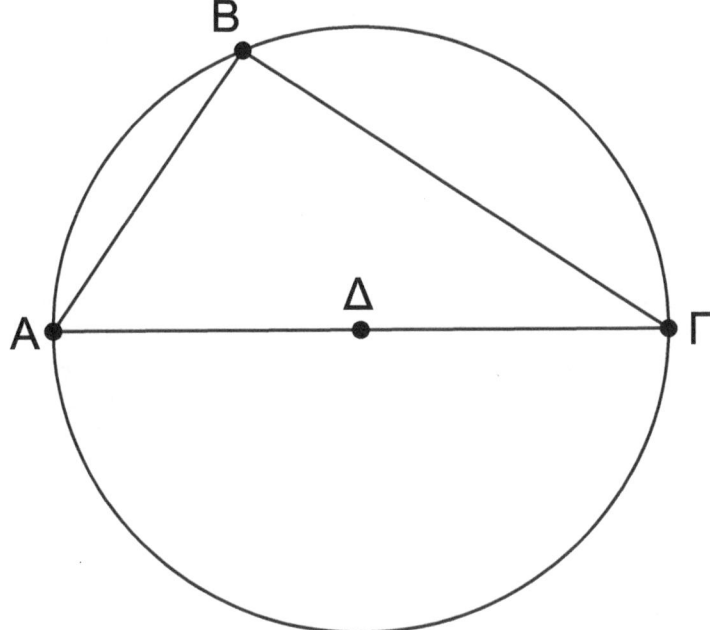

Figure 4.4 If ABΓ is right, then AΓ halves the circle centered at Δ.

then, the hypotenuse AΓ halves the circle centered at Δ. The argument proceeds by *reductio ad absurdum*, showing that ABΓ cannot be greater or smaller than a right angle.

Scholion 52(28)

When developing proposition 7, Aristarchus states that since angle BΔK is 1/30 of a right angle, arc BK will be 1/60 of the complete circle. See the diagram of proposition 7 on p. 99. The scholiast demonstrates that this is so. He shows that the ratio between an angle of a triangle and the three angles of that triangle (that is, two right angles) is the same between an arc and the complete circle. Thus, if angle BΔK is 1/30 of a right angle, it will be 1/60 of two right angles. Accordingly, this same ratio, 1/60, will exist between arc BK and the complete circle.

Scholion 54(29)

In proposition 7, Aristarchus uses again the proto-trigonometric relation that we have called *geometric theorem 4* when he applies it to the proportion between arcs BΛ and BK and their chords. To prove the theorem, the scholiast refers to the demonstration of scholion 32(16). As we have seen, scholion 32(16) demonstrates the proportion between sides and angles. Scholion 54(29) takes an easy step, moving from angles to arcs. And then, through the operation called inversion, the equality is reversed. While scholion 32(16) proves that the lines have a ratio greater than their angles (the greater line and the greater angle being in the denominator), this scholion demonstrates that the ratio between arcs is greater than that between the chords (the greater arc and the greater chord being in the numerator).

Scholion 59

In proposition 10, Aristarchus concludes that the ratio between the volumes of the Sun and the Moon is greater than 5,832 but smaller than 8,000. The scholion compares these values with those of Ptolemy. Once again, Ptolemy's values are within the limits established by Aristarchus. Indeed, according to Ptolemy (*Almagest* V, 16; Toomer 1998: 257), the ratio is 6,644.5. In the scholion, the value appears rounded to 6,644.

Scholion 60(33)

The scholion is very similar to the previous one. It also compares the results of proposition 10 with those of Ptolemy. However, while Ptolemy's value is 6,644.5, that of the scholion is 6,641.5. It is probably a transcription error, confusing a Δ (4) with an A (1).

Scholion 77(47)

Towards the end of proposition 12, Aristarchus once again affirms that the ratio between the chords is greater than that between their arcs (when the greater arc and chord are in the denominator). In this case, the two arcs are not part of a circle subtending a triangle. Instead, there are two parallel lines: ΔΓ, the line dividing the visible and the invisible parts of the Moon, and HZ, representing a great circle parallel to ΔΓ.

See the diagram of proposition 12 on p. 127. The scholiast shows that the relation still holds despite not being part of a triangle simply by transposing one line and its corresponding arc to the end of the other, making two sides of a triangle.

Scholion 83

The scholiast criticizes the first diagram of proposition 13 (see Figure 13a on p. 133), stating that it is inaccurate since the entire circle ΛMN (representing the Moon) must be inside the cone of the shadow. Indeed, in proposition 13, Aristarchus asks that the diagram represent the first instant in which the Moon is totally eclipsed. Therefore, it is natural to think that the circle of the Moon should be wholly immersed in the cone, touching at only one point the edge of the cone. The first thing to point out that this feature is present in almost all manuscripts. Berggren and Sidoli (2007) have shown that the diagram has been modified in some

Arabic manuscripts, making the entire sphere of the Moon remain within the cone. However, this modification produces other inconsistencies. And this is unavoidable because the proposition asks that the end points of the visibility line touch, respectively, the axis of the cone and one of its extremes. This can only be done by leaving a section of the lunar sphere outside the cone. As we know, the visibility line divides what is visible from the Earth and what remains hidden. Now, the question is, the moment in which the Moon is, for the first time, totally eclipsed, is it the moment in which the entire Moon falls under the Earth shadow or the moment in which the whole visible part of the Moon falls under the Earth shadow? The second option is correct. We see the Moon totally eclipsed for the first time when the whole part of the Moon that we see falls under shadow for the first time. It still remains a part of the Moon illuminated, but it is not visible from Earth. Accordingly, Aristarchus considers that the first instant of the eclipse is that in which not the entire Moon but the entire portion of the Moon visible from Earth falls under the shadow cone. That portion is precisely the portion comprised by line ΛN. Thus, the diagram is correct and the scholion is not. See Carman (2014).

Scholion 99(66)

In proposition 13, Aristarchus needs to establish the ratio between two circles belonging to the solar sphere: a solar greatest circle and the circle touching the cone, that is, the visibility circle of the Sun. To do this, he affirms that, since the same cone contains the Sun and the Moon, the ratio between these two circles is the same in both spheres. This equivalence is helpful, for he had already calculated the ratio between the lunar greatest circle and the lunar visibility circle in proposition 12. The scholion demonstrates Aristarchus's assertion. The proof is short but correct. Commandino offers essentially the same proof, but more detailed, in note R.

Scholion 3(5)

Τῶν ὑποθέσεων Ἀριστάρχου, αἱ μέν εἰσιν ἀληθέσταται· αἱ δὲ ψευδέσταται· αἱ δὲ ἐφαπτόμεναι πως τῆς ἀληθείας, καί ἐγγύς ἀληθεῖς.

Καί τὸ μὲν τὴν σελήνην ἀπὸ τοῦ ἡλίου τὸ φῶς λαμβάνειν, ἀλήθες. Δῆλον δὲ τοῦτο ἐκ τῆς σεληνιακῆς ἐκλείψεως· καὶ ἐκ τοῦ, τὸ πεφωτισμένον τῆς σελήνης ἀεὶ πρὸς τὸν ἡλίον νεύειν, καὶ πάσχειν φάσιν.

Τὸ δὲ τὴν γῆν κέντρου λόγον ἔχειν πρὸς τὴν τῆς σελήνης σφαίραν, ψευδές. Ἡ γὰρ πρὸς τὸ κέντρον τῆς γῆς ἐποχὴ τῆς σελήνης ἡμῖν φαινομένης πολὺ διαφέρει· ἣν διαφορὰν καλοῦσι μὲν παράλλαξιν· τοῦτο δὲ οὐκ ἂν ἐγένετο, εἰ καὶ ἡ ὄψις ἡμῖν κέντρον ἦν τῆς σελήνης σφαίρας.

Τὸ δὲ νεύειν εἰς τὴν ἡμετέραν ὄψιν τὸν διορίζοντα τὸ λαμπρὸν καὶ τὸ σκιερὸν τῆς σελήνης, ἀλήθες. Δέδεικται γὰρ ἐν τοῖς ὀπτικοῖς, ὅτι ἐὰν κύκλου περιφέρεια ἐν τῷ αὐτῷ ἐπιπέδῳ τεθῇ, ἐν ᾧ τὸ ὄμμα, εὐθεῖα γραμμὴ ὀφθήσεται.

Ἐπεὶ οὖν ἡ σελήνη ἐν ταῖς διχοτομίαις εὐθεῖα γραμμὴ φαίνεται, ἡ τοῦ κύκλου περιφέρεια ἐν τῷ αὐτῷ ἐστιν ἐπιπέδῳ τῇ ὄψει.

Ἡ δε τέταρτη καί ἡ πέμπτη ὑπόθεσις ἐγγύς ἀληθείας. Δέδεικται γὰρ ἐν τῇ συντάξει, ἐν τῷ πέμπτῳ βιβλίῳ, κατὰ τὸ μέγιστον ἀπόστημα τῆς σελήνης πρὸς τὴν γῆν, διπλασιεπιτρίπεμπτος ἡ διάμετρος τῆς σκιᾶς τῆς διαμέτρου τῆς σελήνης.

Ἡ δε ἔκτη ὑπόθεσις, ψευδές. Ὑπὸ τὴν γὰρ β΄ μόροι ὑποτείνει ἡ σελήνη, ἀλλάξασα λα΄ ἔγγιστα, κατὰ τὸ αὐτὸ μέγιστον ἀπόστημα, ὡς ἐν ἐκείνοις δέδεικται. Δῆλον γὰρ ὅτι οὐ δεῖ ἀδιορίστως λέγειν· κατὰ γὰρ τὰ διάφορα ἀποστήματα, καὶ αὐτὴ μείζων καὶ ἐλάττων φαίνεται· ἐν γὰρ τῷ περιγείῳ ἑαυτῆς, ἡ σελήνη πωλλῷ μείζων φαίνεται ἢ πρὸς τῷ ἀπογείῳ τὴν διάμετρον ἔχουσα.

Translation of the scholia

Scholion 3(5)

Among the hypotheses of Aristarchus, some are wholly true, others are wholly false, and others possess a certain degree of truth, and are almost truth.

For example, that the Moon receives its light from the Sun [is] true. This is manifest from lunar eclipses and from [the fact] that: that the illuminated [part] of the Moon always points towards the Sun, and that it undergoes a phase.

That the Earth has a ratio of a center with respect to the sphere of the Moon [is] false. For the position of the Moon with respect to the center of the Earth differs a lot from [the position of the Moon] as it appears to us. This difference is named "parallax." This would not happen if our eye were also the center of the sphere of the Moon.[1]

That the [circle] delimiting the shaded and the bright parts of the Moon points towards our eye [is] true. For it has been pointed out in *The Optics*,[2] that if an arc of a circle is placed in the same plane as the eye, a straight line will be seen. Since the Moon shows a straight line at dichotomies, the arc of the circle is in the same plane as our eye.

The fourth and fifth hypotheses possess a certain degree of truth. For it has been pointed out in the *Syntaxis*, in the fifth book,[3] that, at the maximum distance of the Moon with respect to the Earth, the diameter of the shadow [of the Earth] is two and three-fifths times the diameter of the Moon.

The sixth hypothesis is false. For according to it, the Moon subtends 2 degrees, in place of approximately $31'$ at the maximum distance, as it is shown in those [books of the *Syntaxis*].[4] For, one should not make a statement [about the size] without qualification.[5] For at different distances, also [the Moon] appears bigger or smaller. For at its perigee, the Moon clearly has a much greater diameter than at its apogee.

---

1 This refers to the center of the orbit of the Moon and not to the center of the Moon itself.

2 Euclid's *Optics*, 22 (Heiberg and Menge 1895: 32–33).

3 *Almagest* (VI,14, Toomer 1998: 254).

4 *Almagest* (VI,15, Toomer 1998: 255).

5 This refers to the fact that it is not convenient to talk about lunar apparent size independently of the distance, because it changes with the former.

Scholion 5(4)

Οἵου ἐστὶν ἡ ἐκ τῆς γῆς α΄, τοιούτου δέδεικται τὸ μὲν τῆς σελήνης ἀπόστημα κατά τινα ἐποχὴν ξδ΄ α΄, τὸ δὲ τοῦ ἡλίου α͵ σι΄, ἐν τοῖς μετὰ τὸ ιβ΄ θεωρήματι τοῦ ε΄ τῆς συντάξεως· Τὸ δὲ ἀπὸ τοῦ κέντρου τῆς γῆς μέχρι τῆς κορυφῆς τοῦ κώνου τῆς σκιᾶς ξη΄. Καὶ οἵου ἐστὶν ἡ ἐκ τοῦ κέντρου τῆς γῆς α΄, τοιούτου ἡ μὲν ἐκ τοῦ κέντρου τῆς σελήνης ιζ΄ λγ΄· ἡ δὲ ἐκ τοῦ κέντρου τοῦ ἡλίου ε΄ λ΄ μόριοι εἰσίν· καί εἰσιν ἀναλόγως, ὡς τὸ ιζ΄ λγ΄ πρὸς ξδ΄ ι΄, οὕτως ε΄ λ΄ μόριοι πρὸς α͵ σι΄.

Scholion 5(4)

In the [passage] following the 12th [proposition] of the fifth [book] of the *Syntaxis*, it has been demonstrated that the distance of the Moon in a certain position is 64;1 [parts],[6] and that of the Sun [is] 1210 [parts] such that the radius of the Earth is 1. And the [distance] from the center of the Earth to the vertex of the shadow cone [is] 68 [parts].[7] And the radius of the Moon is 0;17,33 [parts], and the radius of the Sun is 5;30 parts, such that the radius of the Earth is 1.[8] And, proportionally, as 0;17,33 degrees is to 64;10 [parts], so is 5;30 degrees to 1210 [parts].

---

6 According to Ptolemy (*Almagest*, VI,16, Toomer 1998: 257), they are 64;10 parts and not 64;1 parts.

7 According to Ptolemy (*Almagest*, VI,15, Toomer 1998: 257), they are 268 and not 68 parts.

8 *Almagest*, (VI,16, Toomer 1998: 257).

Scholion 13(8)

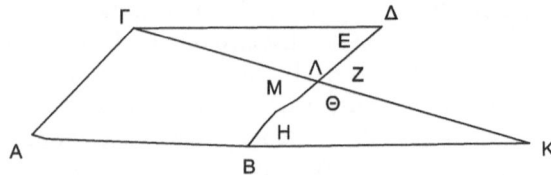

Ἔστω γὰρ ὡς ἡ ΑΚ πρὸς τὴν ΚΒ, οὕτως ἡ ΑΓ πρὸς τὴν ΒΖ· καὶ ἔστω παράλληλος ἡ ΑΓ τῇ ΒΖ. Λέγω ὅτι ἡ ΓΖΚ, εὐθεῖά ἐστιν. Ἤχθω γὰρ διὰ τοῦ Γ τῇ ΑΒ παράλληλος, ἡ ΓΔ· καὶ ἐκβεβλήσθω ἡ ΒΖ ἐπὶ τὸ Δ· παραλληλόγραμμον ἄρα ἐστὶ τὸ ΑΔ· καὶ ἐπεὶ ἐστιν ὡς ἡ ΑΚ πρὸς τὴν ΚΒ, οὕτως ἡ ΓΑ πρὸς τὴν ΖΒ, τουτέστιν ἡ ΔΒ πρὸς τὴν ΖΒ· διελόντι, ἔστιν ὡς ἡ ΑΒ πρὸς τὴν ΒΚ, οὕτως ἡ ΔΖ πρὸς τὴν ΒΖ· καὶ ἐναλλὰξ ἄρα ἐστὶν ὡς ἡ ΑΒ πρὸς τὴν ΔΖ, τουτέστιν ὡς ἡ ΓΔ πρὸς ΔΖ, οὕτως ἡ ΚΒ πρὸς ΒΖ· καὶ ἔστιν ἴση ἡ Ε γωνία, τῇ Η. Δύο δὴ τρίγωνα μίαν γωνίαν μιᾷ γωνίᾳ ἴσην ἔχοντα, καὶ περὶ τὰς ἴσας γωνίας αἱ πλευραὶ ἀνάλογαι, [ἰσογώνια] εἰσιν· ἰσογώνια ἄρα ἔστι τὰ τρίγωνα [ΓΔΖ, ΚΒΖ]. Ἴση ἄρα ἡ Θ τῇ Λ. Κοινὴ προσκείσθω ἡ Μ. Δύο ἄρα αἱ Θ, Μ, δύο ταῖς Μ, Λ, ἴσαι εἰσίν· Αἱ δὲ Λ, Μ, δύο ὀρθαῖς ἴσαι εἰσίν· καὶ αἱ Θ, Μ, ἄρα δύο ὀρθαῖς ἴσαι εἰσίν. Ὥστε ἐπ' εὐθείας ἐστιν ἡ ΓΖ τῇ ΖΚ.

Scholion 13(8)

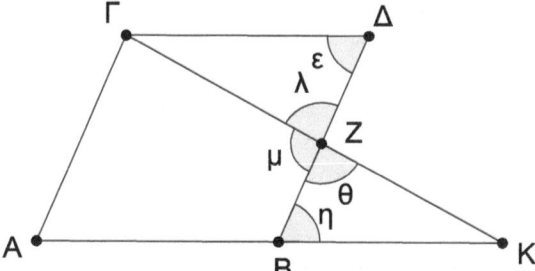

Figure 4.1.1 Diagram of scholion 13(8)

As line AK is to line KB, so let line AΓ be to line BZ. And let line AΓ be parallel to line BZ. I say that line ΓZK is straight. Let line ΓΔ be drawn, through point Γ, parallel to line AB; and let line BZ be produced to point Δ. Therefore, AΔ is a parallelogram. And since, as line AK is to line KB, so is line ΓA to line ZB, that is, line ΔB to line ZB, [therefore] by separation, as line AB is to line BK, so is line ΔZ to line BZ. And, therefore, by alternation, as line AB is to ΔZ, that is, as line ΓΔ is to ΔZ, so is line KB to BZ. And angle E is equal to angle H Now two triangles with an angle [of the first triangle] equal to an angle [of the second triangle], and with the sides of the equal angles, proportional, are equiangular. Therefore, the triangles [ΓΔZ and KBZ] are equiangular. Therefore, angle Θ is equal to angle Λ. Let common angle M be introduced. Therefore, the two angles Θ and M are equal to the two angles M and Λ. But angles Λ and M are equal to two right angles. And therefore, angles Θ and M are equal to two right angles. Thus, line ΓZ is on the [same] straight line than line ZK.

Scholion 14(9)

Δῆλον ὅτι ἡ ΓΛ καὶ ἡ ΖΜ οὔτε κατὰ τῶν κέντρων πίπτουσι τῶν κύκλων, οὔτε ἐπὶ τὰ ΑΞ, καὶ ἐπὶ τὰ ΒΗ μέρη. Εἰ μὲν γὰρ ἐπὶ τῶν κέντρων, οἷον ἐπὶ τοῦ Α· τριγώνου τοῦ ΑΚΓ αἱ δύο γωνίαι ὀρθαὶ ἔσονται, ἥ τε πρὸς τῷ Γ, καὶ ἡ πρὸς τῷ Α. Εἰ δὲ ἐπὶ τὰ ΑΞ μέρη, ἀμβλεῖα μὲν ἔσται ἡ πρὸς τῷ Γ· Ἐπιζευχθείσης γὰρ ἐπὶ τὸ κέντρον τὸ Α εὐθείας, ἐκείνη ἐξώτερον πίπτουσα ἀμβλεῖαν ποιεῖ τὴν πρὸς τῷ Γ κάθετος οὖσα, καὶ ἑτέραν ὀρθὴν τὴν πρὸς ᾧ πίπτει σημεῖον.

Scholion 14(9)

It is evident that lines ΓΛ and ZM fall neither in the center of the circle, nor in sections AΞ and BH.[9] For if they fall at the centers, for example at point A, the two angles of triangle AKΓ will be right, the one at point Γ and the one at point A. And if they fall at section AΞ, the angle at point Γ will be obtuse. For with a straight line to center A having been joined, the [line] falling outside [that line] being a perpendicular [to the axis], produces an obtuse angle at point Γ, and the other angle, the one at the point in which it falls [on the axis], is right.

---

9 This refers to the second diagram of proposition 1 at page 51.

Scholion 19(11)

Ἐπεὶ γὰρ ἐφάπτεται ἡ ΑΘΗ τῶν κύκλων, ὀρθαί εἰσιν αἱ ΒΗΘ, ΑΘΓ γωνίαι. Ὥστε ἐν τῷ τριγώνῳ τῷ ΑΒΗ παρὰ μίαν τὴν ΗΒ παράλληλός ἐστιν ἡ ΘΓ· καὶ ὅμοιον ἔσται τὸ ΗΒΑ τρίγωνον τῷ ΑΘΓ τριγώνῳ· καὶ ἔσται ὡς ἡ ΗΒ πρὸς ΑΒ, οὕτως ἡ ΘΓ πρὸς ΓΑ. Ὥστε καὶ ἐναλλάξ, ἔσται ὡς ἡ ΗΒ πρὸς ΘΓ, οὕτως ἡ ΑΒ πρὸς ΑΓ.

Scholion 19(11)

Since line AΘH is tangent to the circles, angles
BHΘ and AΘΓ are right.[10] And so, in triangle
ABH, line ΘΓ is parallel to one [line, namely] to
line HB. And triangle HBA will be similar to tri-
angle AΘΓ. And, as line HB is to AB, so will line
ΘΓ be to ΓA. And so also, by alternation, as line
HB is to ΘΓ so will line AB be to AΓ.[11]

---

10 This refers to the diagram of proposition 3 in page 65.
11 In most of the manuscripts, it says ΘΓ. However, Fortia (FG:239) is right in changing it by AΓ. In
ms. F (fol. 146) is AΓ.

Scholion 21(13)

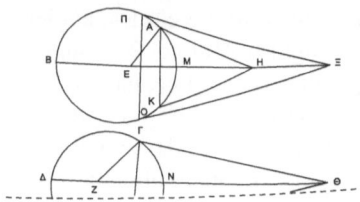

Ὑποκείσθωσαν γὰρ τὰ σχήματα· καὶ ἔστωσαν οἱ μὲν κύκλοι ἴσοι, ἡ δὲ ΖΘ τῆς ΕΞ ἐλάσσων· καὶ ἐφαπτόμεναι αἱ ΞΠ ΞΟ ΘΓ ΘΛ· καὶ ἀπὸ τῶν ἀφῶν εὐθεῖαι αἱ ΟΠ ΓΛ. Λέγω ὅτι καὶ ἡ ΓΛ ἐλάσσων ἐστὶ τῆς ΟΠ. Ἐπεὶ γὰρ ὑπόκειται ἡ ΘΖ τῆς ΕΞ ἐλάσσων, κείσθω τῇ ΘΖ ἴση ἡ ΕΗ, καὶ ἀπὸ τοῦ Η ἐφαπτόμεναι αἱ ΗΑ, ΗΚ· Φανερὸν γὰρ ὅτι ἐν ταῖς ΜΟ, ΜΠ περιφερείαις ἐφάψονται. Καὶ ἐπεζεύχθω ἡ ΑΚ· καὶ ἐπεὶ ἴση ἐστὶν ἡ ΒΗ τῇ ΔΘ, ἡ δὲ ΗΜ τῇ ΘΝ· τὸ ἄρα ὑπὸ τῶν ΒΗ, ΗΜ, ἴσον ἐστὶ τῷ ἀπὸ τῶν ΔΘ, ΘΝ. Καὶ ἔστι τὸ μὲν ὑπὸ τῶν ΒΗ, ΗΜ ἴσον τῷ ἀπὸ τῆς ΑΗ, διὰ τὸ παρατέλευτον τοῦ γ΄ τῶν ἐπιπέδων. Τὸ δὲ ὑπὸ τῶν ΔΘ, ΘΝ, ἴσον τῷ ἀπὸ τῆς ΓΘ. Ὥστε καὶ ἡ ΑΗ ἴση ἐστὶν τῇ ΓΘ. Ἔστι δὲ καὶ ἡ ΕΗ τῇ ΖΘ ἴση· καὶ δύο αἱ ΑΗ, ΗΕ δύο ταῖς ΓΘ, ΘΖ ἴσαι εἰσί, καὶ βάσις ἡ ΑΕ βάσει τῇ ΓΖ ἐστιν ἴση· γωνία ἄρα ἡ ὑπὸ ΑΗΕ γωνίᾳ τῇ ὑπὸ ΓΘΖ γωνίᾳ ἐστὶν ἴση. Ὁμοίως δὴ δειχθήσεται καὶ ἡ μὲν ΗΚ εὐθεῖα, τῇ ΘΛ ἴση, ἡ δὲ ὑπὸ ΕΗΚ γωνία τῇ ὑπὸ ΖΘΛ γωνίᾳ· ὥστε ὅλη ἡ ὑπὸ ΑΗΚ ὅλῃ τῇ ὑπὸ ΓΘΛ ἐστιν ἴση. Δύο δὴ αἱ ΑΗ, ΗΚ, δύο ταῖς ΓΘ, ΘΛ ἴσαι εἰσί. καὶ γωνία ἡ ὑπὸ ΑΗΚ γωνίᾳ τῇ ὑπὸ ΓΘΛ ἐστιν ἴση. Ἡ βάσις ἄρα ἡ ΑΚ βάσει τῇ ΓΛ ἐστιν ἴση. Ἀλλὰ ἡ ΑΚ ἐλάσσων ἐστὶ τῆς ΟΠ καὶ ἡ ΓΛ ἄρα ἐλάσσων ἐστὶ τῆς ΟΠ· διὰ τὸ πορρώτερον εἶναι τοῦ κέντρου.

Scholion 21(13)

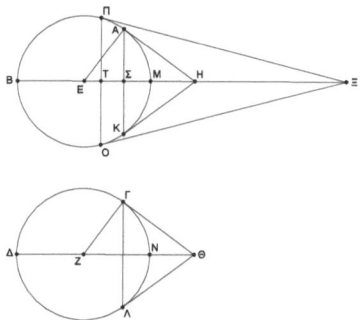

Figure 4.1.2 Diagram of scholion 21(13).

Let the figures be hypothesized. And let the circles be equal, and [let] line ZΘ [be] smaller than EΞ. And [let] lines ΞΠ, ΞO, ΘΓ, and ΘΛ, [be] tangents. And, from the points of contact, [let there be] straight lines ΟΠ and ΓΛ. I say that also line ΓΛ is smaller than line ΟΠ. Since it is hypothesized that line ΘZ is smaller than EΞ, let line EH be equal to line ΘZ, and from point H, [let] tangents, straight lines HΛ and HK [be drawn]. It is evident that they will touch [the circle] at arcs MO and MΠ. And let AK be joined. And, since line BΠ is equal to ΔΘ, and line HM to line ΘN, therefore, the rectangle constructed from lines BH and HM is equal to the rectangle constructed from lines ΔΘ and ΘN. And the rectangle constructed from BH and HM is equal to the square constructed from line AH, because of the penultimate [proposition] of the third book of *On Planar Figures* [of Euclid]. And the rectangle constructed from ΔΘ and ΘN, is equal to the square constructed from line ΓΘ. So that also line AH is equal to line ΓΘ. And also line EH is equal to line ZΘ. And the two lines, AH and HE, are equal to the two lines, ΓΘ and ΘZ; and the base AE is equal to the base ΓZ. Therefore, angle AHE is equal to angle ΓΘZ. In the same manner, it has been demonstrated also that straight line HK is equal to line ΘΛ, and that angle EHK is equal to angle ZΘΛ. So that the whole angle AHK is equal to the whole angle ΓΘΛ. The two lines AH and HK are equal to the two lines ΓΘ and ΘΛ. And angle AHK is equal to angle ΓΘΛ. Therefore, base AK is equal to base ΓΛ. But line AK is smaller than line ΟΠ and, therefore, line ΓΛ is smaller than line ΟΠ, because it is farther from the center.

Scholion 32(16)

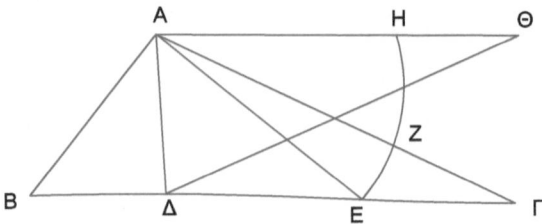

Ἔστω τρίγωνον τὸ ΑΒΓ· καὶ ἐλάττων ἡ ΑΒ τῆς ΑΓ. Λέγω ὅτι ἡ ΑΒ πρὸς τὴν ΑΓ μείζονα λόγον ἔχει, ἤπερ ἡ Γ γωνία πρὸς τὴν Β. Ἤχθω κάθετος ἡ ΑΔ· καὶ τῇ ΒΔ παράλληλος ἡ ΑΘ· καὶ ἐπεὶ μείζων ἡ ΑΓ τῆς ΑΒ, τουτέστιν τὸ ἀπὸ ΑΓ τοῦ ἀπὸ ΑΒ, τουτέστιν τὰ ἀπὸ ΑΔ, ΔΓ, τῶν ἀπὸ τῶν ΑΔ, ΔΒ· κοινοῦ ἄρα ἀφαιρέσει τοῦ ἀπὸ ΑΔ, μείζων ἐστὶν ἡ ΔΓ τῆς ΒΔ. Κείσθω τῇ ΒΔ ἴση, ἡ ΔΕ· καὶ ἐπεζεύχθω οὖν ἡ ΑΕ. Ἴση ἄρα ἡ ΑΕ τῇ ΑΒ·

καὶ κέντρῳ τῷ Α, διαστήματι δὲ τῷ ΑΕ, κύκλος γεγράφθω ὁ ΕΖΗ· καὶ ἐπεζεύχθω ἡ ΕΖ· καὶ ἐκβεβλήσθω. Συμπεσεῖται γὰρ τῇ ΑΘ, ἐπεὶ καὶ τῇ παραλλήλῳ αὐτῆς τῇ ΒΓ. Ἐπεὶ οὖν πάλιν τὸ ΘΑΖ τρίγωνον πρὸς τὴ[ὸ]ν ΗΑΖ τομέα μείζονα λόγον ἔχει ἤπερ τὸ ΖΑΕ τρίγωνον πρὸς τὸν ΖΑΕ τομέα· καὶ ἐναλλὰξ καὶ βάσις καὶ γωνία. Ἡ ΘΖ ἄρα πρὸς τὴν ΖΕ, τουτέστιν, ἡ ΑΖ πρὸς τὴν ΖΓ, διὰ τὴν ὁμοιότητα τῶν ΘΑΖ, ΖΕΓ, τριγώνων, μείζονα λόγον ἔχει, ἤπερ ἡ ὑπὸ ΘΑΖ γωνία, πρὸς τὴν ὑπὸ ΖΑΕ γωνίαν. Συνθέντι καὶ ἀναστρέψαντι, καὶ ἀνάπαλιν, ἡ ΑΖ πρὸς ΑΓ, μείζονα λόγον ἔχει, ἤπερ ἡ ὑπὸ ΗΑΖ πρὸς τὴν ὑπὸ ΗΑΕ. Ἀλλ' ἡ μὲν ΑΖ τῇ ΑΕ ἐστὶν ἴση, τουτέστι τῇ ΑΒ. Ἡ δὲ ὑπὸ ΗΑΓ, τῇ ὑπὸ ΑΓΕ, ἐναλλὰξ γὰρ ἡ δὲ ὑπὸ ΕΑΝ, τῇ ὑπὸ ΑΕΔ, τουτέστι τῇ ΑΒΕ· ἴση γὰρ ἡ ΕΑ τῇ ΑΒ. Ἡ ΒΑ ἄρα πρὸς τὴν ΑΓ, μείζονα λόγον ἔχει, ἤπερ ἡ Γ γωνία, πρὸς τὴν Β.

Scholion 32(16)

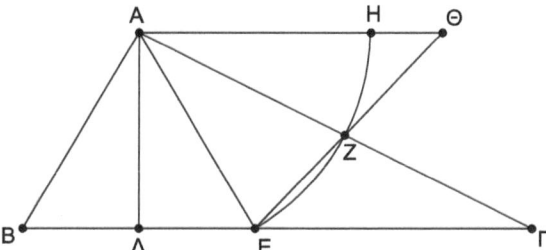

Figure 4.1.3 Diagram of scholion 32(16).

Let there be triangle ABΓ, and [let] line AB [be] smaller than line AΓ. I say that line AB has to line AΓ a greater ratio than angle Γ to angle B. Let the perpendicular line AΔ, and line AΘ, parallel to line BΔ, be drawn. And, since line AΓ is greater than line AB, that is, the square of AΓ [is greater] than the square of AB, that is, the [sum of the] squares of AΔ and ΔΓ than the [sum of the] squares of AΔ and ΔB, therefore, by elimination of the common square of AΔ, line ΔΓ is greater than line BΔ. Let line ΔE be equal to line BΔ. And let AE be joined. Therefore, line AE will be equal to line AB.

And, with center at point A and radius AE, let circle EZH be drawn, and EZ be joined and let it be produced: it will fall upon line AΘ, since it also [falls upon] the parallel to [AΘ], line BΓ. Since, again, triangle ΘAZ has to sector HAZ a greater ratio than triangle ZAE to sector ZAE, by alternation, also the base and the angle [in each triangle have this relation]. Therefore, line ΘZ has to line ZE, that is, line AZ has to line ZΓ – because of the similarity of triangles ΘAZ and ZEΓ – a greater ratio than angle ΘAZ to angle ZAE. By composition and by conversion, and, by inversion, line AZ has to line AΓ a greater ratio than angle HAZ to angle HAE. But line AZ is equal to line AE, that is, to line AB. And angle HAΓ [is equal] to angle AΓE, since by alternation, angle EAN [is equal] to angle AEΔ, that is, to angle ABE. But line EA is equal to line AB. Therefore, line BA has to line AΓ a greater ratio than angle Γ to angle B.

Scholion 37

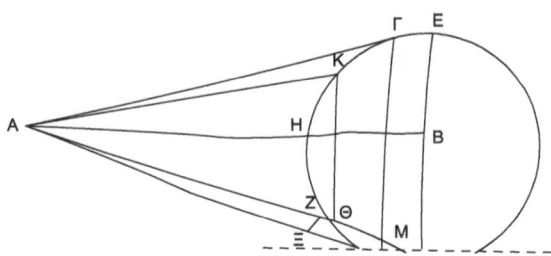

Ἐπεζεύχθω γὰρ ἡ ΑΖ· καὶ φανερὸν ὅτι μείζων ἐστὶν ἡ ΑΖ τῆς ΑΔ. Κείσθω οὖν τῇ ΑΔ ἴση ἡ ΑΜ· καὶ ἐπεζεύχθω ἡ ΔΜ· καὶ ἐπεὶ ἴση ἐστὶν ἡ ΑΜ τῇ ΑΔ, αἱ ἄρα πρὸς τοῖς ΜΔ γωνίαι ὀξεῖαί εἰσιν· ἀμβλεῖα ἄρα ἡ ὑπὸ τῶν ΖΜΔ. Ἐπεζεύχθω ἡ ΔΖ. Μείζων ἄρα ἡ ΔΖ τῆς ΔΜ. Πάλιν δὴ ταῖς ΑΚ, ΑΘ, ἴσαι κείσθωσαν αἱ ΑΝ, ΑΞ. Ἐπεὶ οὖν ἴση ἡ ΜΑ τῇ ΑΔ, ἡ δὲ ΑΝ τῇ ΑΞ, καὶ λοιπὴ ἄρα ἡ ΝΜ λοιπῇ τῇ ΞΔ ἴση ἐστίν, ὡς ἄρα ἡ ΑΝ πρὸς ΝΜ, οὕτως ἡ ΑΞ πρὸς ΞΔ· παράλληλος ἄρα ἡ ΞΝ τῇ ΔΜ· ὅμοιον ἄρα τὸ ΜΑΔ τρίγωνον τῷ ΝΑΞ τριγώνῳ. Ἔστιν ἄρα ὡς ἡ ΔΜ πρὸς ΜΑ, οὕτως ἡ ΞΝ πρὸς ΝΑ· καὶ ἐναλλάξ, μείζων δὲ ἡ ΜΑ τῆς ΑΝ· μείζων ἄρα καὶ ἡ ΜΔ τῆς ΝΞ. Ἀλλὰ τῆς ΜΔ μείζων ἐδείχθη ἡ ΔΖ. Πολλῷ ἄρα μείζων ἡ ΔΖ τῆς ΝΞ. Ἴση δὲ ἡ ΔΖ τῇ ΘΚ, καὶ αἱ περιφέρειαι· καὶ ἡ ΚΘ ἄρα μείζων ἐστὶ τῆς ΝΞ· ἐπεὶ οὖν δύο αἱ ΚΑ, ΑΘ, δύο ταῖς ΝΑ ΑΞ ἴσαι εἰσιν· ἀλλὰ καὶ βάσις ἡ ΚΘ, βάσεως τῆς ΝΞ μείζων ἐστίν· γωνία ἄρα ἡ ὑπὸ ΚΑΘ, γωνίας τῆς ὑπὸ ΝΑΞ μείζων ἐστίν· ὥστε ἐλάττων ἡ ὑπὸ ΖΑΔ τῆς ὑπὸ ΚΑΘ.

Scholion 37[12]

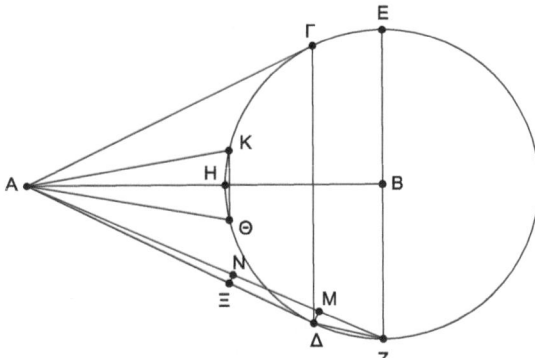

Figure 4.1.4 Diagram of scholion 37.

Let AZ be joined. It is evident that line AZ is greater than line AΔ. Let then line AM be equal to line AΔ. And let ΔM be joined. And, since line AM is equal to line AΔ, the angles at points M and Δ are acute. Therefore, angle ZMΔ is obtuse. Let ΔZ be joined. Therefore, line ΔZ [will be] greater than line ΔM. Again, let lines AN and AΞ be equal to lines AK and AΘ. Since line MA is equal to line AΔ and line AN to line AΞ, also the remaining line NM is equal to the remaining line ΞΔ. Therefore, as line AN is to NM, so is line AΞ to ΞΔ. Therefore, line ΞN is parallel to line ΔM. Therefore, triangle MAΔ is similar to triangle NAΞ. Therefore, as line ΔM is to line MA, so is line ΞN to line NA. And, by alternation, line MA is greater than line AN. Therefore, also line MΔ is greater than line NΞ. But it has been proved that line ΔZ is greater than line MΔ. Therefore, the more so is line ΔZ greater than line NΞ. And line ΔZ is equal to line ΘK, and [also] the arcs. And therefore, line KΘ is greater than line NΞ. Since the two lines KA and AΘ are equal to the two lines NA and AΞ, but also base KΘ is greater than base NΞ, therefore, angle KAΘ is greater than angle NAΞ, and so angle ZAΔ is smaller than angle KAΘ.

---

12 The figure in ms. A is wrong: in it, lines AΘ and AZ collapse into one. In our version, we have corrected it. mss. B, K, and P follow ms. A in the error. ms. Y has a correct, albeit simplified, version.

Scholion 45(23)

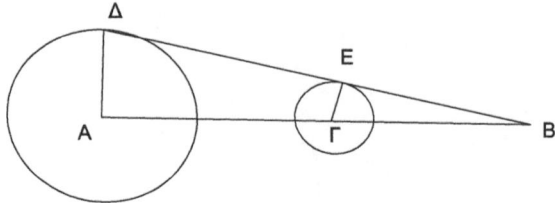

Καὶ τοῦτο σύμφωνον τοῖς τῇ συντάξει δεδειγμένοις. Δέδεικται γὰρ ἐν ἐκείνοις οἵων ἡ ἐκ τοῦ κέντρου τῆς γῆς α΄, τοιούτων ἡ μὲν ἐκ τοῦ κέντρου τῆς σελήνης ο ιζ΄ λγ΄΄· ἡ δὲ ἐκ τοῦ κέντρου τοῦ ἡλίου ε΄ λ΄. ἡ ἄρα ἐκ τοῦ κέντρου τοῦ ἡλίου τῆς ἐκ τοῦ κέντρου τῆς σελήνης, τουτέστι διάστημα τοῦ ἡλίου διαστήματος τῆς σελήνης μείζον μὲν ἢ ιη΄, ἔλαττον δὲ ἢ εἰκοσαπλάσιον. Ὡς δὲ ἡ ἐκ τοῦ κέντρου τοῦ ἡλίου πρὸς τὴν ἐκ τοῦ κέντρου τῆς σελήνης, οὕτω τὸ ἀπόστημα ὃ ἀπέχει ὁ ἥλιος ἀπὸ τῆς γῆς, πρὸς τὸ ἀπόστημα ὃ ἀπέχει ἡ σελήνη ἀπὸ τῆς γῆς. Ἔστιν γὰρ ὡς ἡ ΑΒ πρὸς ΒΓ, οὕτως ἡ ΑΔ πρὸς ΓΕ. Ὄψεως ὑποτίθεται τοῦ Β, καὶ ἐκ τοῦ κέντρου τῆς σελήνης τῆς ΓΕ· καὶ ἐκ τοῦ κέντρου τοῦ ἡλίου τῆς ΑΔ. Καὶ τὸ ἀπόστημα τοῦ ἡλίου ἀπὸ τῆς γῆς τοῦ ἀποστήματος τῆς σελήνης ἀπὸ τῆς γῆς μείζον μὲν ἢ ιη΄, ἔλαττον δὲ ἢ κ΄.

Scholion 45(23)

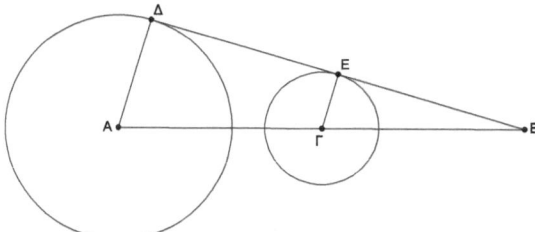

Figure 4.1.5 Diagram of scholion 45(23).

And this agrees with [the propositions] demonstrated in the *Syntaxis*. For it has been demonstrated in them that the radius of the Moon is 0 17′
33″ [parts] and the radius of the Sun, 5 30′ [parts]
such that the radius of the Earth is one. Therefore,
the line departing from the center of the Sun, that
is, the radius of the Sun, is more than 18, but
smaller than 20 times the line departing from the
center of the Moon, that is, the radius of the Moon.
And, as the radius of the Sun is to the radius of the
Moon, so is the distance of the Sun from the Earth
to the distance of the Moon from the Earth. For, as
line AB is to BΓ, so is line AΔ to ΓE. It is supposed
that the eye is at point B, that the radius of the
Moon is line ΓE, and that the radius of the Sun is
line AΔ. And the distance of the Sun from the Earth
is greater than 18 but smaller than 20 times the
distance of the Moon from the Earth.

Scholion 47(30)

Ἐπεὶ γὰρ τριγώνου τοῦ ΖΒΕ ἡ ὑπὸ ΖΒΕ γωνία δίχα τέτμηται ὑπὸ τῆς ΒΗ εὐθείας· αὐτὴ δὲ ἡ ΒΗ τέμνει καὶ τὴν ΖΕ βάσιν· τὰ ἄρα ΖΗ, ΗΕ τμήτατα τὸν αὐτὸν ἕξει λόγον ταῖς λοιπαῖς τοῦ τριγώνου πλευραῖς ταῖς ΖΒ, ΒΕ. Ἔστιν ἄρα ὡς ἡ ΖΒ πρὸς ΒΕ, οὕτως ἡ ΖΗ πρὸς τὴν ΗΕ. Ὥστε καὶ ὡς τὸ ἀπὸ τῆς ΖΕ τετράγωνον, πρὸς τὸ ἀπὸ τῆς ΒΕ τετράγωνον οὕτως τὸ ἀπὸ τῆς ΖΗ τετράγωνον πρὸς τὸ ἀπὸ τῆς ΗΕ τετράγωνον.

Scholion 47(30)

For angle ZBE of triangle ZBE is bisected by straight line BH. And line BH itself also cuts base ZE. Therefore, segments ZH and HE will have the same ratio as the remaining sides, ZB and BE, of the triangle. Therefore, as line ZB is to BE, so is line ZH to line HE. So that also, as the square of line ZB is to the square of line BE, so is the square of line ZH to the square of line HE.

Scholion 50(27)

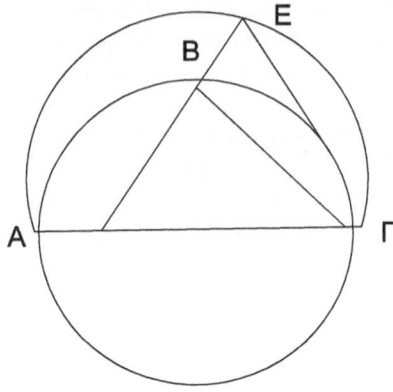

Ἀποδεδειγμένου γὰρ τοῦ τὴν ἐν ἡμικυκλίῳ γωνίαν ὀρθὴν εἶναι· καὶ τὸ ἀντίστροφον ἔσται· ἐὰν ᾖ, ὀρθή, ὅτι ἐν ἡμικυκλίῳ ἐστίν. Ἔστω γὰρ τρίγωνον ὀρθογώνιον τὸ ΑΒΓ, ὀρθὴν ἔχον τὴν πρὸς τῷ Β, καὶ περιγεγράφθω κύκλος ὁ ΑΒΓΔ. Λέγω ὅτι ἡ ΑΒΓ περιφέρεια ἡμικύκλιόν ἐστιν.

Εἰ γὰρ μή, ἢ μείζων ἐστὶν ἢ ἐλάττων. Ἔστω πρότερον ἐλάττων, καὶ περιγεγράφθω ἡ ΑΕΓ περιφέρεια ἡμικυκλίου, καὶ ἐπεζεύχθω ἡ ΕΓ. Ἐπεὶ οὖν ἐν ἡμικυκλίῳ ὀρθή ἐστι ἡ Ε, ὀρθὴ καὶ ἡ πρὸς τῷ Β. Ἴση ἄρα ἡ ἐντὸς τῇ ἐκτός, ὅπερ ἄτοπον· ὁμοίως κἂν μείζων ᾖ ἡ περιφέρεια.

Scholion 50(27)[13]

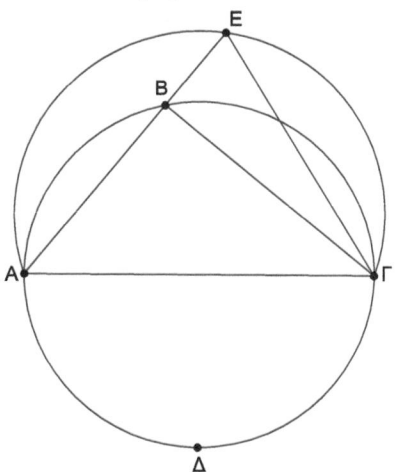

Figure 4.1.6 Diagram of scholion 50(27).

It has been proved that the angle [inscribed] in a semi-circle is right, and the converse will be: if it were right, [it will be so] because of being in a semicircle. Let there be a right-angled triangle, the triangle ABΓ, with the right angle at point B. And let circle ABΓΔ be drawn around it. I say that arc ABΓ is a semicircle.

If not, it is greater or smaller. Let it first be smaller, and let arc AEΓ of the semicircle be drawn around [the triangle], and let EΓ be joined. Since in the semicircle angle E is right, the [angle] at point B is also right. Hence the [angle] that is inside is equal to the [angle] that is outside, which is absurd. The same also [would happen] if the arc were greater [than a semicircle].

13 Point Δ is not in the diagram of several manuscripts, like mss. A, C, and D, but it is in others (like in mss. B, Y, and Z). We decided to include it because it is explicitly mentioned in the text of the scholion.

Scholion 52

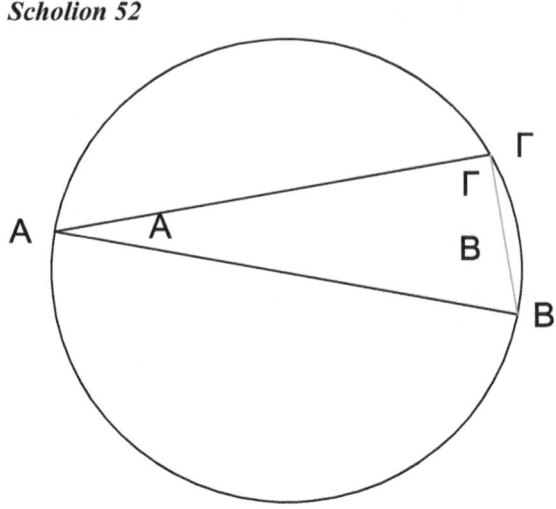

Ἐκκείσθω γὰρ ὅλος ὁ κύκλος, καὶ εἰς αὐτὸν ἡ τὸ
λ΄ τῆς ὀρθῆς ἔχουσα μέρος ἡ ὑπὸ ΒΑΓ. Λέγω ὅτι ἡ
ΒΓ περιφέρεια ἑξηκοστόν ἐστιν ὅλου τοῦ κύκλου.
Ἐπεὶ γὰρ ἐν κύκλῳ αἱ γωνίαι τὸν αὐτὸν ἔχουσιν
λόγον ταῖς περιφερείας, ἔστιν ἄρα, ὡς ἡ Α γωνία
πρὸς τὰς Α, Β, Γ γωνίας, τουτέστιν ὡς ἡ Α πρὸς β΄
ὀρθάς, οὕτως ἡ ΒΓ περιφέρεια πρὸς ὅλον τὸν ΑΒΓ
κύκλον. Ἀλλὰ ἡ Α γωνία τῶν Α, Β, Γ, τουτέστι, δύο
ὀρθῶν, ἑξηκοστὸν μέρος ἐστίν. Εἰ γὰρ τῆς α΄
τριακοστὸν τῶν δύο ἑξηκοστόν· Καὶ ἡ ΒΓ ἄρα
ὅλου τοῦ κύκλου ἑξηκοστὸν μέρος ἐστίν.

Scholion 52

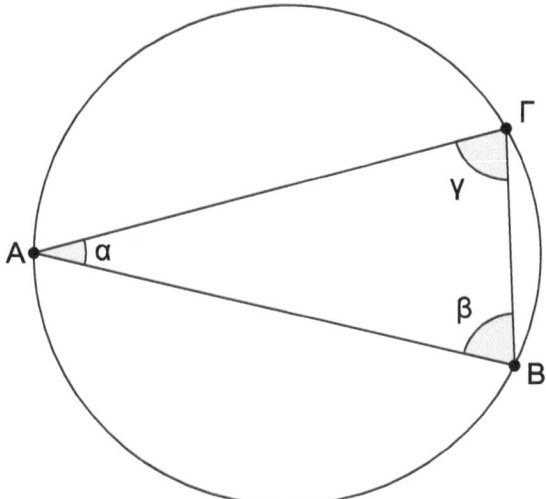

Figure 4.1.7 Diagram of scholion 52.

Let the whole circle be drawn, and in it, angle BAΓ [which is] the thirtieth part of a right angle. I say that arc BΓ is the sixtieth [part] of the whole circle. Since, in a circle, the angles have the same ratio as the arcs, therefore, as angle A is to angles A, B, and Γ, that is, as angle A is to two right angles, so is arc BΓ to the whole circle ABΓ. But angle A is the sixtieth part of angles A, B, and Γ, that is, of two right angles. If it is the thirtieth [part] of 1, it will be the sixtieth of two. And therefore, [arc] BΓ is the sixtieth part of the whole circle.

Scholion 54

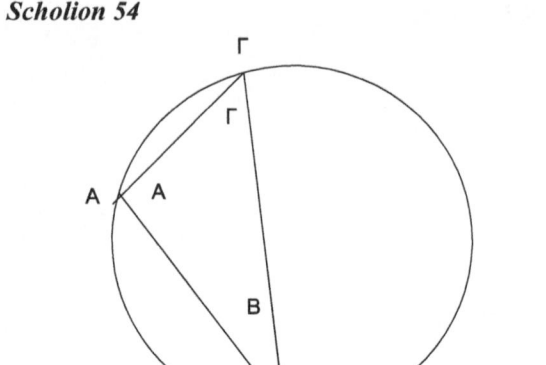

Ἐκκείσθω γὰρ ὁ ΑΒΓ, καὶ μείζων ἡ ΑΒ τῆς ΑΓ περιφερείας· ὥστε ἡ ΓΑ εὐθεῖα, ἐλάττων τῆς ΑΒ· καὶ ἐπεζεύχθω ἡ ΒΓ. Διὰ δὴ τὸ λῆμμα, τὸ ἐν τῷ δ᾽, ἡ ΑΓ πρὸς τὴν ΑΒ μείζονα λόγον ἔχει, ἤπερ ἡ πρὸς τῷ Β γωνία πρὸς τὴν Γ. Ὡς δὲ ἡ Β πρὸς τὴν Γ, οὕτως ἡ ΓΑ περιφέρεια πρὸς τὴν ΑΒ· καὶ ἡ ΑΓ ἄρα εὐθεῖα πρὸς τὴν ΑΒ μείζονα λόγον ἔχει, ἤπερ ἡ ΓΑ περιφέρεια, πρὸς τὴν ΑΒ. Ἀνάπαλιν, ἄρα ἡ ΑΒ εὐθεῖα πρὸς ΑΓ ἐλάττονα λόγον ἔχει, ἤπερ ἡ ΑΒ περιφέρεια πρὸς τὴν ΑΓ περιφέρειαν. Ὥστε ἡ ΑΒ περιφέρεια πρὸς τὴν ΑΓ περιφέρειαν μείζονα λόγον ἔχει, ἤπερ ἡ ΑΒ εὐθεῖα πρὸς ΑΓ εὐθεῖαν.

Scholion 54

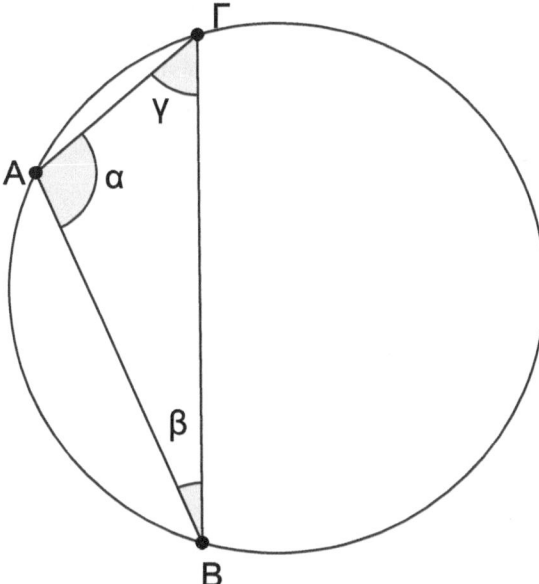

Figure 4.1.8 Diagram of scholion 54.

Let [circle] ABΓ and arc AB, greater than arc AΓ, be drawn, so that straight line ΓA is smaller than line AB. And let BΓ be joined. Now because of the lemma in the fourth [book],[14] line AΓ has to line AB a greater ratio than angle at point B [has] to angle at point Γ. And, as angle B is to angle Γ, so is arc ΓA to arc AB. And, therefore, the straight line AΓ has to line AB a greater ratio than arc ΓA [has] to arc AB. By inversion, therefore, straight line AB has to AΓ a smaller ratio than arc AB [has] to arc AΓ. So that arc AB has to arc AΓ a greater ratio than straight line AB [has] to straight line AΓ.

14 This refers to scholion 32(16). See p. 269.

Scholion 59

Κατὰ Πτολεμαῖον ἡ διάμετρος τοῦ ἡλίου ἔστι τμημάτων ιη´ καὶ δ᾽ πέμπτων, οἵων ἡ τῆς σελήνης διάμετρος ἑνός· γίνεται οὖν ἡ σφαῖρα τοῦ ἡλίου, ϛχμδ´, οἵων ἡ σφαῖρα τῆς σελήνης ἑνός. Hoc scholion pertinet ad 10 theorema.

Scholion 59

According to Ptolemy, the diameter of the Sun is 18 and 4 fifths parts[15] such that the diameter of the Moon is one. Then the sphere of the Sun will be 6644[16] parts such that sphere of the Moon is one. *This scholion belongs to theorem 10.*

15 It is very unusual to find expressions like "four fifth-parts" to represent fractions in Greek. This kind of expressions seems to be restricted to astronomical and mathematical authors and is particularly characteristic of Hipparchus.

16 In Noack's translation, it appears 6,644 $^1/_2$, which is correct, but it is not in his edition of the Greek text. The text in manuscript R, the only one in which this scholion appears, actually reads, ϛυηγ' νδ' λϛ" (6443, 54′ 36″). She has corrected it based on the value of *Almagest* V, 16 (Toomer 1998: 257), but it would be interesting to analyze where this value comes from.

Scholion 60(33)

Ἐν τῇ Συντάξει δέδεικται ὁ ἥλιος πρὸς τὴν σελήνην λόγον ἔχων ὅν τὰ ϛ,χμα΄ S΄ πρὸς α΄. Συμφωνεῖ οὖν ἐκεῖνο τούτῳ.

Scholion 60(33)

In the *Syntaxis* it is demonstrated that the Sun has
to the Moon the ratio that 6641 1/2 has to 1.[17]
Then that is in agreement with this.

---

17 According to Ptolemy (*Almagest*, VI,15, Toomer 1998: 257), they are 6,644 and ½ and not 6,641 and ½.

Scholion 77(47)

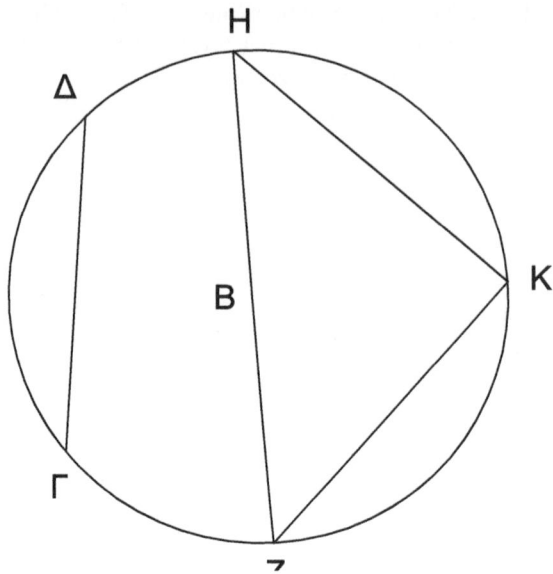

Ἐπεὶ γὰρ μείζων ἐστὶν περιφέρεια ἡ ΗΚΖ τῆς ΓΕΔ περιφερείας· κείσθω γὰρ τῇ ΓΕΔ ἴση ἡ ΖΚ, καὶ ἐπεζεύχθωσαν αἱ ΗΚ, ΗΖ· ἴση ἄρα ἡ ΖΚ τῇ ΔΓ. Καὶ ἐπεὶ ἡ ΖΗ τῆς ΖΚ μείζων ἐστὶν, ἡ ΖΚ ἄρα πρὸς ΖΗ μείζονα λόγον ἔχει ἤπερ ἡ Η γωνία πρὸς τὴν Κ γωνίαν. Ἴση δὲ ἡ ΖΚ τῇ ΓΔ. Καὶ ἡ ΓΔ ἄρα πρὸς τὴν ΖΗ μείζονα λόγον ἔχει ἤπερ ἡ Η γωνία πρὸς τὴν Κ. Ὡς δὲ ἡ Η πρὸς Κ, οὕτως ἡ ΚΖ περιφέρεια πρὸς τὴν ΗΕΖ περιφερείαν, καὶ ἡ ΓΔ ἄρα πρὸς τὴν ΖΗ, καὶ λόγον ἔχει ἤπερ ἡ ΖΚ περιφέρεια, τουτέστιν ἡ ΔΕΓ περιφέρεια, πρὸς τὴν ΗΕΖ περιφερείαν.

Scholion 77(47)[18]

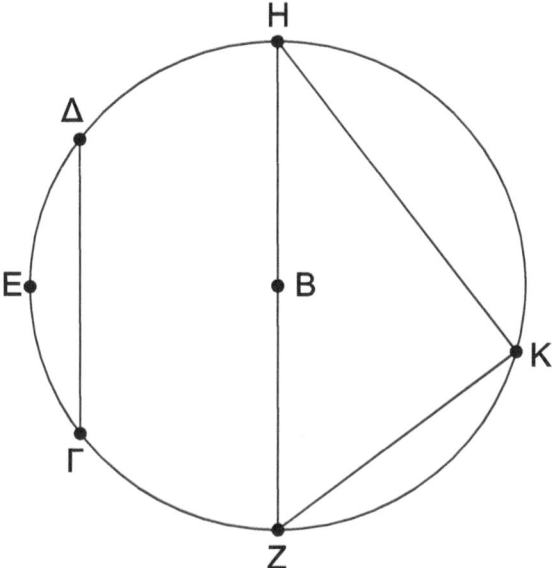

Figure 4.1.9 Diagram of scholion 77(47).

For arc HKZ is greater than arc ΓΕΔ. Let [arc] ZK, be equal to [arc] ΓΕΔ, and let HK and HZ be joined. Therefore, line ZK [will be] equal to line ΔΓ. And, since line ZH is greater than line ZK, line ZK has to ZH a greater ratio than angle H to angle K. And line ZK is equal to line ΓΔ. And, therefore, line ΓΔ has to line ZH a greater ratio than angle H has to angle K. And as angle H is to [angle] K, so is arc KZ to arc HEZ, and, therefore, line ΓΔ to line ZH, and it has the ratio that arc ZK, that is, arc ΔΕΓ, has to arc HEZ.

18 Point E is not in the diagrams of mss. A, B, C, and D, but it is in those of mss. L, M. N, and O (in ms. K, it is not possible to know because it is damaged in that part). We decided to include it since it is explicitly mentioned in the text of the scholion.

Scholion 83

Ἔσφαλται ἡ καταγραφή· ὅλος γὰρ ὁ ΛΜΝ κύκλος ἐντὸς τοῦ κώνου ὀφείλει εἶναι.

Scholion 83

The diagram[19] is wrong, since the whole circle
ΛMN should be inside the cone.

---

19 This refers to the first diagram of proposition 13.

Scholion 99(66)

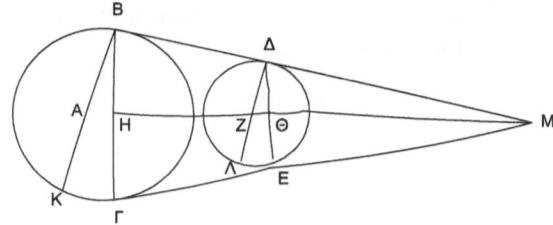

Ἔστω γὰρ τοῦ ἡλίου κύκλος ὁ ΒΓΚ, οὗ κέντρον τὸ Α· σελήνης δὲ κύκλος ὁ ΔΕΛ, κέντρον δὲ αὐτῆς τὸ Ζ· αἱ δὲ πλευραὶ τοῦ κώνου τοῦ περιλαμβάνοντος τόν τε ἡλίον καὶ τὴν σελήνην, αἱ ΒΔΜ, ΓΕΜ· αἱ δὲ διορίζουσαι αἱ ΒΓ, ΔΕ· καὶ ἀπὸ τῶν Β, Δ, ἐπὶ τὰ Α, Ζ ἐπιζευχθεῖσαι ἐκβεβλήσθωσαν ἐπὶ τὰ Κ, Λ.

Ἐπειδὴ ὀρθαί εἰσιν αἱ Η, Θ, παράλλελός ἐστιν ἡ ΒΗ τῇ ΔΘ. Ὥς τε ὅμοιόν ἐστι τὸ ΒΗΜ τρίγωνον, τῷ ΔΘΜ τριγώνῳ· ἔστιν ἄρα ὡς ἡ ΒΗ πρὸς ΔΘ, οὕτως ἡ ΒΜ πρὸς ΜΔ.

Ὡς δὲ ἡ ΒΗ πρὸς ΔΘ, οὕτως ἡ ΒΓ πρὸς ΔΕ· καὶ ὡς ἄρα ἡ ΒΜ πρὸς ΜΔ, οὕτως ἡ ΒΓ πρὸς ΔΕ.

Ὁμοίως δὴ δείξομεν ὅτι καὶ ὡς ἡ ΒΚ πρὸς ΔΛ, οὕτως ἡ ΒΜ πρὸς ΜΔ.

Παράλληλος γὰρ ἐστὶν ἡ ΑΒ τῇ ΔΖ. Οὐκοῦν ἔσται καὶ ὡς ἡ ΔΕ πρὸς ΒΓ, οὕτως ἡ ΔΛ πρὸς ΒΚ· καὶ ἐναλλὰξ ὡς ἡ ΔΕ πρὸς ΔΛ, οὕτως ἡ ΒΓ πρὸς ΒΚ, ὅπερ ἔδει δεῖξαι.

Scholion 99(66)

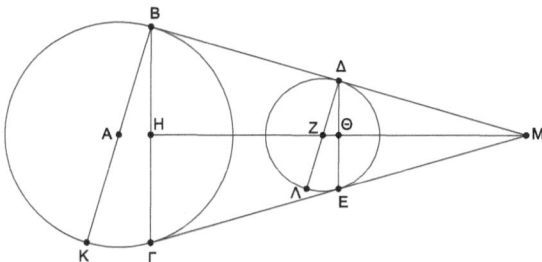

Figure 4.1.10 Diagram of scholion 99(66).

Let the circle of the Sun be BΓK, the center of which is point A, the circle of the Moon, [be] ΔEΛ, and its center, point Z. And [let] the sides of the cone enveloping the Sun and the Moon, [be] lines BΔM and ΓEM. And [let] those dividing [lines], [be] lines BΓ and ΔE. And let [the lines] be joined from points B and Δ, and to points A and Z; and let they also be produced to points K and Λ.

Since angles H and Θ are rights, line BH is parallel to line ΔΘ. And so, triangle BHM is similar to triangle ΔΘM. Therefore, as line BH is to line ΔΘ, so is line BM to MΔ.

And as line BH is to ΔΘ, so is line BΓ to ΔE. And, therefore, as line BM is to MΔ, so is line BΓ to ΔE.

Certainly, we will prove in the same manner that, as line BK is to ΔΛ, so is line BM to MΔ.

For line AB is parallel to line ΔZ. Accordingly, also, as line ΔE is to BΓ, so will line ΔΛ be to BK. And, by alternation, as line ΔE is to ΔΛ, so [will] line BΓ [be] to BK, what had to be demonstrated.

Appendix 1

Mathematical tools

In this appendix, we will briefly explain the main mathematical tools used by Aristarchus in the treatise. These can be divided into two groups. On the one hand, he uses a series of pseudo- or pre-trigonometric relationships, that is, functions allowing him to move from ratios between angles to ratios between sides or vice versa. Unlike trigonometry, the two ratios will not be equal, but one greater or smaller than the other. It is precisely the impossibility of establishing equalities that explains why, for each required value, Aristarchus has to propose an upper and a lower limit. On the other hand, Aristarchus uses certain relationships between inequalities, allowing him to operate between them. Let us first look at the pseudo- or pre-trigonometric relationships and then the operations between inequalities.[1]

Proto-trigonometric relationships

In *On Sizes*, Aristarchus uses four proto-trigonometric relationships. Note that he uses the relationships without justifying them. This clearly shows that they were propositions already known to his potential readers.

Geometric theorem 1

The first theorem (see Figure 5.1) establishes that given two right triangles with a leg in common, the inverse of the ratio between the angles opposite to the common leg is smaller than the ratio between the unequal legs. In the figure:

$$\frac{\beta}{\alpha} < \frac{BD}{BC}$$

Aristarchus uses this theorem, for example, in propositions 11a.

An implicit application of this theorem can be found in proposition 4.[2]

---

1 In this section, we follow Berggren and Sidoli (2007: 223–227).
2 Cfr. Sidoli and Berggren (2007: 228, note 38).

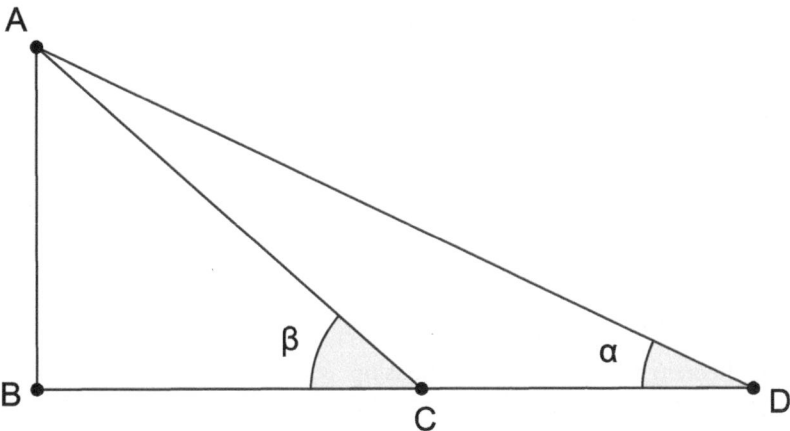

Figure 5.1 Geometric theorem 1.

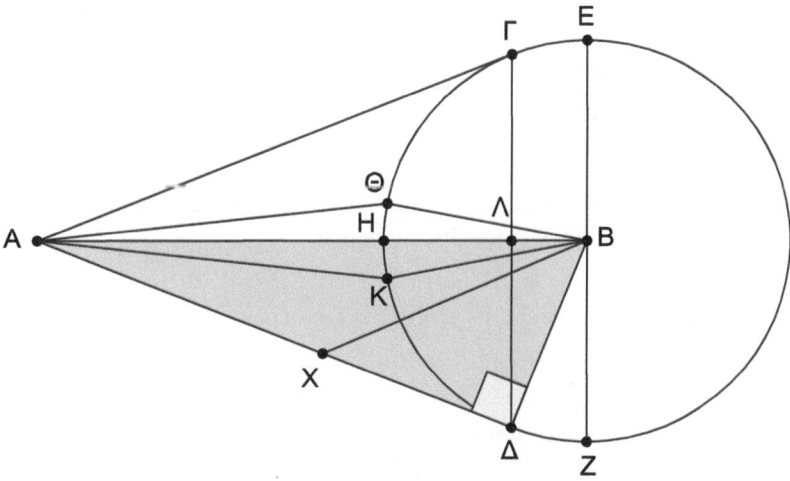

Figure 5.2 Application of geometrical theorem 1 to proposition 4.

Refer to Figure 5.2. In triangle BΔA, the angle at *A* is equal to 1° (half the apparent size of the Moon), while the angle at *Δ* is right. Aristarchus establishes the ratio between sides AΔ and ΔB from the relationship between angles at A and Δ. To make the use of the theorem explicit, let us draw line BX in such a way that line ΔX is equal to BΔ. Thus, the angle BXΔ will be 45°. Now we can apply the theorem:

$$\frac{x}{a} < \frac{\Delta A}{\Delta X}$$

Where x is the angle at X and a, the angle at A. Now, we know that ΔX is equal to BΔ, and that x is $45°$ and a is $1°$. Therefore, we know that:

$$\frac{45°}{1°} < \frac{\Delta A}{B\Delta}$$

That is, the ratio between BΔ and ΔA is smaller than the ratio between 1 and 45. Therefore, BΔ is smaller than 1/45 times ΔA.

Geometric theorem 2

The second theorem holds a similar relationship, but with the hypotenuses of the respective triangles. In this case, the inverse of the ratio of the angles opposite the common leg is greater than the ratio of the hypotenuses.

$$\frac{\beta}{\alpha} > \frac{AD}{AC}$$

This theorem is used in proposition 4.

Geometric theorem 3

The third geometrical theorem establishes a relationship between the complements of those angles with the unequal legs. It asserts that:

$$\frac{90-\alpha}{90-\beta} < \frac{BD}{BC}$$

This theorem is used in proposition 7.

Geometric theorem 4

This last theorem establishes a relationship between the arcs and their chords. See Figure 5.3. It holds that the ratio between the arcs is greater than the ratio between their chords (when the major arc is on the numerator):

$$\frac{arc\ DC}{arc\ CB} > \frac{chord\ DC}{chord\ CB}$$

This is the most used theorem. It appears, for example, in propositions 7b, 11b, and 12.

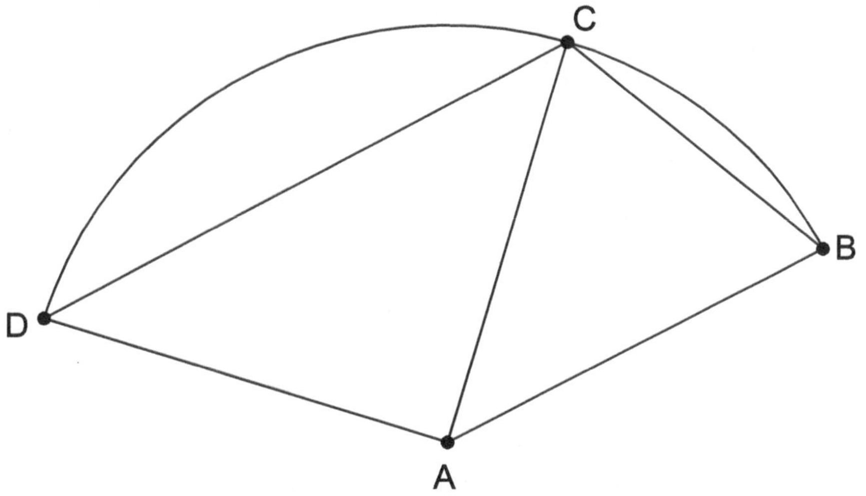

Figure 5.3 Geometrical theorem 4.

Operations with quotients

Aristarchus also makes certain operations on inequalities whose terms are quotients. Here we list the ones that he uses in the treatise. In Euclid's *Elements*, they are expressed as operations between equalities, but Aristarchus uses them with inequalities (Heath 1913: 328). In each one, we add the reference to Netz 1999 for further discussion.

Alternation (καὶ ἐναλλάξ). Netz (1999: 139) (f.56).

$$\frac{A}{B} \gtrless \frac{C}{D} \Rightarrow \frac{A}{C} \gtrless \frac{B}{D}$$

Inversion (ἀνάπαλιν). Netz (1999: 140) (f.57).

$$\frac{A}{B} \gtrless \frac{C}{D} \Rightarrow \frac{B}{A} \lessgtr \frac{D}{C}$$

Composition (συνθέντι). Netz (1999: 140) (f.58).

$$\frac{A}{B} \gtrless \frac{C}{D} \Rightarrow \frac{(A+B)}{B} \gtrless \frac{(C+D)}{D}$$

Separation (διελόντι). Netz (1999: 140) (f.59).

$$\frac{A}{B} \gtrless \frac{C}{D} \Rightarrow \frac{(A-B)}{B} \gtrless \frac{(C-D)}{D}$$

Conversion (ἀναστρέψαντι). Netz (1999: 140) (f.60).

$$\frac{A}{B} \gtrless \frac{C}{D} \Rightarrow \frac{A}{(A-B)} \lessgtr \frac{C}{(C-D)}$$

Equality of Terms (δι' ἴσου). Netz (1999: 157) (f.102).

$$\frac{A}{B} \gtrless \frac{C}{D} \text{ and } \frac{B}{E} \gtrless \frac{D}{F} \Rightarrow \frac{A}{E} \gtrless \frac{C}{F}$$

Appendix 2

List of manuscripts

Id.	Manuscript	Epoch
A	Vat. Gr. 204	Late 9th century
B	Par. Gr. 2342	14th century
C	Par. Gr. 2488	16th century
D	Vat. Gr. 1055	16th century
E	Ricc. 1192	16th century
F	Vat. Gr. 192	End of 11th century, beginnings of 12th century
G	Magl. 11	16th century
H	Par. Suppl. Gr. 12	16th century
J	Ambr. J 84 inf.	16th century
K	Vat. Gr. 203	13th century
L	Vat. Barb. Gr. 186	15th century
M	Vat. Ross. 978	16th century
N	Esc. y-I-7	16th century (1548 approximately)
O	Ricc. 38	16th century
P	Vat. Gr. 202	13th century
Q	Vat. Gr. 2472	Mid-14th century
R	Vat. Gr. 2363	Mid-15th century
S	Marc. Gr. 304	Mid-15th century
T	Esc. X-I-4	16th century (dated November 1, 1542)
U	Ambr. A 101 sup.	16th century
V	Upps. UB Gr. 53	Beginnings of century XVII
W	Savil. 10	End of 16th century, beginnings of 17th century
X	Cant. Trinit. O.5.12	17th century
Y	Par. Gr. 2364	End of 15th century, beginnings of 16th century
Z	Par. Gr. 2386	Mid-16th century
1	Vat. Gr. 191	13th–14th centuries
2	Vat. Gr. 1346	16th century
3	Ambr. C 263 inf.	Mid-16th century
4	Vindob Suppl. Gr. 9	Mid-16th century
5	Par. Gr. 2366	16th century (before 1550)

References

Editions of *On Sizes*

Carman, C. and Buzón, R. (2020). *Aristarco de Samos, Acerca de los tamaños y las distancias del Sol y de la Luna. Estudio preliminar, revisión del texto griego y traducción al castellano*. Barcelona: Universitat de Barcelona Edicions.

Commandino, F. (1572). *Aristarchi de magnitudinibus et distantiis solis et lunae liber cum Pappi Alexandrini explicantionibus quibusdam a Federico Commandino Urbinate in latinum conversus et commentariis illustratus*. Cum Priuilegio Pont. Max. in annos X. Pesaro: Camillo Francischini.

Fortia, Comte de (1810). *Histoire d'Aristarque de Samos, suivie de la traduction de son ouvrage sur les distances du Soleil et de la Lune, de l'histoire de ceux qui ont porté le nom d'Aristarque avant Aristarque de Samos, et le commencement de celle des Philosophes qui ont paru avant ce même Aristarque. Par M. de F*****. Paris: Duminil-Leuseur.

Fortia, Comte de (1823). *Traité d'Aristarque de Samos sur les grandeurs et les distances du Soleil et de la Lune, traduit en français pour la première fois, par M. le Comte de Fortia d'Urban*. Paris: Firmin Didot Père.

Gioé, A. (2007). *Aristarque de Samos. Sur les dimensions et les distances du Soleil et de la Lune. Édition critique, présentation, traduction et notes*. Ph.D. Thesis. Paris: Université Paris IV-Sorbonne.

Heath, Th. (1913). *Aristarchus of Samos. The Ancient Copernicus. A History of Greek Astronomy to Aristarchus Together with Aristarchus' Treatise on the Sizes and Distances of the Sun and Moon. A New Greek Text with Translation and Notes by sir Thomas Heath*. Oxford: Oxford University Press.

Massa Esteve, M. R. (2007). *Aristarco de Samos sobre los tamaños y las distancias del Sol y la Luna, texto latino de Federico Commandino. Introducción, traducción y notas de María Rosa Massa Esteve*. Cádiz: Universidad de Cádiz.

Nizze (1856). *Ἀριστάρχου Σαμίου βιβλίον περὶ μεγεθῶν καὶ ἀποστημάτων ἡλίου καὶ σελήνης, mit kritischen Berichtigungen von E. Nizze. Mit zwei Figuren – Tafeln*. Stralsund: Koeniglichen Regierungs Buchdruckerei.

Nokk (1854). *Aristarchos über die Grössen und Entfernungen der Sonne und des Mondes. Uebersetzt und erläutert von A. Nokk. Als Beilage zu dem Freiburger Lyceums-Programme von 1854*. Friburgo.

Spandagos, E. (2001). *Περὶ μεγεθῶν καὶ ἀποστημάτων ἡλίου καὶ σελήνης τοῦ Ἀριστάρχου τοῦ Σαμίου. Εἰσαγωγή, Ἀρχαῖο κείμενο. Μετάφραση, Ἐπεξηγήσεις, Σχόλια Ἱστορικά Στοιχεία*. Ahtens: Αἴθρα.

Stamatis, E. (1980). Ἀριστάρχου Σαμίου περὶ μεγεθῶν καὶ ἀποστημάτων ἡλίου καὶ σελήνης. *2300 τῇ ἐπέτειος τῆς γεννήσεώς του 320 π.Χ. -1980*. Athens. Private edition.

Thomas, I. (1941). *Selections Illustrating the History of Greek Mathematics with an English Translation by Ivor Thomas in Two Volumes. Volume II: From Aristarchus to Pappus*. Cambridge, MA: Harvard University Press.

Valla, G. (1498). *Giorgio Valla Placentino Interprete. Hoc in volumine hec continetur: Nicephori Logica, Georgij Valle libellus de argumentis, Euclidis quartusdecimus Elementorum, Hypsiclis interpretatio eiusdem libri Euclidis, Nicephorus de astrolabo, Proclus De astrolabo, Aristarchi Samij De magnitudinibus et distantijs solis et lune, Timeus De mundo, Cleonidis musica, Eusebii Pamphili De quibusdam theologicis ambiguitatibus, Cleomedes De mundo, Athenagore philosophi De resurrectione, Aristotelis De celo, Aristotelis Magna ethica, Aristotelis Ars poetica, Rhazes De pestilentia, Galenus De in equali distemperantia, Galenus De bono corporis habitu, Galenus De confirmatione corporis humani, Galenus De presagitura, Galenus De presagio, Galeni introductorium, Galenus De succidaneis, Alexander aphroditeus De causis febrium, Pselus De victu humano*. Venecia: Simon Bevilacqua de Pavia.

Vera Fernández de Córdoba, F. (1956–1969). *Historia de la Cultura Científica*. Buenos Aires: Ediar.

Wallis, J. (1688). *Aristarchi Samii De Magnitudinibus et distantiis Solis et Lunae, liber. Nunc primum Graeche editus cum Federici Commandini versiones Latina notisque illius et Editoris. Pappi Alexandrini Secundi Libri Mathematicae Collectionis, fragmentum, hactenus desideratum. E. Codice MS. Edidit, latinum fecit, Notisque illustravit*. Johannes Wallis, S.T.D. Geometriae Professor Savilianus; et Regalis Societetis Londini, Sodalis. Oxoniae, et Theatro Sheldoniano, 1688. Re-edited in Johannis Wallis Opera Mathematica, 1693–1699, Vol. 3, pp. 565–594. We indicate the pages of this last edition.

General bibliography

Abers, E. y Ch. Kennel (1975). "Commentary: The Role of Error in Ancient Methods for Determining the Solar Distance". In R. Westman (ed.), *The Copernican Achievement*. Berkeley: University of California Press, pp. 130–137.

Adler, A. (1928–1938). *Suidae Lexicon*, 5 vol. (Lexicographi Graeci). Stuttgart: Teubner.

Al-Bīrūnī (1954–1956). *Al-Qānūn al-Mas'ūdi*, 3 vols. Hyderabad-Deccan: Dā'irat al-ma'ārif al-'Uthmāniyya.

Al-Farghani (1910). *Elementa astronomica (liber de aggregatione scientiae stellarum)*. Ed. R. Campani. Firenze: Passerini.

Aristotle (1942). *Meteorologica: With an English Translation by H. D. P. Lee*. The Loeb Classical Library. Cambridge, MA: Harvard University Press.

Bacon, R. (1928). *The Opus Maius of Roger Bacon. A Translation of Robert Belle Burke*, 2. vols. Philadelphia: University of Pennsylvania Press; London: Oxford University Press.

Baldi, B. (1714). "Vita Federici Commandini". *Giornale de'letterati d'Italia*, 19, pp. 140–185. Reprinted in G. Ugolini y F. L. Polidori (eds.) (1859). *Versi e presse scelete di Bernardino Baldi*. Firenze: Felice Le Monnier, pp. 513–537 (We follow this pagination).

Berggren, J. L. and A. Jones (2000). *Ptolemy's Geography. An Annotated Translation of the Theoretical Chapters*. Princeton: Princeton University Press.

Berggren, J. L. and N. Sidoli (2007). "Aristarchus's On the Sizes and Distances of the Sun and the Moon: Greek and Arabic Texts". *Archive for History of Exact Sciences*, 61, pp. 213–254.

Berggren, J. L. and R. Thomas (2006). *Euclid's Phaenomena. A Translation and Study of a Hellenistic Treatise in Spherical Astronomy*. Winnipeg: University of Manitoba.

Blass, F. (1883). "Der Vater des Archimedes". *Astronomische Nachrichten*, 104, pp. 255–256.

Boll, F. (1894). "Studien über Claudius Ptolemäus". *Jahrbücher für classische Philologie*, Supl. Bd. 21, Lepizig, pp. 51–243.

Boulliau, I. (1645). *Astronomia Philolaica*. Paris: Simeon Piget.

Bowen, A. and B. Goldstein. (1994). "Aristarchus, Thales, and Heraclitus on Solar Eclipses: An Astronomical Commentary on P. Oxy. 53.3710 Cols. 2.33–3.19". *Physics*, 31, pp. 689–729.

Bowen, A. and R. Tood (2004). *Cleomedes' Lectures on Astronomy. A Translation of The Heavens with an Introduction and Commentary by Bowen and Todd*. Berkeley: University of California Press.

Britton, J. P. (1992). *Models and Precision: The Quality of Ptolemy's Observations and Parameters*. New York: Garland.

Carman, C. (2009). "Rounding Numbers: Ptolemy's Calculation of the Earth-Sun Distance". *Archive for History of Exact Sciences*, 63 (2), pp. 205–242.

Carman, C. (2014). "Two Problems in Aristarchus's Treatise On the Sizes and Distances of the Sun and Moon". *Archive for History of Exact Sciences*, 68, pp. 35–65.

Carman, C. (2015). "The Planetary Increase of Brightness During Retrograde Motion: An *explanandum* Constructed *ad explanantem*". *Studies in History and Philosophy of Science*, 54, pp. 90–101.

Carman, C. (2017). "Martianus Capella's Calculation of the Size of the Moon". *Archive for History of Exact Sciences*, 71, pp. 193–210.

Carman, C. (2018a). "The First Copernican Was Copernicus: The Difference between Pre-Copernican and Copernican Heliocentrism". *Archive for History of Exact Sciences*, 72 (1), pp. 1–20.

Carman, C. (2018b). "Accounting for Overspecification and Indifference to Visual Accuracy in Manuscript Diagrams: A Tentative Explanation Based on Transmission". *Historia Mathematica*, 45, pp. 217–236.

Carman, C. (2020a). "On the Distances of the Sun and Moon According to Hipparchus". In A. Jones and C. Carman (eds.), *Instruments – Observations – Theories: Studies in the History of Astronomy in Honor of James Evans*, pp. 177–203. http://doi.org/10.5281/zenodo.3928498. Chapter DOI: 10.5281/zenodo.3975739. Open access distribution under a Creative Commons Attribution 4.0 International (CC-BY) license.

Carman, C. (2020b). "Vestiges of the Emergence of Overspecification and Indifference to Visual Accuracy in the Mathematical Diagrams of Medieval Manuscripts". *Centaurus*, 62, pp. 141–157.

Carman, C. and J. Evans (2015). "The Two Earths of Eratosthenes". *ISIS*, 106 (1), pp. 1–16.

Carmody, F. J. (1947). *Leopold of Austria "Li Compilacions de le Science des Estoilles"*. Berkeley: University of Chicago Press.

Carmody, F. J. (1960). *The Astronomical Works of Thabit B. Qurra*. Berkeley: The University of California Press.

Christianidis, J. P. and N. Kastanis (1992). "In Memoriam: Evangelos S. Stamatis". *Historia Mathematica*, 19, pp. 99–105.

Ciocci, A. (2022). "Federico Commandino and his Latin edition of Aristarchus's On the Sizes and Distances of the Sun and the Moon". *Archives for History of Exact Sciences*. https://doi.org/10.1007/s00407-022-00294-7. Online First.

Cioni, A. (1998). "GABI, Simone, detto Bevilacqua". In: *Dizionario Biografico degli Italiani*, Vol. 51. Roma: Istituto della Enciclopedia Italina.

Cobos Bueno, J. M. and M. Pecellin Lancharro (1997). "Francisco Vera Fernández de Córdoba, historiador de las ideas científicas". *Llull*, 20, pp. 507–528.

Commandinus, F. (1558). *Archimedis Opera non nulla a Federico Commandino Urbinate nuper in Latinum conversa, et commentariis illustrata*. Venezia: Paulum Manutium, Aldi F.

Commandinus, F. (1572). *Euclidis Elementorum libri XV. Una cum Scholiis antiquis*. Pesaro: Camillum Franceschinum

CTC (1960–2003). *Catalogus translationum et commentariorum. Mediaeval and Renaissance Latin Translations and Commentaries: Annotated Lists and Guides*, 8 vol. Washington, DC: The Catholic University of America Press.

Dammert, F. L. (1870). *Anton Nokk, Doktor der Philosophie, Großherzoglich Badischer Geheimer Hofrat, Professor und Lyceumsdirektor, Ritter des Ordens vom Zähringer Löwen. Ein Lebensbild*. Freiburg Br.: Wangler.

De Goeje, M. J. (1892). *Ibn Rustah, Kitāb al-A'lāk an-Nafīsa*. Leiden: E. J. Brill.

Derenzini, G. (1973), "Per la tradizione manoscritta del testo e degli scolii del «Περὶ μεγεθῶν καί ἀποστημάτων ἡλίου καὶ σελήνης". *Physis. Rivista internazionale di storia della scienza*, 15, pp. 329–332.

Dicks, D. R. (1960). *The Geographical Fragments of Hipparchus*. London: Athlone Press.

Diels, H. (1879). *Doxographi Graeci*. Berlin: G. Reimer.

Dilke, O. A. W. (1985). *Greek and Roman Maps*. Ithaca: Cornell University Press.

Dillon, J. (2003). *The Heirs of Plato. A Study of the Old Academy (347–274 BC)*. Oxford: Clarendon Press.

Diogenes Laërtius (1853). *Lives and Opinions of Eminent Philosophers*. London: Haddon and Son.

Donahue, W. H. (2015). *Johannes Kepler, Astronomia Nova*. New Revised Edition. Trans. Donahue. Santa Fe, NM: Green Lion.

Dreyer, J. L. (1913–1929). *Tychonis Brahe Opera Omnia*, 15 vol. Copenhague: Libraria Gyldendaliana.

Dreyer, J. L. (1953). *A History of Astronomy from Thales to Kepler*, 2nd ed., originally published as *History of the Planetary Systems from Thales to Kepler*. 1905. New York: Dover.

Evans, J. (1998). *The History and Practice of Ancient Astronomy*. Oxford: Oxford University Press.

Evans, J. and J. L. Berggren (2004). *Gemino's Introduction to Phenomena: A Translation and Study of a Hellenistic Survey of Astronomy*. Princeton: Princeton University Press.

Fabricius, J. A. (1795). *Bibliotheca Graeca. Editio nova curante Gottlieb Christophoro Harles, Hamburgi apud Carolum Ernestum Bohn*. Hamburg.

Festugière, A. J. (1968). *Proclus, Commentaire sur le Timée*, Vol. 4, book IV. Paris: J. Vrin.

Fisher, I. (1932). *Claudii Ptolemaei Geographiae, Codex Urbinas Graecus 82, Tomus Prodromus, Pars Priori*, Leiden and Leipzig: Bibliotheca Apostolica Vaticana.

Fowler, D. (1999). *The Mathematics of Plato's Academy. A New Reconstruction*. Oxford: Clarendon Press.

Frisch, C. (1858–1870). *Joannis Kepleri astronomi opera omnia*, 8 vols. Francfurt: Heyder & Zimmer.

Galilei, G. (1953). "Dialogo dei massimi sistemi". In *La Letteratura italiana, storia e testi, Vol. 34: Galileo e gli scienziati del Seicento, T. 1, Opere di Galileo Galilei*. Verona: Valdonega.

Gardenal, G. (1981). "Giorgio Valla e le scienze esatte". In V. Branca (ed.), *Giorgio Valla tra scienza e sapienza*. Firenze: Olschki, pp. 9–54.

Gingerich, O. (1975). "The Origins of Kepler's Third Law". *Vistas in Astronomy*, 18, pp. 595–601.

Gingerich, O. (1980). "Was Ptolemy a Fraud?". *Quarterly Journal of the Royal Astronomical Society*, 21, pp. 253–266.

Gingerich, O. (1981). "Ptolemy Revisited: A Reply to R. R. Newton". *Quarterly Journal of the Royal Astronomical Society*, 22, pp. 40–44.

Goldstein, B. R. (1967). "The Arabic Version of Ptolemy's Planetary Hypotheses". *Transactions of the American Philosophical Society, New Series*, 57, Part. 4.

Goldstein, B. R. and N. Swerdlow (1970). "Planetary Distances and Sizes in an Anonymous Arabic Treatise Preserved in Bodleian Ms. Marsh 621". *Centaurus*, 15 (2), pp. 135–170.

Häckermann, A. (1886). "Nizze, Johann Ernst". En *Allgemeine Deutsche Biographie (ADB)*, Band 23. Leipzig: Duncker & Humblot, pp. 744–745.

Hall, A. and M. Hall (1953). *The Correspondence of Henry Oldenburg*. Vol. 8. Madison: University of Wisconsin Press.

Halley, E. (1679). *Catalogus Stellaroum Australium Sive Supplementum Catalogi Tychonici*. London: Thomas James.

Hamilton, N., N. Swerdlow and G. Toomer (1987). "The Canobic Inscription: Ptolemy's Earliest Work". In J. L. Berggren y B. R. Goldstein (eds.), *From Ancient Omens to Statistical Mechanics*. Copenague: University Library, pp. 55–73.

Hartner, W. (1964). "Medieval Views on Cosmic Dimensions and Ptolemy's Kitab al. Manshurat". In A. Koyré (ed.), *Mélanges Alexandre Koyré, publiés à l'occasion de son soixante-dixième anniversaire*, Vol. 1. Paris: Hermann, pp. 254–282.

Hartner, W. (1980). "Ptolemy and Ibn Yûnus on Solar Parallax". *Archives Internationales d'Histoire des Sciences*, 30, pp. 5–26.

Hartner, W. (2008). "Al-Battānī, Abū 'Abd Allāh Muhammad Ibn Jābir Ibn Sinān Al-Raqqī Al-Harrānī Al-Sābi'". In C. C. Gllspie (ed.), *Complete Dictionary of Scientific Biography*. New York: Scribner.

Heath, T. L. (1897). *The Works of Archimedes, Edited in Modern Notation with Introductory Chapters*. Cambridge: Cambridge University Press.

Heath, T. L. (1912). *The Method of Archimedes: Recently Discovered by Heiberg*. Cambridge: Cambridge University Press.

Heath, T. L. (1921). *A History of Greek Mathematics. Vol. I: From Thales to Euclid*. Oxford: Clarendon Press.

Heath, T. L. (1931). *A History of Greek Mathematics. Vol. II: From Aristarchus to Diophantus*. Oxford: Clarendon Press.

Heath, T. L. (1932). *Greek Astronomy*. London: J.M. Dent & Sons.

Heath, T. L. (2002). *Euclids Elements: All Thirteen Books Complete in One Volume*. Santa Fe: Green Lion Press.

Heiberg, J. L. (ed.) (1880). *Archimedis Opera omnia: cum commentariis Eutocii*, Vol. 1. Leipzig: Teubner.

Heiberg, J. L. (ed.) (1894). *Simplicii in Aristotelis De Caelo Commentaria*. Leipzig: Teubner.

Heiberg, J. L. (1896). "Beiträge zur Geschichte Georg Vallas und seiner Bibliothek". *Centralblatt für Bibliothekswese*, 16, pp. 2–6 and 54–103.

Heiberg, J. L. (ed.) (1898–1903). *Claudii Ptolemaei Opera quae exstant omnia. Vol. I: Syntaxis Mathematica*, 2 vols. Leipzig: Teubner.

Heiberg, J. L. (ed.) (1907). *Claudii Ptolemaei Opera quae exstant omnia. Vol. II: Opera Astronomica Minora*. Leipzig: Teubner.

Heiberg, J. L. (ed.) (1913). *Archimedis Opera omnia: cum commentariis Eutocii*, Vol. 2. Leipzig: Teubner.

Heiberg, J. L. and H. Menge (eds.) (1895). *Euclidis Opera Omnia. Euclidis Optica, Opticorum Recensio Theonis, Catoptrica, cum scholiis antiquis*, Vol. 7. Leipzig: Teubner.

Henderson, J. (1975). "Erasmus Rheinhold's Determination of the Distance of the Sun from the Earth". En R. Westman (ed.), *The Copernican Achievement*. Berkeley: University of California Press, pp. 108–129.

Henderson, J. (1991). *On the Distances between Sun, Moon & Earth According to Ptolemy, Copernicus & Reinhold*. Leiden: E. J. Brill.

Hiller, E. (1878). *Theonis Smyrnaei, philosophi platonici, Expositio rerum mathematicarum ad legendum Platonem utilium*. Leipzig: Teubner.

Horrocks, J. (1662). "Venus in Sole visa". In J. Hevelius (ed.), *Mercurius in Sole visus*. Gdansk: Reiniger, pp. 111–145.

Horrocks, J. (1673). *Jeremiae Horrocci Opera Posthuma*. London: John Wallis.

Horrocks, J. (1859). *Memoir of the Life and Labors of the Rev. Jeremiah Horrox to Which is Appended a Translation of the Celebrated Discourse upon the Transit of Venus Across the Sun*. London: Wertheim, Macinthosh and Hunt.

Hultsch, F. (1867). *Censorini de Die Natali, recensuit Fridericus Hultsch*. Leipzig: Teubner.

Hultsch, F. (1877). *Pappi Alexandrini Collectionis quae supersunt e libris manu scriptis edidit, latina interpretatione et commentariis, instruxit Fridericus Hultsch*, Vol. 2. Berlin: Weidman.

Hultsch, F. (1897). *Poseidonios über die Grösse und Entfernung der Sonne*. Berlin: Weidman.

Hultsch, F. (1900). "Hipparchos über die Grösse und Entfernung der Sonne". In *Berichte über die Verhandlungen der kgl. Sächsischen Ges. d. Wissenschaften*, Vol. 52. Leipzig: Teubner, pp. 169–200.

Huygens, C. (1888–1950). *Oeuvres complètes de Christiaan Huygens. Observations astronomiques. Systèmes de Saturne. Travaux astronomiques, 1658–1666*. La Haya: Société hollandaise des sciences.

Ibn Rustah (1982). "Kitab al-Aᶜlaq an-Nafisa VII". In M. J. de Goeje (ed.), *Bibliotheca Geographorum Arabicorum*, Vol. VII. Leiden: E. J. Brill.

Jones, A. (2005). "Ptolemy's *Canobic Inscription* and Heliodorus' Observation Reports". *SCIAMVS*, 6, pp. 53–97.

Jones, A. (2006). "The Keskintos Astronomical Inscription: Text and Interpretations". *SCIAMVS*, 7, pp. 3–41.

Jones, A. (2008). "Eratosthenes of Kurene". In P. T. Keyser and G. L. Irby-Massie (eds.), *The Encyclopedia of Ancient Natural Scientists*. London: Routledge, pp. 297–300.

Jones, H. L. (1960). *The Geography of Strabo*, Vol. 1. Cambridge, MA: Harvard University Press.

Kepler, J. (1609). *Astronomia nova, seu physica coelestis, tradita commentariis de motibus stellae martis*. Heidelberg: Voegelin.

Kepler, J. (1620). *Epitome Astronomiae Copernicanae*. Linz: Johann Planck.

Kepler, J. (1621). *Prodromus dissertationvm cosmographicarvm: continens mysterivm cosmographicvm, de admirabili proportione orbivm coelestivm, de qve cavsis cælorum numeri, magnitudinis, motuumque periodicorum genuinis & proprijs, demonstratvm, per qvinqve regularia corpora geometrica*. Francfurt: Kempfer.

Kepler, J. (1937–). *Johannes Kepler Gesammelte Werke*. Munich: C. H. Beck.

Kidd, I. G. (1988). *Posidonius, Vol II: The Commentary*. Cambridge: Cambridge University Press.

Kieffer, J. S. (1964). *Galen's Institutio Logica. English Translation, Introduction and Commentary*. Baltimore: Johns Hopkins Press.

Krössler, F. (2008). "Nizze, Ernst". In *Personenlexikon von Lehrern des 19. Jahrhunderts. Berufsbiographien aus Schul-Jahresberichten und Schulprogrammen 1825–1918, vo. "Naarmann – Nymbach"*. Gießen: Universität Gießen.

Koyré, A. (1964). *Mélanges Alexandre Koyré, publiés à l'occasion de son soixante-dixième anniversaire, 2 vols. Vol I: L'aventure de la science*. Paris: Hermann.

Kuhn, T. (1957). *The Copernican Revolution. Planetary Astronomy in the Development of Western Thought*. Cambridge, MA: Harvard University Press.

Kunitzsch, P. (1974). *Der Almagest. Die Syntaxis Mathematica des Claudius Ptolemäus in arabisch-lateinischer Überlieferung*. Wiesbaden: Harrassowitz.

Kūshyār ibn Labbān (1948). "Al-ab'ād wa-'l-ajrām" (Distances and sizes). In *Rasā'il mutafarriqa fī al-hay'a li-'l-mutaqaddimīn wa-mu'āsirī al-Bīrūnī*. Hyderabad-Deccan: Dā'irat al-ma'ārif al-'Uthmāniyya.

Lasbergius, J. P. (1631). *Uranometriae Libri tres. In quibus Lunae, Solis et reliquorum Planetarum, et inerrantium Stellarum distantiae a Terra, et magnitudines, hactenus ignoratae perspicue demostrantur*. Midleburg: Zacharias Roman.

Lebedev, A. (1990). "Aristarchus of Samos on Thales Theory of Eclipses". *Apeiron*, XXIII, pp. 77–85.

Lusa Monforte, G. (2008). "Midiendo los cielos". *Quaderns d'Historia de L'Engynyeria*, IX, pp. 325–327.

Maar, D. (1982). *Vision. A Computational Investigation into the Human Representation and Processing of Visual Information*. New York: Freeman.

Marcovich, M. (1986). *Hippolytus Refutatio Omnium Heresium*. Berlin: Walter de Gruyter.

Massa Esteve, M. (1999). *Estudis matemàtics de Pietro Mengoli (1625–1686): Taules triangulars i quasi proporcions com a desenvolupament de l'àlgebra de Viète*. Barcelona: Servei de Publicacions de la UAB.

Massa Esteve, M. (2006). *L'algebrització de les matemàtiques. Pietro Mengoli (1625–1686)*. Barcelona: Servei de Publicacions de la UAB.

McColley, G. (1941). "George Valla: An Unnoted Advocate of the Geo-Heliocentric Theory". *Isis*, 33 (3), pp. 312–314.

Nallino, C. A. (1899–1903). *Al-Battānï sive Alhutenii Opus astronomicum, ad fidem codicis escurialensis arabice editum, latine versum, adnotationibus instructum a Carolo Alphonso Nallino*. Milán: Reale Osservatorio di Brera.

Netz, R. (1999). *The Shaping of Deduction in Greek Mathematics. A Study in Cognitive History*. Cambridge: Cambridge University Press.

Netz, R. (2004). *The Works of Archimedes. Translation and Commentary. Vol. 1: The Two Books on the Sphere and the Cylinder*. Cambridge: Cambridge University Press.

Netz, R., W. Noel, N. Tchernetska and N. Wilson (2011a). *The Archimedes Palimpsest. Vol. I: Catalogue and Commentary*. Cambridge: Cambridge University Press.

Netz, R., W. Noel, N. Tchernetska and N. Wilson (2011b). *The Archimedes Palimpsest. Vol. II: Images and Transcriptions*. Cambridge: Cambridge University Press.

Neugebauer, O. (1957). *The Exact Sciences in Antiquity*, 2nd ed. New York: Dover.

Neugebauer, O. (1968). "On the Planetary Theory of Copernicus". *Vistas in Astronomy*, 10, pp. 89–103.

Neugebauer, O. (1974). "Notes on Autolycus". *Centaurus*, 18, pp. 66–69.

Neugebauer, O. (1975). *A History of Ancient Mathematical Astronomy*, 3 vols. Berlin: Springer.

Newton, R. (1977). *The Crime of Claudius Ptolemy*. Baltimore and London: John Hopkins University Press.

Nizze, H. (1907). *Dr. Johann Ernst Nizze: Professor und Direktor am Gymnasium zu Stralsund – Ein Lebensbild eines Lützower Jägers nach alten Papieren von seiner ältesten Tochter Hedwig Nizze*. Stralsund: Zemsch.

Noack, B. (1992). *Aristarch von Samos. Untersuchungen zur Überlieferungsgeschichte der Schrift περὶ μεγεθῶν καὶ ἀποστημάτων ἡλίου καὶ σελήνης*. Wiesbaden: Ludwig Reichert.

Öttinger, L. (1850). *Die Vorstellungen der alten Griechen und Römer über die Erde als Himmelskörper*. Freiburg: Diernfellner.

Pedersen, O. (1974). *A Survey of the Almagest*. Odense: Odense University Press.

Pérez Sedeño, E. (1987). *Las Hipótesis de los Planetas*. Madrid: Alianza.

Pliny (1962). *Natural History in Ten Volumes. Vol. I: Praefatio, Libri I, II with an English Translation by H. Rarckham*. Cambridge, MA: Harvard University Press.

Plutarch (1957). *Plutarch's Moralia. With an English Translation by Harold Cherniss and William C. Helmbold*, Vol. XII. Cambridge, MA: Harvard University Press.

Prell, H. (1959). "Die Vorstellung des Altertums von ther Erdumfangslänge". *Abhandlungen der Sächsischen Akademie der Wissenschaften*, 16 (1).

Proclus (1906). *In Platonis Timaeum Commentaria*, 3 vols. Leipzig: Diehl.

Proclus (1909). *Hypotyposis astronomicarum positionum*. Leipzig: Teubner.

Ptolemy, C. (1843–1845). *Claudii Ptolemaei Geographia*, 3 vols., ed. C. F. A. Nobbe. Lepzig: Karl Tauchnitz.

Ptolemy, C. *The Planetary Hipotheses*. See Goldstein (1969), Heiberg (1907) and Pérez Sedeño (1987).

Reeve, C. D. C. (1997). *Plato, Cratylus: Translated with Introduction and Notes*. Indianapolis and Cambridge: Hackett.

Riccioli, G. B. (1651). *Almagestum Novum*, 2 vols. Bologna: Vittorio Benacci.

Rome, A. (1931). *Commentaires de Pappus et de Théon d'Alexandrie sur l'Almageste, 3 vols. Vol I: Pappus d'Alexandrie: Commentaire sur les livres 5 et 6 de l'Almageste)*. Roma: Biblioteca Apostólica Vaticana.

Rosen, E. (1995). *Copernicus and His Successors*. London: The Hambledon Press.

Saito, K. and N. Sidoli (2012). "Diagrams and Arguments in Ancient Greek Mathematics: Lessons Drawn from Comparisons of the Manuscript Diagrams with Those in Modern Critical Editions". In K. Chemla (ed.), *The History of Mathematical Proof in Ancient Traditions*. Cambridge: Cambridge University Press, pp. 135–162.

Saliba, G. (1979). "The First Non-Ptolemaic Astronomy at the Maraghah School". *Isis*, 70, pp. 571–576.

Saunders, M. (1993). "Ivor Bulmer-Thomas" (Obituary). *The Independent*, 8 October 1993.

Schiaparelli, G. V. (1898). "Origine del sistema planetario eliocentrico presso I Greci". *Memorie del R. Istituto Lombardo di scienze e lettere*, Vol. 18, Part 5. Milano: U. Hoepli.

Scriba, C. (1970). "The Autobiography of John Wallis, F.R.S.". *Notes and Records of the Royal Society of London*, 25 (1), pp. 17–46.

Sextus Empiricus (2012). *Against the Physicists* (Adversus Mathematicos IX and X), trans. Richard Bett. Cambridge: Cambridge University Press.

Sidoli, N. (2007). "What We Can Learn from a Diagram: The Case of Aristarchus's *On the Sizes and Distances of the Sun and Moon*". *Annals of Science*, 64 (4), pp. 525–547.

Simplicius (1984). *Aristotelis De Caelo Commentaria*, ed. Johan Ludvig Heiberg. Berlin: Reimer.

Smith, D. E. (1917). "John Wallis as a Cryptographer". *Bulletin of the American Mathematical Society*, 24 (2), pp. 82–96.

Stamatis, E. S. (1972). *Archimedis Opera Omnia*, 3 vols. Leipzig: Teubner.

Steele, J. M. (2000). "A Re-analysis of the Eclipse Observations in Ptolemy's *Almagest*". *Centaurus*, 42, pp. 89–108.

Swerdlow, N. M. (1968). *Ptolemy's Theory of the Distances and Sizes of the Planets: A Study in the Scientific Foundation of Medieval Cosmology*. Ph.D. Thesis. Yale University.

Swerdlow, N. M. (1969). "Hipparchus on the Distance of the Sun". *Centaurus*, 14 (1), pp. 287–305.

Swerdlow, N. M. (1973). "Al Battānī's Determination of the Solar Distance". *Centaurus*, 17, pp. 97–105.

Swerdlow, N. M. (1979). "Ptolemy on Trial". *American Scholar*, 48, pp. 523–531.

Swerdlow, N. M. (1992). "Shadow Measurement: The Sciametria from Kepler's Hypparchus: A Translation with Commentary". In P. M. Harman and Alan E. Shapiro (eds.), *An Investigation of Difficult Things: Essays on Newton and the History of the Exact Sciences*. Cambridge: Cambridge University Press, pp. 19–70.

Tannery, P. (1893). "Aristarque de Samos". In P. Tannery (ed.), *Recherches sur l'histoire de l'astronomie ancienne*. Paris: Gauthier-Villars.

Tannery, P. (1912). "Aristarque de Samos". In. J. L. Heiberg and H.-G. Zeuthen (eds.), *Mémories Scientifiques, vol. I: Sciences exactes dans L'antiquité 1876–1884*. Paris: Gauthier-Villars.

Tarán, L. (1975). *Academica: Plato, Philip of Opus and the Pseudo-Platonic Epinomis*. Philadelphia: American Philosophical Society.

Tassinari, P. (ed.) (1994). *Ps. Alessandro d'Afrodisia, Trattato sulla febbre*. Alessandria: Edizioni dell'Orso.

Taub, L. C. (1993). *Ptolemy's Universe. The Natural Philosophical and Ethical Foundations of Ptolemy's Astronomy*. Chicago and La Salle: Open Court.

Taylor, A. E. (1929). "Plato and the Origins of the Epinomis". *Proceedings of the British Academy*, 15, pp. 235–317.

Thomas, I. (1957). *Selections Illustrating the History of Greek Mathematics with an English Translation of Ivor Thomas. 2 vols. Vol. 2: From Aristarchus to Pappus*. Cambridge, MA: Harvard University Press.

Thorndike, L. (1949). *The Sphere of Sacrobosco and Its Commentators*. Chicago: University of Chicago Press.

Toomer, G. J. (1967). "The Size of the Lunar Epicycle According to Hipparchus". *Centaurus*, 12 (3), pp. 145–150.

Toomer, G. J. (1974). "Hipparchus on the Distances of the Sun and Moon". *Archive for History of Exact Sciences*, 14, pp. 126–142.

Toomer, G. J. (1998). *Ptolemy's Almagest*. Princeton: Princeton University Press.

Toomer, G. J. (2005). "Philippus of Opus". In *The Oxford Classical Dictionary*, 3rd ed. Oxford: Oxford University Press, p. 250.

Valyus, N. A. (1962). *Stereoscopy*. New York: The Focal Press.

Van Brummelen, G. (2009). *The Mathematics of the Heavens and the Earth. The Early History of Trigonometry*. Princeton: Princeton University Press.

van Helden, A. (1986). *Measuring the Universe. Cosmic Dimensions form Aristarchus to Halley*. Chicago: University of Chicago Press.

Ver Eecke, P. (1933). *Pappus d'Alexandrie, la Collection Mathématique*, 2 vols. Paris: Librairie Scientifique et Technique Albert Blanchard.

Vera Fernández de Córdoba, F. (1970). *Científicos Griegos. Recopilación, estudio preliminar, preámbulos y notas por Francisco Vera*, 2 vols. Madrid: Aguilar.

Vitruvius (1999). *Ten Books on Architecture*. Cambridge: Cambridge University Press.

Wall, B. E. (1975). "Anatomy of a Precursor: The Historiography of Aristarchos of Samos". *Studies in History and Philosophy of Science*, 6 (3), pp. 201–228.

Waszink, E. J. (1962). *Timaeus a Calcidio Translatus Commentarioque Instructus*. London and Leiden: E. J. Brill.

Wegman, E. J. (1995). "Huge Data Sets and the Frontiers of Computational Feasibility". *Journal of Computational and Graphical Statistics*, 4 (4), pp. 281–295.

Wendelin, G. (1626). *Loxias seu de Obliquitate Solis Diatriba*. Amberes: Jeronimum Verdessium.

Wendelin, G. (1644). *Eclipses lunares ab anno 1573 ad 1643 observatae*. Amberes: Jeronimum Verdessium.

Wright, R. R. (1934). *The Book of Instruction in the Elements of the Art of Astrology, by Abū 'r-Raihān Muhammad ibn Ahmad al-Bīrūnī*, translated with a facsimile of Brit. Mus. Ms Or. 3849 (Arabic text), by R. Ramsey Wright. London: Luzac.